Superplasticity: 60 Years after Pearson

Superplasticity:
60 Years after Pearson

Proceedings of the Conference organised on behalf of
the Superplastic Forming Committee of the Manufacturing Division
of The Institute of Materials, and held at the
University of Manchester Institute of Science and Technology
(UMIST) on 7–8 December 1994

Edited by
NORMAN RIDLEY

THE INSTITUTE OF MATERIALS

Book 618
First published in 1995 by
The Institute of Materials
1 Carlton House Terrace
London SW1Y 5DB

© The Institute of Materials 1995
All rights reserved

ISBN 0 901716 77 4

Printed and finished in the UK by
Bourne Press, Bournemouth

Table of Contents

Foreword ix

Introductory paper

C.E. Pearson and his Observations of Superplasticity 1
N. Ridley

Fundamental Aspects of Superplasticity

Mechanisms of Superplastic Flow 9
T. G. Langdon

Grain Boundary Sliding – the Main Mechanism of Superplastic Flow 25
O. A. Kaibyshev

Grain Boundary Structure and Superplastic Behaviour 33
R. I. Todd

Physical Model for Superplastic Flow Rheology in a Wide Strain Rate Range 43
S. A. Larin and V. N. Perevezentsev

The Theory of the Evolution of the Microstructure of
Superplastic Alloys and Ceramics 51
V. N. Perevezentsev

A Composite Model for Superplasticity 60
B. Baudelet and J. Lian

Cavitation during Superplastic Flow

Cavitation in Superplastic Materials 63
N. Ridley and Z. C. Wang

A Comparative Study of Cavitation Characteristics of a Superplastic
Aluminium Composite and an Aluminium Alloy 75
H. Iwasaki, M. Mabuchi, K. Higashi and T. Mori

Cavitation in Superplastically Deformed Ceramic Materials   85
Z. C. Wang, T. J. Davies and N. Ridley

Superplastic Flow at High Strain Rates

Positive Exponent Superplasticity in Metallic Alloys and Composites   93
K. Higashi.

Microstructural Design for Superplastic Metal Matrix Composites   103
M. Mabuchi and K. Higashi.

Superplasticity in Mg–Al–Ga Alloys produced by
Rapid Solidification Processing   113
T. Shibata, K. Higashi, Y. Yamaguchi, A. Inoue and T. Masumoto

Superplasticity in Ceramics and Intermetallics

Superplastic Ceramics and Intermetallics and their Potential Applications   121
J. Wadsworth and T. G. Nieh

Superplasticity in Al_2O_3–ZrO_2–Al_2TiO_5 Ceramics   133
J. Pilling and J. Payne

Electronic Contributions to the Superplastic Partition in Ceramics   139
T. J. Davies, A. A. Ogwu, N. Ridley and Z. C. Wang

Effect of Fe and Co Additions on the Superplastic Behaviour in the
NiAl–Ni_3Al Two-Phase Alloy   151
S. Ochiai and M. Kobayashi

Microstructural Evolution during Superplastic Flow

Microstructure and Textural Changes during the Superplastic Deformation
of Modified IMI550 Titanium Alloy (Ti–4Al–3Mo–2Sn–lFe–0.5Si)   161
M. Tuffs and C. Hammond

Microstructural Evolution of Al–Cu–Zr Alloys during
Thermomechanical Processing   173
E. Cullen, F. J. Humphreys and N. Ridley

Macro and Microtexture Development during the Superplastic
Deformation of AA 8090 183
P. L. BLACKWELL and P. S. BATE

Superplasticity in Aluminium Alloys in Recrystallised
and Unrecrystallised Conditions 193
A. A. ALALYKIN, I. I. NOVIKOV, V. K. PORTNOY and V. I. PAVLOV

Superplastic Forming – Modelling

Finite Element Modelling of Superplastic Forming with Precise Dies 201
R. SADEGHI and Z. PURSELL

Aspects of the Numerical Simulation of SPF including
Material Parameter Evaluation 206
R. D. WOOD and J. BONET

Numerical Simulation of Superplastic Forming/Diffusion
Bonding Processes 218
O. BALDO and J. DIAZ

Simulation of Uni-axial Superplasticity Specimen Testing in Ti–6Al–4V 235
F. P. E. DUNNE and I. KATRAMADOS

Superplastic Forming – Equipment Alloys, Forming and Applications

Design and Manufacture of Hydraulic Presses for Superplastic Forming 253
R. WHITTINGHAM

Estimation of Strain Rate Sensitivity in Superplastic Compression Test 260
M. YOSHIZAWA and H. OHSAWA

Superplastic Sheet Aluminium Alloys: Their Forming, Application
and Future 273
R. G. BUTLER and R. J. STRACEY

Properties of Superplastic 5083 alloy and its Applications 277
M. MATSUO

New Market Areas for Superplastic Aluminium　284
R. J. STRACEY and R. G. BUTLER

Roll Bonding/Superplastic Forming (RB/SPF) of
Superplastic Aluminium Alloys　296
S. FURIHATA and H. OHSAWA

Superplastic Forming and Diffusion Bonding – An Overview　304
D. STEPHEN

Titanium Alloys for Superplastic Forming　305
A. WISBEY and M. W. KEARNS

Superplasticity in the Titanium Alloy SP700 with Low SPF Temperature　324
A. WISBEY, B. C.WILLIAMS, H. S. UBHI, B. GEARY,
C. M. WARD-CLOSE and A. W. BOWEN

Superplastic Forming (SPF) at AWE – Forming Processes and
Post-SPF Characterisation of Ti–6Al–4V Plate　338
F. J. MORAN

The Exploitation of Superplasticity for Rolls-Royce's Wide
Chord Fan Blade　347
G. A. FITZPATRICK

Use of Superplastic Forming for the Production of Turboprop Nacelle and
Large Ti–6Al–4V Components　351
W. SWALE

Application of SPF/DB Titanium Technology to Large Commercial Aircraft　359
C. F. DRESSEL

Industrialisation of SPF within BAe Military Aircraft　377
A. D. COLLIER and N. JACKSON

Closing Remarks

Where is Superplasticity 60 Years after Pearson?　387
M. J. STOWELL

Foreword

The meeting was organised on behalf of the Superplastic Forming Committee of the Manufacturing Division of the Institute of Materials to commemorate the 60th Anniversary of the publication by the former Institute of Metals of the classic paper by C.E. Pearson entitled "The Viscous Properties of Extruded Eutectic Alloys of Lead-Tin and Bismuth-Tin." Not only did Pearson observe exceptionally high tensile strains in metallic materials, he also identified important microstructural aspects of the phenomenon, which subsequently became known as superplasticity. In addition, he demonstrated the feasibility of superplastic bulge-forming using gas pressure, a procedure which is the basis of forming techniques employed commercially today.

A major objective of the meeting was to determine the current status of the science of superplasticity, the practice of superplastic forming, and the likely directions of future developments. To this end, important areas were addressed in both keynote presentations and contributed papers which dealt with fundamental aspects of superplastic flow, high strain rate superplasticity, cavitation, microstructural evolution, superplasticity in ceramics and intermetallics, and the modelling and practice of superplastic forming.

The meeting, which also included a poster session, was well supported, with almost half of those present travelling from abroad. The scope of the topics covered at the meeting ensured a good interaction of workers having a wide range of interests in superplasticity.

Thanks are due to fellow members of the Organising Committee: Mr. M.J. Ball, Dr. R. Grimes, Dr. C. Hammond, Mr. D.M. Ward and Dr. A. Wisbey, for their help in putting the programme together; to the Conference Department of the Institute of Materials and, in particular, to Teresa Davies for her efficient organisation; and to Mr. Peter Danckwerts of the Institute for his rapid production of the proceedings.

Norman Ridley
Manchester Materials Science Centre
February, 1995.

C.E. Pearson and his Observations of Superplasticity

N. Ridley
University of Manchester/UMIST, Materials Science Centre,
Grosvenor Street,
Manchester M1 7HS, U.K.

Claude E. Pearson: A Short Biography

Professor C.E. Pearson

     Claude Edmund Pearson was born in 1903 in Saltburn-by-Sea in the North Riding of Yorkshire. He studied metallurgy at the University of Sheffield where he graduated with a B. Met. degree in 1923. On leaving Sheffield, he spent 2 years with the Skinningrove Iron and Steel Company which was located near his home town in North Yorkshire. He was subsequently (1927) awarded the M. Met. degree for his research studies on "The Growth of Commercial Grey Cast Iron".

     In 1925, Claude Pearson was appointed Lecturer in Metallurgy at Armstrong College, Newcastle-upon-Tyne, then part of the University of Durham, now the University of Newcastle-upon-Tyne. He was promoted to a Readership in 1939 and was made Head of Department in 1944. He became the first Professor of Metallurgy being appointed when the William Cochrane Chair was created in 1946. Pearson resigned his chair in 1948 and joined Durham Chemicals at Chester-le-Street, Co. Durham, as Technical Manager, subsequently holding the title of Technical Director. At this company he developed interests in metal powders and devised a process for producing zinc dust, which is widely used as a component of various anti-corrosion products. The process was of considerable commercial significance and was operated under licence in a number of countries. Pearson remained with Durham Chemicals until 1968 when he retired.

     Throughout his academic career, Pearson maintained an interest in the extrusion of metals and in the microstructure and mechanical properties of the extruded products. He was the author of a book on "The Extrusion of Metals", which was published by Chapman and Hall in 1944. The book was substantially revised and a second edition, co-authored with

Redvers N. Parkins, was published in 1960. It was not known at the time of his death in Woolsington, near Newcastle-upon-Tyne in 1971 whether Pearson was aware of the developing commercial interest in the phenomenon, then known as superplasticity, which he had so spectacularly demonstrated in 1934.

Acknowledgments

The author is grateful to Emeritus Professor R.N. Parkins, Materials Division, University of Newcastle-upon-Tyne, the Alumni Office, University of Sheffield, and Mr. R. Haddick of Durham Chemicals for their help in supplying biographical information.

Pearson's Observations

In 1934, C.E. Pearson published a paper in the Journal of the Institute of Metals entitled "The Viscous Properties of Extruded Eutectic Alloys of Lead-Tin and Bismuth-Tin" [1]. In this paper he included the celebrated photograph of an extruded Bi-Sn eutectic which had been stretched at room temperature under a constant stress of 250 p.s.i. ($\sim$ 1.7 MNm^{-2}) to an elongation of 1950%. As the initial specimen had a gauge length of 4" ($\sim$ 100 mm), the elongated specimen was 82.1" ($\sim$ 2.1 metres) long, and had to be coiled for photographic convenience [Fig. 1]. Elongations of up to 1500% were also recorded for extruded Pb-Sn eutectic alloy.

Pearson noted that the deformation of the alloys resembled viscous flow of substances such as glass and further demonstrated this by the expansion of extruded tubes under internal pressure. Fig. 2 shows tubing of initial o.d. $\sim$ 10 mm extruded from a billet of Pb-Sn eutectic, which had been subjected to a pressure of 300 p.s.i. (2.1 MPa). The "bulge-formed" tube (12" long, $\sim$ 300 mm) expanded uniformly at first but eventually burst after 50 minutes when its internal volume was 9.1 times that of the starting material.

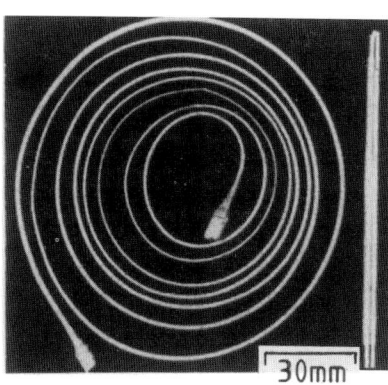

Fig.1. Extruded Bi-Sn eutectic alloy.
Elongation 1950%.

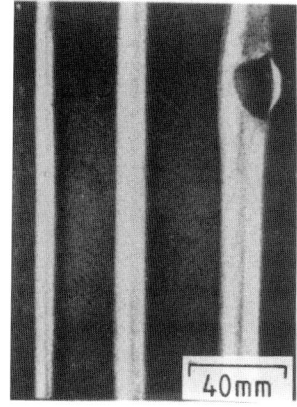

Fig.2. Extruded Pb-Sn eutectic alloy.
Deformed under internal pressure.

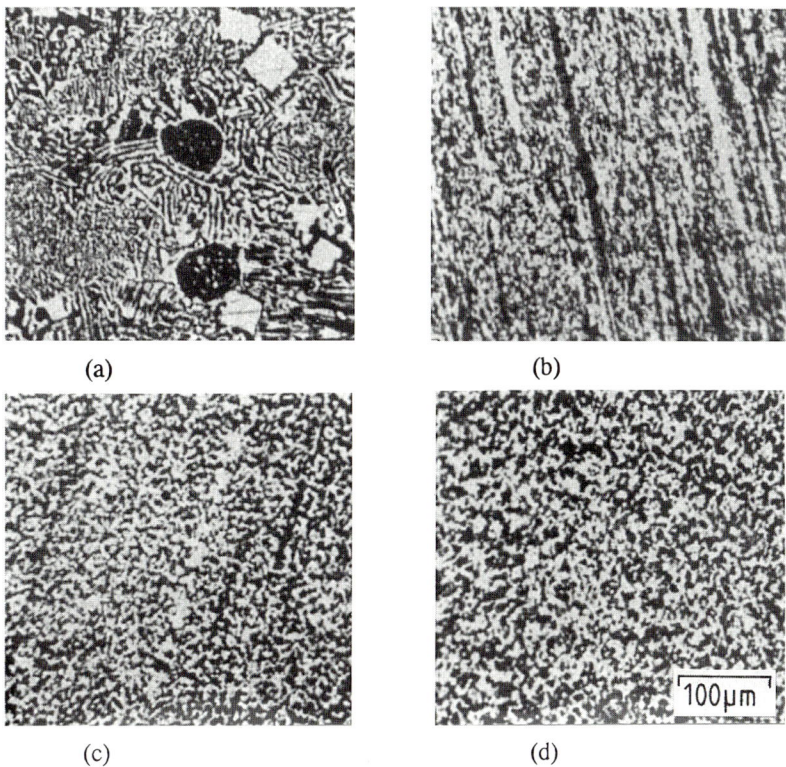

Fig.3. Bi-Sn eutectic alloy. (a) as-cast, (b) extruded-longitudinal section, (c) elongated 150%, (d) elongated 600%.

Pearson's work was not the earliest to demonstrate superplastic behaviour. The first observation was probably that by Bengough [2] when in 1912 he reported an elongation of 162% at 700°C for an extruded α/β brass containing small concentrations of alloying additions. Further high tensile strains were recorded by Jenkins [3] in 1928 working on cold rolled Cd-Zn and Pb-Sn eutectics. For the former alloy, an elongation of 410% was observed at room temperature, while for the latter, elongations of 361% and 405% were observed on testing at room temperature and 117-120°C, respectively.

However, the studies of Pearson in 1934 were certainly the most convincing of the earlier demonstrations of superplasticity. Not only did Pearson record spectacularly high elongations, but he was also aware of many aspects of the relationship of microstructure and mechanical behaviour. He observed that the directionality of microstructure (banding) present in the extruded material was progressively removed during slow straining until the structure ceased to show directionality (Fig. 3a, b, c, d), although if the alloy was deformed rapidly it failed at a low tensile strain and the grains became elongated. Microscopic observations made on smooth surfaces of specimens at intervals during their deformation showed the development of a granular appearance and then the tilting of grains relative to the original surface [Fig. 4]. The shapes of the grains remained unchanged and they were free from slip-bands. These observations were interpreted as showing that the type of

Fig.4. Extruded Pb-Sn eutectic alloy Annealed 5 hours at 100°C, elongated 50%.

deformation shown by the alloys was due to flow taking place at the boundaries of the grains, i.e. by grain boundary sliding.

For a given stress it was noted that the rate of deformation of Pb-Sn eutectic was highest when the grain size was small (Fig. 5). If the grain size was increased by natural ageing or low temperature annealing, then for a given strain rate the flow stress was increased. Low flow stresses were associated with high elongations and after deformation, the original grain structure existed; no recrystallisation had occurred. The readiness and extent of deformation in the extruded alloys was attributed to the persistence of extremely small grain sizes, stabilised by having two closely intermingled phases (i.e. a microduplex structure), at a temperature sufficiently high for intergranular flow.

Data given by Pearson for the relationship between flow stress, σ, and strain rate, $\dot{\varepsilon}$, for Pb-Sn eutectic specimens having had various natural ageing or low temperature annealing

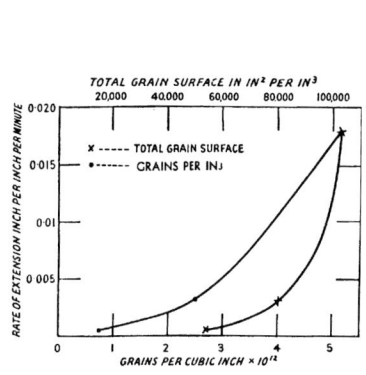

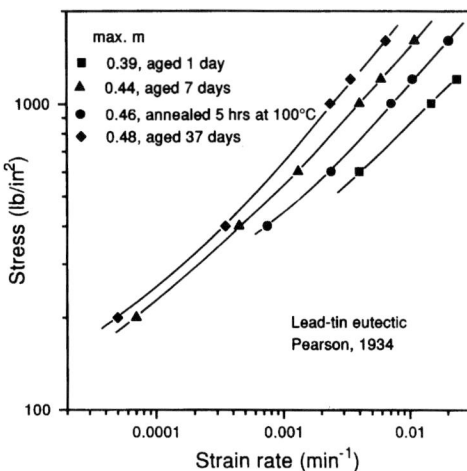

Fig.5. Extruded Pb-Sn eutectic. Effect of grain size on strain rate at 1000 psi (7 MNm^{-2}).

Fig.6 Extruded Pb-Sn eutectic. Logarithmic plot of stress versus strain rate.

treatments, are shown plotted logarithmically in Fig. 6. The maximum gradients of the plots give strain rate sensitivity of flow stress, or m, values of 0.4-0.5 which are fully in accord with superplastic behaviour. The relative positions of the stress-strain curves, although not their relative maximum slopes, are consistent with the anticipated variation in grain size according to the ageing/annealing treatments given to the material.

Summary

The studies of Pearson identified many of the important features of superplastic flow but of particular significance was his recognition of the major role that intergranular displacement, or grain boundary sliding, was playing in the deformation process.

References

1. C.E. Pearson, "The Viscous Properties of Extruded Eutectic Alloys of Lead-Tin and Bismuth-Tin", J. Inst. Metals, 54, 1934, 111-124.

2. G.D. Bengough, "A Study of the Properties of Alloys at High Temperatures", J. Inst. Metals, 7 1912, 123-174.

3. C.M.H. Jenkins, "The Strength of a Cadmium-Zinc and of a Lead-Tin Alloy Solder", J. Inst. Metals, 40, 1928, 21-39.

Fundamental Aspects of Superplasticity

Mechanisms of Superplastic Flow

Terence G. Langdon

Departments of Materials Science and Mechanical Engineering
University of Southern California
Los Angeles, CA 90089-1453, U.S.A.

Abstract

Polycrystalline materials with very small grain sizes are capable of exhibiting high tensile ductility over a limited range of imposed strain rates at elevated temperatures. In metallic alloys, detailed measurements and microstructural observations show that superplastic flow occurs by the process of grain boundary sliding through the relative displacement of individual grains within the crystalline matrix. This process is limited by intragranular dislocation creep at the faster strain rates and by impurity effects in the grain boundaries at the slower strain rates. In ceramic materials such as yttria-stabilized zirconia, grain boundary sliding is again an important deformation mechanism but the mechanical behavior depends critically upon whether there is an amorphous phase at the grain boundaries. This paper examines the nature of the deformation mechanisms occurring during superplastic flow in metals and ceramics.

1. Introduction

Superplasticity is defined as **"the ability of a polycrystalline material to exhibit, in a generally isotropic manner, very high tensile elongations prior to failure"** [1]. Although superplastic elongations are generally of the order of some hundreds or even a few thousands of per cent, it is possible to achieve remarkably high tensile elongations in some materials under carefully selected testing conditions: for example, an elongation of >7000% was attained in a Pb-62% Sn eutectic alloy [2].

To date, the occurrence of superplasticity has been reported in a very wide range of metallic alloys, intermetallics, ceramics, and metal and ceramic matrix composites [3]. From an historical perspective, however, the precise origin of the discovery of superplasticity is not clearly established. Although there is some evidence for the possible utilization of the superplastic effect in the fabrication of ancient steels and bronzes dating back to a period from two to four thousand years ago [4,5], the first laboratory report of superplastic-like flow appeared in the scientific literature in the early part of this century [6]. Nevertheless, the early reports documented rather limited tensile elongations, of the order of some hundreds of per cent, and it was not until 1934 that Pearson [7] published a detailed description of very high tensile ductilities in the Pb-Sn and Bi-Sn eutectic alloys, including the classic demonstration of an elongation of 1950% in an extruded and aged Bi-Sn eutectic alloy which was essentially brittle in the as-cast condition. This definitive report clearly served to establish superplasticity as a scientific phenomenon.

Pearson's work received very little or no attention in the west, but extensive investigations of superplasticity were conducted in the Soviet Union about ten years later by Bochvar and Sviderskaya [8,9]. It is not clear whether this Russian work was prompted by knowledge of Pearson's experiments. In the English language translation of the first book devoted exclusively to superplasticity by Presnyakov [10], it states that the experiments of Pearson "were continued" by Bochvar and Sviderskaya. However, Novikov [11] has claimed very recently that the Russians were unaware of Pearson's work. In the circumstances, it is perhaps more prudent to recognize the considerable contributions made by all of these early pioneers to our understanding of superplasticity and to state simply that Bochvar and co-workers rediscovered the superplastic effect.

Over the last two decades, much attention has been devoted to examining and attempting to explain the physical mechanisms which occur during superplastic deformation. Clearly, a detailed understanding of these mechanisms is a prerequisite for any meaningful extension of superplasticity either to new advanced materials or to new testing conditions which are more favorable for the successful utilization of superplasticity in industrial forming operations (for example, by extending the superplastic effect to faster strain rates). The purpose of this paper is therefore to present an overview of the current understanding of the mechanisms of superplasticity in both metals and ceramics.

2. Superplasticity in Metals

2.1 Characteristics of Flow

When specimens of a superplastic metal are pulled in tension at a constant temperature, there is generally a sigmoidal or S-shaped relationship between the flow stress, σ, and the imposed strain rate, $\dot{\varepsilon}$. An experimental example is shown in the lower half of Fig. 1 for a Zn-22% Al eutectoid alloy tested with a grain size, d, of 2.5 μm at absolute temperatures, T, from 423 to 503 K [12]: the slope of this type of plot is termed the strain rate sensitivity, m ($= \partial \ln \sigma / \partial \ln \dot{\varepsilon}$). It should be noted that a very small grain size is a requirement for superplasticity. Inspection of Fig. 1 shows that, at each testing temperature, the experimental datum points fall along a single line which may be divided into three distinct regions: the value of m is ~ 0.22 at low strain rates in region I, it increases to ~ 0.50 at intermediate strain rates in the vicinity of 10^{-2} s^{-1} in region II, and it decreases again, probably to ~ 0.2, at high strain rates in region III.

A high tensile ductility requires a high value for m [13]. This may be demonstrated by pulling to failure, at a constant displacement rate, each of the specimens used to construct the plot of σ versus $\dot{\varepsilon}$. The result is shown in the upper part of Fig. 1, where the elongation to failure, $\Delta L/L_o \%$, is plotted against $\dot{\varepsilon}$, where ΔL is the total increase in length at fracture and L_o is the initial gauge length. Thus, maximum ductility is achieved in region II and there is a decrease in the total elongations to failure in both regions I and III. In the example shown in Fig. 1, the elongations to failure are in excess of 2000% at the two highest testing temperatures.

The experimental characteristics of regions I-III have been documented elsewhere [14,15]. Briefly, superplasticity in region II is associated with a high value of m (typically of the order of ~ 0.5), a low activation energy which appears to be associated with grain boundary diffusion and an inverse dependence of strain rate upon the grain size raised to a power of ~ 2.

The behavior in region III is similar to normal high temperature creep, with a low value of m (typically ~ 0.2) and a high activation energy of the order of the value for lattice self-diffusion. As in the creep of materials with large grain sizes, deformation in this region arises from the glide and climb of dislocations within the grains. The transition from region III to region II is attributed to a breakdown in the ability to form subgrains within the grains. Subgrains are formed in region III but the equilibrium subgrain size, λ, varies inversely with the applied stress, and a reduction in the stress level therefore leads to a condition where $d \leq \lambda$ and subgrains cannot form [16].

There is direct experimental evidence, from tests on the Pb-62% Sn and Zn-22% Al alloys [17], that the transition from region III to region II occurs at approximately the level of stress where $d \simeq \lambda$.

As described in more detail in the following section, deformation in the superplastic region II arises from grain boundary sliding and the movement of dislocations along the grain boundaries. The transition from region II to region I is associated with the segregation of impurities to the grain boundaries and the consequent effect of this segregation on dislocation movement [18]. Careful experiments on the Zn-22% Al eutectoid [19,20] and the Pb-62% Sn eutectic [21] have established that region I is not observed when using alloys of very high purity. An experimental example is shown in Fig. 2 where the tests were conducted under shear conditions and the plot shows the shear strain rate, $\dot{\gamma}$, versus the shear stress, τ, for Zn-22% Al of two different purity levels [19]: the slope of this plot gives the stress exponent, n ($= 1/m$). It is apparent from Fig. 2 that region I is present only in the alloy containing 180 ppm of impurities and it is absent, so that region II extends to the lowest experimental strain rates, in the material of very high purity.

2.2 The Mechanism of Superplasticity in Metals

When a metal is deformed in the superplastic region II, high tensile elongations are achieved with the grains becoming displaced with respect to each other but with little or no concomitant grain elongation [15]. Nevertheless, there is very extensive grain rotation during superplastic flow and, as shown for three Sn grains in a Pb-62% Sn alloy in Fig. 3(a) where the rotation angle, ϕ, is plotted against the total strain, ε_t, this rotation occurs up to a maximum angle of the order of $\pm 30°$ [22]. These measurements demonstrated that there is no net build-up of rotation but rather the values of the rotation angles increase or decrease within any grain as deformation continues. Experiments using electron microscopy have established that there is only a relatively minor contribution from intragranular dislocation movement in region II [23,24]. Furthermore, Fig. 3(b) shows that the intragranular strain, ε'_{ls}, is non-uniform in the Pb-62% Sn alloy [25], and it occurs in an oscillatory manner with changes from positive to negative contributions as the total elongation increases. From these measurements, it was concluded that intragranular deformation makes no net contribution to the total strain in the superplastic region II. Although the experimental results presented in Fig. 3 (a) and (b) relate to the Sn phase, essentially identical data were obtained also for the Pb phase.

There are experimental observations showing that diffusion creep makes a negligible contribution during superplastic deformation [26] and this is reasonable because of the lack of any significant grain elongation. Therefore, it is concluded that grain boundary sliding is the dominant deformation process in superplasticity. Several investigations have been conducted to determine the contribution from grain boundary sliding to the total strain in a wide range of superplastic alloys but tabulations of the results show that the sliding contributions generally lie in the range of ~ 50-70% [27]. Similar contributions are recorded also if the specimens are repolished during testing [28], thereby serving to demonstrate that grain boundary sliding remains important even when the elongation is very high.

If the sliding contribution is ~ 50-70%, it leaves open the problem of accounting for a "missing strain" of the order of ~ 30-50%. This missing strain cannot be attributed to diffusion creep because, as already noted, the contribution from diffusion creep is essentially negligible; and it cannot be attributed to the movement of dislocations within the grains because, as demonstrated by the measurements associated with the results shown in Fig. 3(b), the intragranular strain makes no net contribution to the total strain. This apparent contradiction led to a re-appraisal of the experimental method used to measure the sliding contributions in superplasticity, and it was demonstrated that the method is in error and will lead to erroneously low values for the sliding contributions, in the range of ~45-90%, even when essentially all of the deformation is due to grain boundary sliding [29].

From these analyses, it is concluded that sliding accounts for all of the strain in superplastic flow. However, grain boundary sliding cannot occur without some accommodation mechanism at the periphery of the grains, and therefore it appears that sliding is accommodated by some limited intragranular dislocation movement which serves to facilitate both the sliding and the re-arrangement of the individual grains [25].

The precise mechanism for grain boundary sliding and the concomitant intragranular accommodation is illustrated schematically in Fig. 4 for (a) large and (b) small grain sizes, respectively. Figure 4(a) represents grain sizes in the mesoscopic grain size range where subgrains are able to form and Fig. 4(b) represents grain sizes in the microscopic grain size range where subgrains are not formed: the precise definitions of these different grain size ranges were given earlier [18]. Under conditions where subgrains form within the grains, as at large grain sizes or at high stress levels in region III for superplastic materials, extrinsic dislocations move along the grain boundaries, pile up at obstacles such as the ledge at A in Fig. 4(a), and then accommodation occurs by dislocations moving across the grain to pile up at the first low angle subgrain boundary. When subgrains are no longer formed, as in superplastic metals tested in region II, the extrinsic dislocations again move along the boundary and pile up at obstacles such as the triple point D when sliding occurs between grains B and C in Fig. 4(b) but, in the absence of any obstacles, the accommodating intragranular slip now moves across the grain to pile up at the opposite grain boundary.

It is possible to model the mechanisms depicted schematically in Fig. 4 by first noting that, at elevated temperatures, the rate of grain boundary sliding, $\dot{\varepsilon}_{gbs}$, may be expressed through a general relationship of the form

$$\dot{\varepsilon}_{gbs} = \frac{A_{gbs} D_{gbs} Gb}{kT} \left(\frac{b}{d}\right)^{p_{gbs}} \left(\frac{\sigma}{G}\right)^{n_{gbs}} \quad (1)$$

where D_{gbs} is the diffusion coefficient for grain boundary sliding [= $D_{o(gbs)}$ exp $(-Q_{gbs}/RT)$, where $D_{(gbs)}$ is a frequency factor, Q_{gbs} is the activation energy for the appropriate diffusion process and R is the gas constant], G is the shear modulus, b is the Burgers vector, k is Boltzmann's constant, p_{gbs} and n_{gbs} are the exponents of the inverse grain size and the stress, respectively, and A_{gbs} is a dimensionless constant. For the situation depicted in Fig. 4, the rate of sliding is controlled by the rate of removal by climb of the dislocations from the head of the intragranular pile up. The expression for the macroscopic sliding rate, $\dot{\varepsilon}_{gbs}$, is then of the form [30]

$$\dot{\varepsilon}_{gbs} \propto \frac{2DL\sigma^2 b^3}{3\sqrt{3} h^2 GkTd} \quad (2)$$

where D is the appropriate diffusion coefficient, L is the length of the intragranular pile up and h is the climb distance.

Using the model depicted in Fig. 4(a) for a mesoscopic grain size with $d > \lambda$, it can be shown that, since $L \propto \lambda$ and $h \propto \sigma^{-1}$, the rate equation for sliding at large grain sizes is given by equation (1) with $n_{gbs} = 3$, $p_{gbs} = 1$ and $D_{gbs} = D_\ell$, where D_ℓ is the coefficient for lattice self-diffusion [30].

Conversely, when $d < \lambda$ in the superplastic region II, L and h are proportional to d and it can be shown that the rate of sliding, which is equivalent to the superplastic strain rate, $\dot{\varepsilon}_{sp}$, is given by

$$\dot{\varepsilon}_{sp} \simeq \frac{A_{sp}D_{gb}Gb}{kT}\left(\frac{b}{d}\right)^{2.0}\left(\frac{\sigma}{G}\right)^{2.0} \tag{3}$$

where D_{gb} is the coefficient for grain boundary diffusion and A_{sp} is a dimensionless constant having a value of ~ 10 [30].

Equation (3) predicts that the activation energy for superplastic flow is equal to the value for grain boundary diffusion, and the strain rate in the superplastic region II varies with the reciprocal of the grain size raised to a power of 2 and with the stress raised to a power of 2 (so that $m = 0.5$). These predictions are reasonably consistent with many of the experimental results currently available for superplastic metallic alloys [15].

3. Superplasticity in Ceramics

Unlike superplastic metals, the observations of superplastic-like flow in ceramic materials are very recent and date only from the report, published in 1986 by Wakai and co-workers [31], describing tensile elongations of more than 100% in a polycrystalline yttria-stabilized tetragonal zirconia, generally termed Y-TZP. As summarized in several recent reviews [32-34], superplasticity has now been documented in a number of different monolithic ceramics and ceramic composites but, nevertheless, most attention has been devoted to the Y-TZP system and there are reports of elongations of up to 800% in Y-TZP [35] or even as high as 1038% in a Y-TZP material containing 5 wt % of SiO_2 [36]. This section will therefore examine the current understanding of flow mechanisms in Y-TZP ceramics.

3.1 The Nature of the Grain Boundaries

It is now firmly established that the grain boundaries play a major role in the superplastic flow of polycrystalline materials. Furthermore, it is reasonable to anticipate that this role will be equally important in Y-TZP ceramics where the grain sizes are often at the submicrometer level.

Experiments have shown that Al_2O_3 and SiO_2 may segregate to the grain boundaries in Y-TZP materials, thereby forming an amorphous phase which becomes liquid at the elevated temperatures used for tensile testing. There are several reports which clearly establish the presence of a grain boundary phase in materials where the Al_2O_3 and/or SiO_2 contents are fairly high: for example, in 3Y-TZP containing 0.092 wt % Al_2O_3 and 0.011 wt % SiO_2 [37] and in 3Y-TZP containing 0.065 wt % Al_2O_3 and 0.002 wt % SiO_2 [38], where the number preceding Y denotes the mol % of Y_2O_3. By contrast, there is no evidence for any amorphous phase in 3Y-TZP samples containing 0.005 wt % Al_2O_3 and 0.002 wt % SiO_2 [35,39,40]. Figure 5 shows the microstructure in a representative high purity 3Y-TZP sample containing 0.005 wt % Al_2O_3 and 0.002 wt % SiO_2: the average grain size of this material was ~ 0.5 μm [40]. The grains visible in Fig. 5 are reasonably equiaxed, the grain boundaries are well-defined and it was shown using high resolution electron microscopy that the crystalline lattices of adjacent grains were continuous up to their common boundaries. However, there was experimental evidence for some yttria segregation to the grain boundary regions.

From these observations, it is apparent that there are two types of superplastic Y-TZP materials depending upon whether an amorphous phase is absent or present at the grain boundaries [40].

These two possibilities are illustrated schematically in Fig. 6, where Fig. 6(a) shows a boundary without an amorphous phase but with grain boundary dislocations and Fig. 6(b) shows a boundary with a thin liquid or amorphous phase and with the presence of steps at the interfaces between the liquid and the crystalline matrix [41]. These two types of boundary are considered separately in the following two sections.

3.2 The Mechanism of Superplasticity in Y-TZP Without a Grain Boundary Amorphous Phase

When tests are conducted on high purity Y-TZP samples where no amorphous phase is present at the grain boundaries, the stress exponent, n, is close to 3 over a limited range of strain rates in the vicinity of $\sim 10^{-4}$ s^{-1} [35,42-44]. However, when experiments are conducted over a wider range of testing conditions, there is evidence for a change in the value of n with an increase in either the stress level [45-47] or the grain size [46,48].

An experimental example is shown in Fig. 7 for 3Y-TZP tested at 1723 K with grain sizes of 0.41 and 1.3 μm [45,46]. For both of these grain sizes, there is evidence for a transition from $n \simeq 3.0$ to $n \simeq 2.0$ with increasing stress. Furthermore, inspection of Fig. 7 shows that the transition stress marking the change between the two regimes decreases when the grain size is increased: thus, the transition stresses are ~ 40 and ~ 10 MPa for the grain sizes of 0.41 and 1.3 μm, respectively. This trend reveals an important difference between Y-TZP ceramics and superplastic metals because the metals exhibit transition stresses between regions I and II which are essentially independent of the specimen grain size [49,50].

The tests conducted on Y-TZP samples involve very small grain sizes where $d < \lambda$. For example, the subgrain size in polycrystalline ceramics is given by the relationship [51]

$$\lambda \simeq \zeta b \left(\frac{\sigma}{G}\right)^{-1} \tag{4}$$

where ζ is a constant having a value of ~ 20. Taking $b = 3.6 \times 10^{-10}$ m and $G = 7 \times 10^4$ MPa for Y-TZP ceramics [52], and putting λ equal to the grain size, the transition stresses marking the condition of $d \simeq \lambda$ are ~ 1230 and ~ 390 MPa for the grain sizes used in Fig. 7 of 0.41 and 1.3 μm, respectively.

In the absence of an amorphous phase, the grain boundaries remain crystalline and therefore it is probable that deformation takes place by a mechanism similar to that occurring in superplastic metals. Thus, the results at the higher stress levels in Fig. 7, where $n \simeq 2.0$, probably represent the superplastic mechanism, and this is supported by experimental results in this region showing a dependence of strain rate on the reciprocal of the grain size raised to a power, p, of ~ 3 [46] and ~ 2.1 [48], respectively. At the lower stresses, however, where $n \simeq 2.0$, the value of p decreases to ~ 1 [46].

Figure 8 gives a schematic illustration of the predicted variation of strain rate with stress for two different grain sizes, d_1 and d_2 (where $d_1 < d_2$), where it is assumed that deformation at the higher stresses occurs by the superplastic mechanism as given by equation (3) and deformation at the lower stresses is due to impurity-controlled creep such as the accommodation of grain boundary sliding by a viscous drag process [53]. A stress exponent of $n = 3$ is characteristic of viscous glide [54] and, since the strain is accumulated by sliding at the boundaries, this process will lead to $p = 1$. In addition, it is anticipated that $D = \tilde{D}$ in the impurity-controlled regime, where $\tilde{D}$ is the diffusion coefficient for the appropriate solute which serves to exert a drag effect on the dislocations.

Although the precise nature of this impurity-controlled creep is not understood, it may arise from the presence of Y^{3+} cations which preferentially segregate to the boundary regions [35,40]. Inspection shows that the predictions of Fig. 8 are consistent with the experimental trends recorded in Fig. 7.

3.3 The Mechanism of Superplasticity in Y-TZP With a Grain Boundary Amorphous Phase

When lower purity Y-TZP samples are tested, containing an amorphous phase at the grain boundaries, there is a significant decrease in the value of the stress exponent, n [33]. An experimental example is shown in Fig. 9 for 2Y-TZP tested at 1673 K with grain sizes of 0.27 and 0.66 μm [55]. These results were obtained in tension and they show a transition with increasing stress from $n \simeq 2.5$ to $n \simeq 1.7$. Experiments conducted in compression on 2Y-TZP at 1673 K indicate a similar transition with increasing stress from $n \simeq 2.3$ to $n \simeq 1.3$, and these data also suggest values of p of ~ 1 and ~ 3 in the low and high stress regions, respectively [56].

If an extensive amorphous phase is present at the grain boundaries, superplastic deformation probably occurs through a diffusion-controlled solution-precipitation mechanism [41,57,58], where $n = 2$, $p = 1$ and $D = D_{liq}$, where D_{liq} is the diffusion coefficient in the liquid. An interphase reaction-controlled creep then becomes rate-controlling as the stress level is reduced so that, typically, $n = 2$, $p = 1$ and $D = D_{ad}$, where D_{ad} is the diffusion coefficient in the adsorption layer between the liquid film and the crystalline matrix. The predictions arising from these mechanisms are illustrated schematically in Fig. 10 for two different grain sizes [53]. Again, they are reasonably consistent with the experimental data.

4. Concluding Remarks

Superplastic flow in metals is now understood reasonably well. The mechanism of flow in the superplastic region II is grain boundary sliding accommodated by limited intragranular dislocation movement. Superplastic flow in yttria-stabilized tetragonal zirconia (Y-TZP) ceramics is more complicated because the behavior depends upon whether there is an amorphous phase at the grain boundaries. Possible deformation mechanisms have been developed for Y-TZP but more experimental data are needed to obtain a complete understanding of the flow characteristics.

Acknowledgement

This work was supported by the United States Department of Energy under Grant No. DE-FG03-92ER45472.

16 *Superplasticity: 60 Years after Pearson*

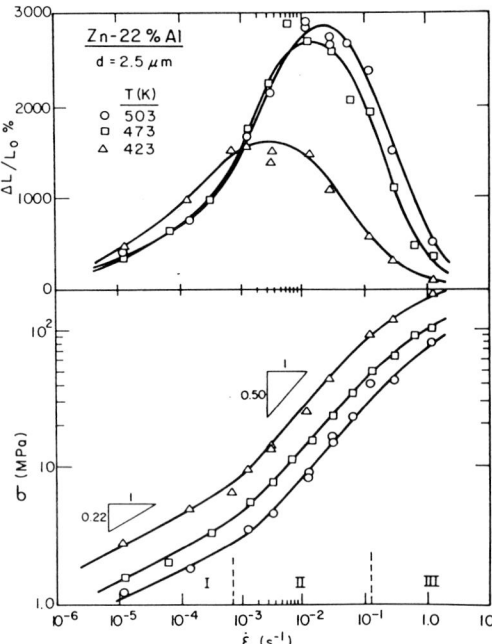

Fig. 1 Elongation to failure (upper) and flow stress (lower) versus imposed strain rate for Zn-22% Al over a range of temperatures [12].

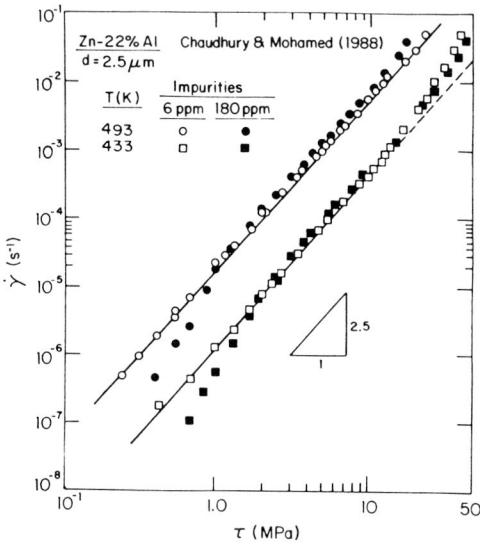

Fig. 2 Shear strain rate versus shear stress for Zn-22% Al showing the absence of region I in an alloy containing only 6 ppm of impurities [19].

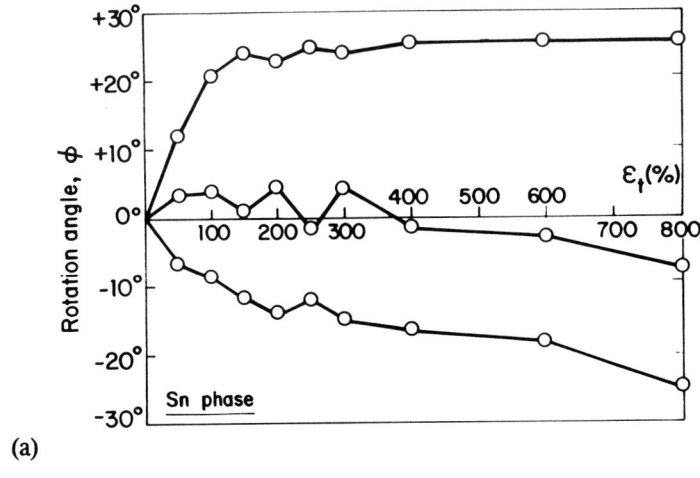

(a)

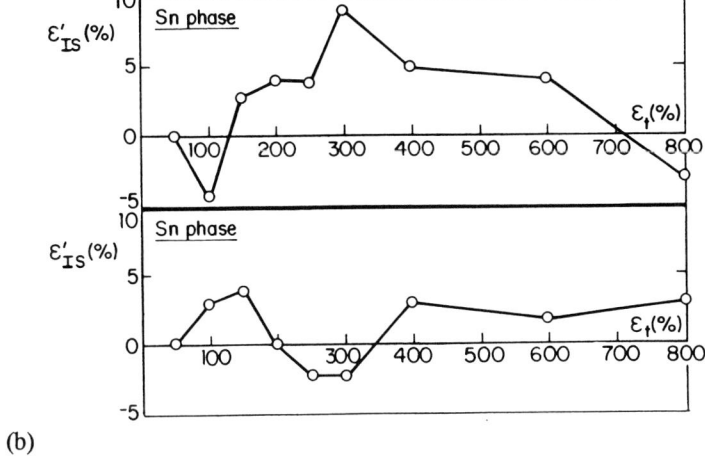

(b)

Fig. 3 (a) Rotation angle [22] and (b) intragranular strain [25] versus total strain for selected grains in a Pb-62% Sn alloy.

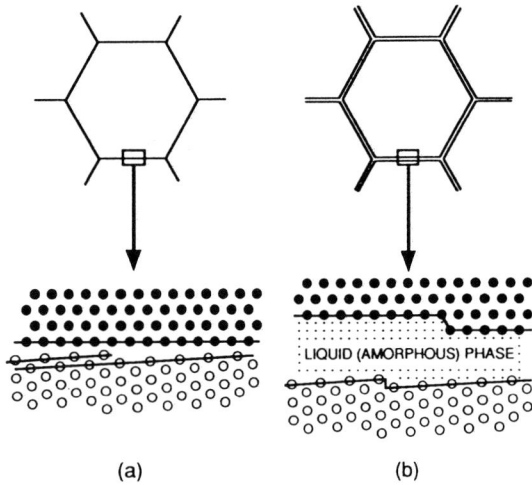

Fig. 6 Grain boundary structure (a) without and (b) with an amorphous boundary phase [41].

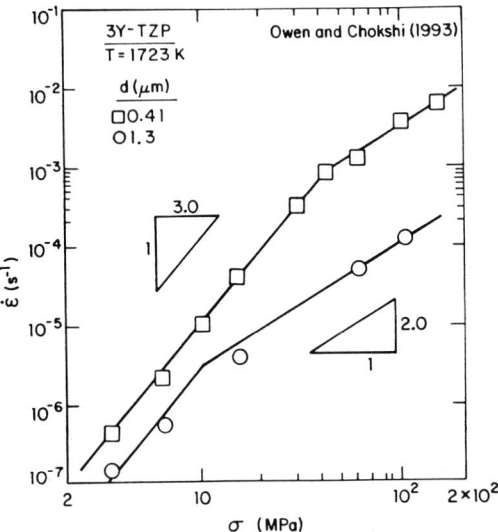

Fig. 7 Strain rate versus stress for 3Y-TZP without a grain boundary amorphous phase [45,46].

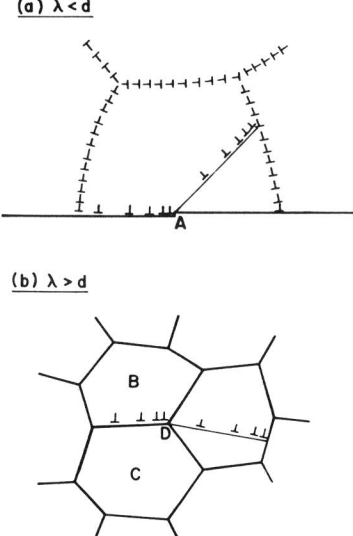

Fig. 4 Grain boundary sliding under conditions where the grain size, d, is (a) larger than or (b) smaller than the equilibrium subgrain size, λ [30]: condition (b) corresponds to deformation in the superplastic region II.

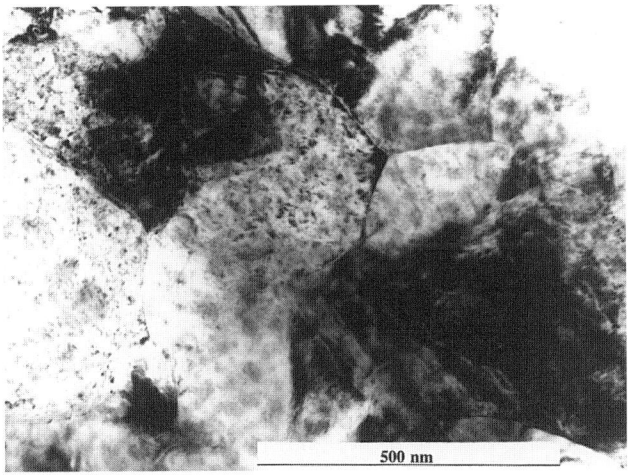

Fig. 5 Microstructure in a high purity 3Y-TZP sample [40].

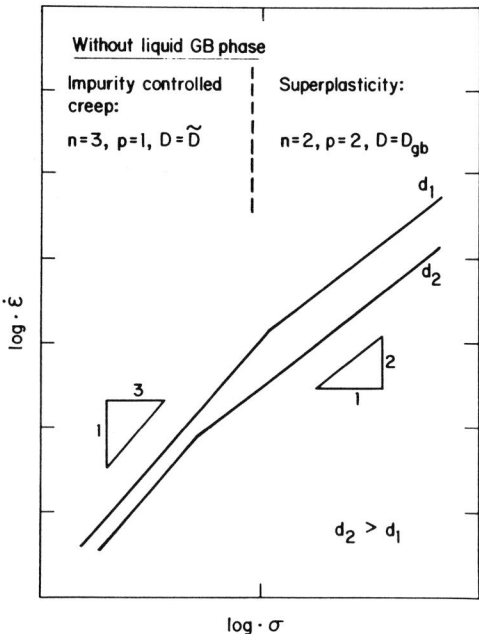

Fig. 8 Predicted variation of strain rate with stress when there is no liquid grain boundary phase [53].

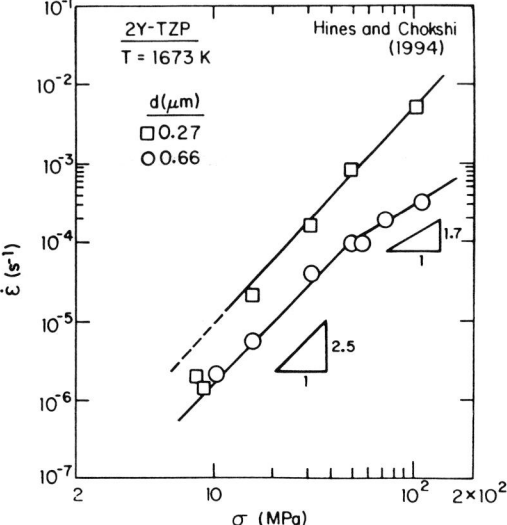

Fig. 9 Strain rate versus stress for 2Y-TZP with a grain boundary amorphous phase [55].

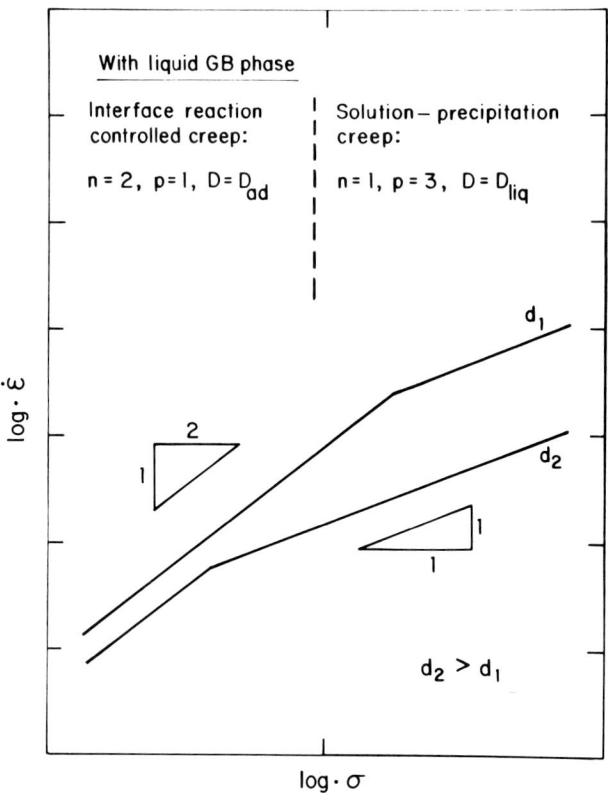

Fig. 10 Predicted variation of strain rate with stress when there is a liquid grain boundary phase [53].

References

1. T.G. Langdon and J. Wadsworth, in "Superplasticity in Advanced Materials" (Ed. S. Hori, M. Tokizane and N. Furushiro), 847; 1991. The Japan Society for Research on Superplasticity, Osaka, Japan.

2. Y. Ma and T.G. Langdon, Metall. Trans. 25A, 1994, 2309.

3. A.H. Chokshi, A.K. Mukherjee and T.G. Langdon, Mater. Sci. Eng. R10, 1993, 237.

4. O.D. Sherby and J. Wadsworth, Prog. Mater. Sci. 33, 1989, 169.

5. O.D. Sherby and J. Wadsworth, Mater. Res. Soc. Symp. Proc. 196, 1990, 3.

6. G.D. Bengough, J. Inst. Metals 7, 1912, 123.

7. C.E. Pearson, J. Inst. Metals 54, 1934, 111.

8. A.A. Bochvar and Z.A. Sviderskaya, Izvest. Akad. Nauk SSSR, Otdel. Tekh. Nauk (9), 1945, 821.

9. A.A. Bochvar and Z.A. Sviderskaya, Izvest. Akad. Nauk SSSR, Otdel. Tekh. Nauk (10), 1946, 1001.

10. A.A. Presnyakov, "Sverkhplastichnost' Metallov i Splavov," 1969, Nauka, Alma-Ata, U.S.S.R.: "Superplasticity of Metals and Alloys," English translation by C.B. Marinkov, 1976, The British Library, Wetherby, England.

11. I.I. Novikov, in "Superplasticity in Advanced Materials: ICSAM-94" (Ed. T.G. Langdon), 3; 1994. Trans Tech Publications, Aedermannsdorf, Switzerland.

12. F.A. Mohamed, M.M.I. Ahmed and T.G. Langdon, Metall. Trans. 8A, 1977, 933.

13. D.A. Woodford, Trans. ASM 62, 1969, 291.

14. T.G. Langdon, Metall. Trans. 13A, 1982, 689.

15. T.G. Langdon, in "Superplastic Forming of Structural Alloys" (Ed. N.E. Paton and C.H. Hamilton), 27; 1982. The Metallurgical Society of AIME, Warrendale, PA, U.S.A.

16. T.G. Langdon, Mater. Sci. Eng. A137, 1991, 1.

17. F.A. Mohamed and T.G. Langdon, Scripta Metall. 10, 1976, 759.

18. T.G. Langdon, Mater. Sci. Eng. A166, 1993, 67.

19. P.K. Chaudhury and F.A. Mohamed, Acta Metall. 36, 1988, 1099.

20. P.K. Chaudhury, V. Sivaramakrishnan and F.A. Mohamed, Metall. Trans. 19A, 1988, 2741.

21. S. Yan, J.C. Earthman and F.A. Mohamed, Phil. Mag. A 69, 1994, 1017.

22. A. Veveçka and T.G. Langdon, Mater. Sci. Eng. A187, 1994, 161.

23. S.-A. Shei and T.G. Langdon, J. Mater. Sci. 16, 1981, 2988.

24. L.K.L. Falk, P.R. Howell, G.L. Dunlop and T.G. Langdon, Acta Metall. 34, 1986, 1203.

25. R.Z. Valiev and T.G. Langdon, Acta Metall. Mater. 41, 1993, 949.

26. R.Z. Valiev and O.A. Kaibyshev, Acta Metall. 31, 1983, 2121.

27. T.G. Langdon, J. Mater. Sci. 16, 1981, 2613.

28. Z.-R. Lin, A.H. Chokshi and T.G. Langdon, J. Mater. Sci. 23, 1988, 2712.

29. T.G. Langdon, Mater. Sci. Eng. A174, 1994, 225.

30. T.G. Langdon, Acta Metall. Mater. 42, 1994, 2437.

31. F. Wakai, S. Sakaguchi and Y. Matsuno, Adv. Ceram. Mater. 1, 1986, 259.

32. Y. Maehara and T.G. Langdon, J. Mater. Sci. 25, 1990, 2275.

33. T.G. Langdon, in "Superplasticity in Aerospace II" (Ed. T.R. McNelley and H.C. Heikkenen), 3; 1990. The Minerals, Metals and Materials Society, Warrendale, PA, U.S.A.

34. T.G. Nieh and J. Wadsworth, in "Superplasticity in Advanced Materials: ICSAM-94" (Ed. T.G. Langdon), 359; 1994. Trans Tech Publications, Aedermannsdorf, Switzerland.

35. T.G. Nieh and J. Wadsworth, Acta Metall. Mater. 38, 1990, 1121.

36. K. Kajihara. Y. Yoshizawa and T. Sakuma, Scripta Metall. Mater. 28, 1993, 559.

37. T. Hermansson, H. Swan and G.L. Dunlop, in "Euro-Ceramics" (Ed. G. de With, R.A. Terpstra and R. Metselaar), 3, 3.329; 1989. Elsevier Applied Science, London, England.

38. T. Stoto, M. Nauer and C. Carry, J. Amer. Ceram. Soc. 74, 1991, 2615.

39. T.G. Nieh, D.L. Yaney and J. Wadsworth, Scripta Metall. 23, 1989, 2007.

40. S. Primdahl, A. Thölén and T.G. Langdon, Acta Metall. Mater. (in press).

41. F. Wakai. Acta Metall. Mater. 42, 1994, 1163.

42. T.G. Nieh, C.M. McNally and J. Wadsworth, Scripta Metall. 22, 1988, 1297.

43. M. Nauer and C. Carry, in "Euro-Ceramics" (Ed. G. de With, R.A. Terpstra and R. Metselaar), 3, 3.323; 1989. Elsevier Applied Science, London, England.

44. Y. Ma and T.G. Langdon, Acta Metall. Mater. 42, 1994, 2753.

45. D.M. Owen and A.H. Chokshi, in "Science and Technology of Zirconia" (Ed. S.P.S. Badwal, M.J. Bannister and R.H.J. Hannink), 432; 1993. Technomic Publishing, Lancaster, PA, U.S.A.

46. A.H. Chokshi, Mater. Sci. Eng. A166, 1993, 119.

47. A. Bravo-León, M. Jiménez-Melendo, A. Domínguez-Rodríguez and A.H. Chokshi, in "Third Euro-Ceramics" (Ed. P. Durán and J.F. Fernández), 3, 877; 1993. Faenza Editrice Ibérica, Castellón de la Plana, Spain.

48. J. Ye, A. Domínguez-Rodríguez, R.E. Medrano and O.A. Ruano, in "Third Euro-Ceramics" (Ed. P. Durán and J.F. Fernández), 3, 525; 1993. Faenza Editrice Ibérica, Castellón de la Plana, Spain.

49. F.A. Mohamed and T.G. Langdon, Acta Metall. 23, 1975, 117.

50. F.A. Mohamed and T.G. Langdon, Phil. Mag. 32, 1975, 697.

51. W.R. Cannon and T.G. Langdon, J. Mater. Sci. 23, 1988, 1.

52. A. Bravo-León, M. Jiménez-Melendo and A. Domínguez-Rodríguez, Acta Metall. Mater. 40, 1992, 2717.

53. T.G. Langdon, in "Plastic Deformation of Ceramics" (in press).

54. F.A. Mohamed and T.G. Langdon, Acta Metall. 22, 1974, 779.

55. J.A. Hines and A.H. Chokshi, in "Superplasticity in Advanced Materials: ICSAM-94" (Ed. T.G. Langdon), 421; 1994. Trans Tech Publications, Aedermannsdorf, Switzerland.

56. M. Nauer and C. Carry, Scripta Metall. 24, 1990, 1459.

57. R. Raj and C.H. Chyung, Acta Metall. 29, 1981, 159.

58. R. Raj, J. Geophys. Res. 87, 1982, 4731.

Grain Boundary Sliding - the Main Mechanism of Superplastic Flow

O. A. Kaibyshev

Institute for Metals Superplasticity Problems
Russian Academy of Sciences Khalturina 39, Ufa, 450001, Russia.
Telephone: (3472) 253717, Telex: 412543 Super SU;
E-mail: vast@ipsm.bashkiria.su

Abstract

Since the publication of the fundamental paper of Pearson it has become clear that Grain Boundary Sliding (GBS) is the main deformation mechanism of superplastic (SP) flow. However, the nature of GBS itself has not been studied. The results of studying the nature of GBS in zinc bicrystals are presented. The most significant result of these investigations is establishing both the existence of stimulated GBS and considerable GBS acceleration on the interaction of dislocations with grain boundaries. Hence it follows that grain boundary structure affects significantly the development of GBS.

In a real polycrystal, GBS takes place in a localized form along the common surface of many grains - cooperative grain boundary sliding. The evidence for the operation of this mechanism of SP flow and for its interaction with the changes of the mechanical properties are given.

Introduction

In the classic work of Pearson [1] two basic facts have been established: the possibility of obtaining enormous deformations reaching 2000%; grain shape change proved to be insignificant. It followed from these facts that the specimen deformation as a whole and the deformation of separate grains were in apparent contradiction: though the specimen elongation was thousands of percent, the grains remained practically equiaxed. Pearson suggested that the deformation was due to the development of grain boundary sliding (GBS). It is clear that the nature of superplastic deformation (SPD) can not be understood without a systematic study of GBS.

Models of GBS development of superplastic (SP) deformation have been proposed [2-6]. It was assumed that for the deformation realization, GBS must be accompanied by grain accommodation due to either diffusion creep or dislocation slip. The hypothetical models made it possible to obtain good enough agreement with the experiment; however, similar results were derived from quite different physical concepts.

All the above observations result in the need to conduct systematic experimental investigations of the nature of GBS.

The aim of the present review is the systematic study of the nature of GBS in different materials [3-6]

1. Stimulated Grain Boundary Sliding

Important results in understanding the nature of GBS have been obtained during experiments on zinc bicrystals [7,8]. The choice of this metal as a model material was due to the following reasons:

Room temperature for zinc is 0.43 T_{melt}, which corresponds to the temperature of transition to the SP state.

In zinc, slip along the basal plane at room temperature is characterized by considerably lower

τ_{cr} than for slip along non-basal planes, which allows control of the influence of dislocation slip on GBS development.

In Zn bicrystals high-angle boundaries of a general type can be easily obtained, these boundaries being characteristic of fine grained materials. The zinc bicrystal was grown so that a high-angle 90° boundary with $\sum > 25$, i.e. a random-type boundary could be obtained in it. Bicrystals of two types were cut from this large bicrystal.

In type I bicrystal the boundary plane is orientated at an angle of 45° to applied stresses and, therefore, the shear stresses are maximum (Fig.1). In this case the shear stresses for slip are equal to 0 for both the crystals. One can easily guess that for type I bicrystals the sliding must develop in a "pure" form, i.e. under the influence of applied stresses while intragranular slip is absent.

In type II bicrystals, dislocation slip in the basal plane of both crystals is possible, the operating shear stresses in the boundary plane being significant enough though not having maximum values, i.e. in this case the sliding occurs under the conditions when the development of intragranular slip in both the crystals is possible.

The operation of GBS in two types of bicrystals proved to be significantly different. The rate of sliding accompanied by intragranular slip increases nearly 50 times as compared with "pure" sliding; that is why it is called stimulated GBS (Fig.2). Such a considerable difference in the GBS development in bicrystals with one and the same boundary but with various possibilities of intragranular slip development is unambiguously indicative of the interrelation of these processes. At the micro-level it seems to be connected with the interaction of lattice dislocations with the grain boundary (GB) [9].

A lattice dislocation entering a GB is known to relax at definite temperatures, not only the dislocation itself but also the boundary structure changing in this case. Evidently, lattice dislocations in a GB dissociate into grain boundary dislocations with small Burgers vector. The motion of these excess (relative to equilibrium concentration) grain boundary dislocations seems to result in GBS acceleration and the development of stimulated GBS.

The question arises: which type of GBS predominantly operates under SPD conditions and controls the flow stress?

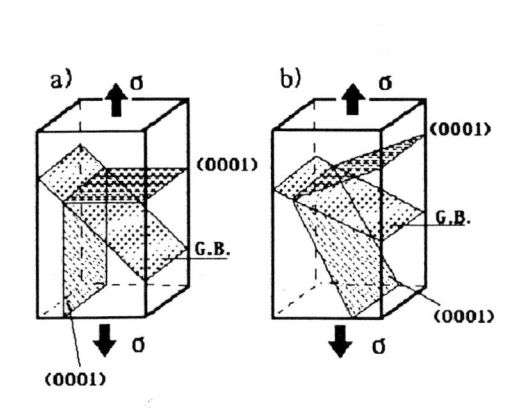

Fig.1 Two types of the zinc bicrystals with 90° tilt boundary: a) for studying "pure" GBS; b) for studying GBS concurrent with intragranular slip (stimulated GBS).

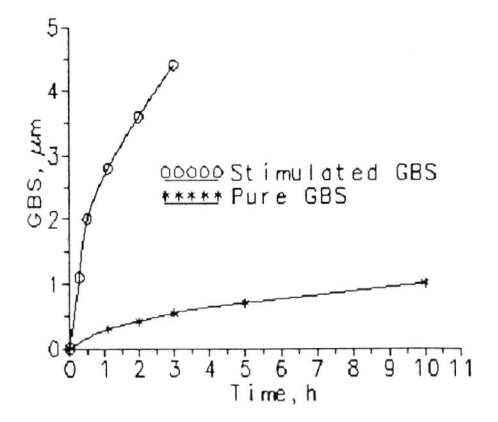

Fig.2 Sliding vs time curves for bicrystals of types a) and b) during testing under similar condition. $\tau=1.4$ MPa.

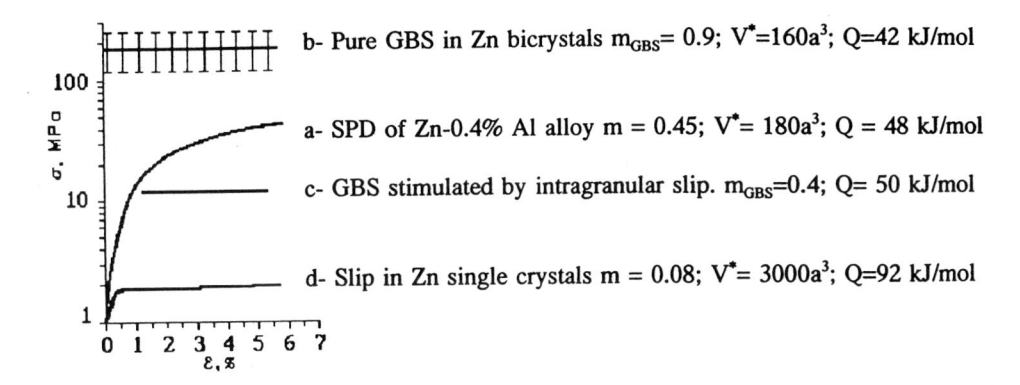

```
100 ┤                  b- Pure GBS in Zn bicrystals $m_{GBS}= 0.9$; $V^*=160a^3$; $Q=42$ kJ/mol

 σ, MPa              a- SPD of Zn-0.4% Al alloy m = 0.45; $V^*= 180a^3$; $Q = 48$ kJ/mol

 10                  c- GBS stimulated by intragranular slip. $m_{GBS}=0.4$; $Q= 50$ kJ/mol

                     d- Slip in Zn single crystals m = 0.08; $V^*= 3000a^3$; $Q=92$ kJ/mol
  1
     0 1 2 3 4 5 6 7
          ε, %
```

Fig.3 Scheme illustrating the ratio of stress levels and m values providing a GBS rate of 6×10^{-10} m/s realized during SPD of Zn-0.4%Al alloy (a) and of bicrystals (b and c).

To answer this question let us correlate the flow stress during the SPD of the Zn-0.4%Al alloy at room temperature and the flow stress necessary for the development of intragranular slip and different types of GBS under the same conditions. At the same time we shall correlate the thermoactivation parameters of these deformation processes (Fig.3). One can see that the thermoactivation parameters of SPD of the polycrystalline alloy Zn-0.4%Al are close to those of stimulated GBS and differ considerably from the corresponding parameters characteristic of "pure" GBS and dislocation slip.

Hence we can conclude that the deformation micromechanism controlling the flow stress during SPD is stimulated grain boundary sliding, while the interaction of lattice dislocations with the GB is the process determining the possibility of the development of this deformation mechanism. The data on the change of the flow stress at changing strain rates [6] testify in favour of this conclusion. Under usual conditions the increase of the strain rate results in an increase of the flow stress while under SPD conditions it leads, on the contrary, to a decrease of the flow stress.

This fact can be understood if we assume that stimulated GBS is the main mechanism of SPD. The increase of the density of lattice dislocations entering a GB at the instantaneous increase of the strain rate promotes the development of the stimulated GBS and reduces the flow stress. Under usual conditions the instantaneous increase of the density of lattice dislocations results, on the contrary, in strain hardening [6].

Special experiments were conducted to study the interaction of lattice dislocations with grain boundaries [9].

The dissociation (spreading) of lattice dislocations in a GB proved to be dependent on temperature and GB structure. Dislocation spreading has been studied in a number of materials including the intermetallic TiAl. After cold deformation of the alloy with d= 10μm to ε=2.5% at room temperature, practically all dislocations are GB. On heating the foil in the microscope column one can see the disappearance of the contrast of lattice dislocations, while defects in the GB structure remain. This effect is likely to be connected with the dissociation of lattice dislocations into a complex of excess grain boundary dislocations. The possible model of the relaxation of lattice dislocations is described in [9].

Excess grain boundary dislocations can result in GBS acceleration, i.e. to the development of stimulated GBS. It is very important to determine the spreading temperature, t_s, and to correlate it with the temperature of the transition to the SP state. The example of determining the t_s is given in Fig.4. The usual transition to SP flow is observed at $t > t_s$. The spreading temperature differs considerably for special GB's and GB's of the general type. For special boundaries the spreading temperature is significantly higher [6].

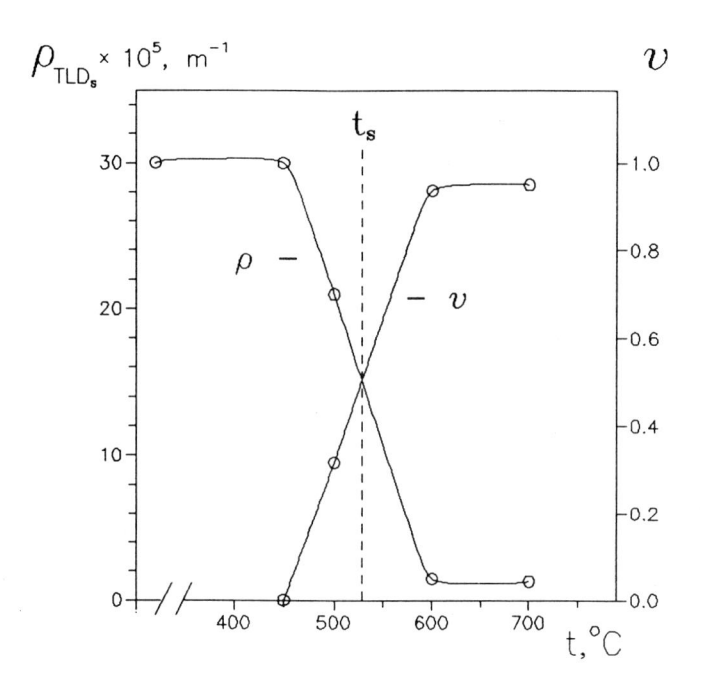

Fig.4 Dependence of trapped lattice dislocations density (ρ) and the proportion of relaxed grains (υ) on annealing temperature (t) in the intermetallic TiAl.

Hence, the following experimental result becomes clear - in a Mg alloy grain refinement was realized due to the introduction of deformation twins. The grain size in this case was reduced 3 times from 10 μm to 3 μm. However, the SP characteristics did not increase; on the contrary, σ increased while the parameters δ and m were reduced.

These results can be understood only if we assume that t_s for special GB's is higher than the temperature of SP deformation and the development of stimulated GBS along these boundaries becomes impossible.

The interconnection of the temperature of spreading with that of transition into the SP state is most conveniently revealed in the intermetallic TiAl [10,11].

The peculiarity of this material is that it contains a considerable proportion of special GB's of the twin type in the structure. The refinement of this material up to d=0.4μm results not only in the decrease of the grain size but also in the rise in the fraction of general type GB's. The temperature of the transition into the SP state is sharply reduced in this case. The independent measurement of t_s using electron microscopic data shows that the transition to SP flow is observed at $t_1 > t_s$.

Thus, stimulated GBS is the main micromechanism of SP flow.

To explain the nature of SPD many hypotheses have been suggested. But the true theory must account not only for the SPD phenomenology but also the structural changes which take place during this deformation. Moreover, the physical model must be based on the experimentally confirmed facts: the existence of stimulated GBS and the decisive influence of the interaction of dislocations with GB's during SPD.

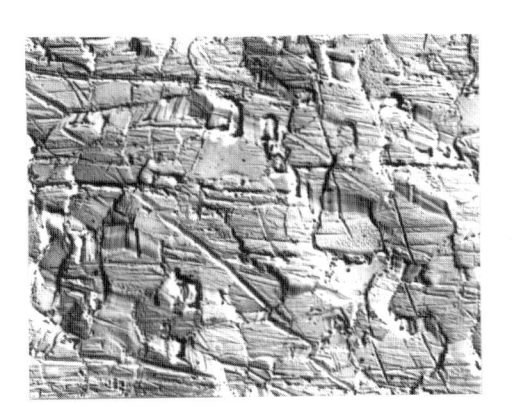

Fig.5 Bands of local deformation during SPD. Deformation relief on the surface of Zn-22%Al alloy. Replica, x 8000.

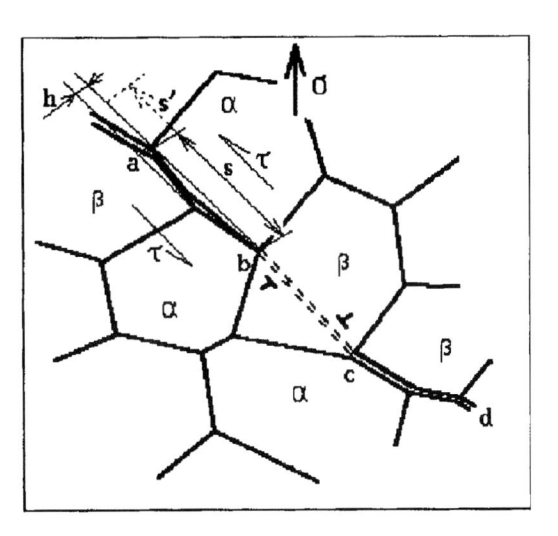

Fig.6 Scheme of cooperative grain boundary sliding (CGBS) interface development in the initial stage ($\varepsilon=3\%$) of SP deformation Zn-22%Al alloy.

Only two of all the suggested models meet these requirements: the model proposed by us [12] in 1984 in which we assume that stimulated GBS is the result of the motion of grain boundary dislocations, and the model suggested by Perevezentsev [13], in which stimulated GBS is considered to be the result of the change of grain boundary structure and the formation of the liquid-like state. At present it is difficult to give preference to either of the two models of the micromechanism of SP deformation.

At the same time, it is quite clear already that even knowing the main deformation micromechanism controlling SP flow, one cannot explain some experimental facts. In particular, it is not quite clear how SP deformation develops at the macroscale. For example, it was demonstrated that in the process of SPD some grains can shift relative to neighbouring grains for considerable distances without changing their shape.

Some macroscopic models [14] of the grain displacement are given. We can single out two groups of models. In the models of the first group the displacement of neighbouring grains and the possibility of operation of different accommodation processes are considered. In other models the macroscopic displacement of grains is presented as the result of the operation of cooperative shear processes. However, these mechanical models treat SP flow as the result of successive macroscopic shears whose physical sense and micromechanism are still vague.

2. Cooperative Grain Boundary Sliding

The importance of macroscopic processes for understanding the SPD nature is clear at least from the fact that mechanical properties and the ultimate deformation of specimens depend on their geometry [15]. The data show that the extent of the stage I of deformation, the maximum elongation to failure and the m value depend on the specimen thickness. When the specimen thickness is reduced from 2 to 0.06mm, δ decreases from 540% to 57% and m from 0.45 to 0.26, while the grain size and temperature-rate deformation conditions remain the same.

These new data demonstrate that in order to understand the nature of SPD special experiments on the study of the mechanism of SP deformation at the meso-level must be conducted.

A number of special techniques have been used to solve this experimental task [16,17].

Fig.5 shows the microstructure of the specimen surface after deformation. The bands of localized deformation can be seen. Their distribution throughout the bulk has been fixed as well. The nature of the bands of localized deformation is of interest - it is the result of the cooperative grain boundary sliding of one group of grains relative to another over some common surface.

The electron micrograph shows an example of the deformation transmission from one pair of grains to another, as well as the passage of the shear through the grain whose boundary arrangement is unfavourable for GBS realization (Fig.6).

It is evident that the transmission of GBS from one pair of grains to another is easily realized in the region where the planes of neighbouring grains have close space orientation throughout the specimen bulk and are orientated along τ_{max}. In the case of the more complex relief appearing in the path of the developing band there arises the need for increasing the stress for initiating additional accommodation mechanisms. The simultaneous shear along several boundaries allows concentration of the stress in the locking point (b). The concentration of stresses can cause the generation of dislocations in the b-c region, grain rotation or intensive diffusion flow which provide further accommodation. On the other hand, lattice dislocations can stimulate GBS in the next region (c-d), i.e. the continuation of the band a-b-c-d in which the shear occurs along α-α, α-β, β-β boundaries and crystallographic planes of several grains simultaneously. Separate regions of localized deformation are united into a band of intensive deformation of the Luders type. In the process of elongation of the specimen there appear new analogous bands which in the case of the optimum strain rate are arranged on the specimen surface at an angle close to 45° to the extension axis. While connecting with each other the bands form the net which separates the whole bulk of the material into groups of grains.

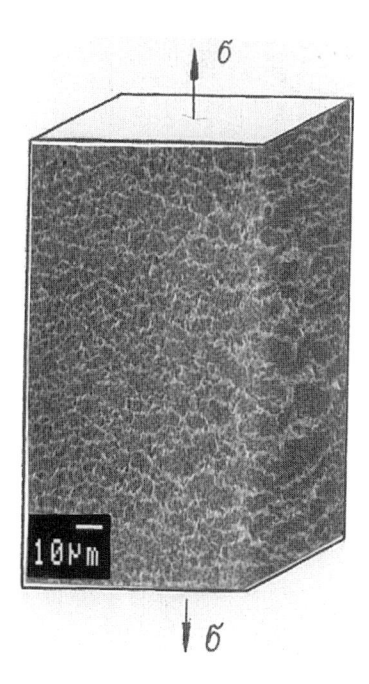

Fig.7 Bands of local deformation in the bulk specimen after 30% SPD Zn-22%Al.

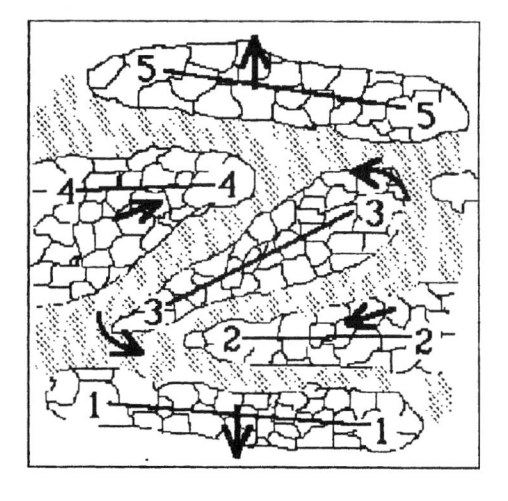

Fig.8 The scheme of the grain groups displacement during SP deformation.

The observation of the specimen front and side planes allows to see that the bands are traces of spatially tilted planes of localized deformation. In this case groups of grains separated by bands of localized deformation as volume conglomerates are clearly seen at the edge (Fig.7).

Let us follow the interrelation of mechanical properties with the appearance of bands of localized deformation on the alloy Zn-22%Al. In the diagrams of the extension of this alloy as well as in the case of most SP materials, two regions are clearly seen: 1 - where σ sharply depends on ϵ (the initial stage of extension) and 2 - where σ - ϵ dependence is weak - the stage of steady-state flow. The degree of deformation and the stress for the transition from stage I to stage II depend on the strain rate. At low strain rates the transition takes place at $\epsilon=2\%$ and $\sigma=1$ MPa. At increasing strain rate both values increase as well.

It turns out that the mechanical behaviour of the material correlates with the peculiarities of the formation of the deformation relief. At $\epsilon=3\%$, along grain boundaries there begin to appear thin (0.1μm), winding deformation bands 15-20 μm long at a distance of 5-8 μm from each other. The bands continuing one after another are arranged across the specimen from one edge to another. Hence, it is demonstrated that the transition from region I of deformation to the region of steady-state flow is connected with formation of localized deformation bands.

The change of the character of σ-ϵ curves depending on strain rate correlates as well with the appearance of deformation bands: the higher the strain rate, the higher is the density of deformation bands and the fewer is the number of grains in groups separated by bands of localized deformation. Experimentally, we managed not only to show the shear deformation along the bands of cooperative GBS but also to evaluate quantitatively the contribution of this process to the total deformation.

In the region ϵ_{opt}, up to 70 - 90% of the total deformation is realized due to cooperative GBS.

Thus, the appearance of localized deformation bands on SPD is closely connected with the change of mechanical properties, and there is the interrelation between these physical values.

As a result, one more specific peculiarity of GBS on SPD can be established - at the meso-level this deformation is realized as cooperative GBS.

The technique of precision observation of one and the same area of the specimen surface before deformation and after it, made it possible to obtain more detailed information concerning the displacement of grain groups and other structural changes during SPD, including the data on the redistribution of grains in the bulk, i.e. the increase of their number in the extension direction and the decrease in the transverse one; both translational and rotational components of the motion of grain conglomerates have been revealed.

Fig.8, which is reproduced from an experimental micrograph demonstrates that groups 1-1 and 5-5 move apart from each other. Groups 2-2 and 4-4 are separated in the extension direction and at the same time approach each other in the transverse direction and are turned at an angle of about 4°. Due to such motion, the groups are arranged along the extension axis. At last, the central group 3-3 mainly performs rotational movements with the orientation in the direction of the extension axis. This rather complicated motion of grain conglomerates is realized by the simultaneous development of plastic flow along the deformation bands of different orientations.

Conclusions

1. Grain boundary sliding can be an independent deformation mechanism ('Pure GBS') and can be stimulated by intragranular slip. The parameters of stimulated GBS meet the requirements of SP flow best of all.

2. During SPD the interaction of lattice dislocations with grain boundaries plays a significant role. SPD is realized at a temperature not lower than the spreading temperature of trapped dislocations.

3. Under SP conditions, GBS is a cooperative process and develops along shear bands. The correlation between the change of mechanical properties of SP materials and the development of cooperative GBS has been shown.

4. Large plastic deformations on SPD are realized due to cooperative GBS which form bands of the Luders type, the band structure being complex: the main element is cooperative GBS accompanied by intragranular slip, GB migration and local diffusion processes.

5. The model of displacement of groups of grains at SPD has been experimentally obtained for a real material.

References

1. C.E. Pearson, J.Inst.Metals 54, 1934, 111-124.

2. J.W. Edington, K.N. Melton, C.P.Cutler, Progr. in Mater. Sci., 21, 1976, 61-158.

3. O.A. Kaibyshev, Plastichnost' i Sverkhplastichnost Metallov, Metallurgia, Moscow (1975).

4. O.A. Kaibyshev, Sverkhplastichnost' Promyshlennykh Splavov, Metallurgia, Moscow (1984).

5. J. Pilling and N. Ridley, Superplasticity in Crystalline Solids, The Institute of Metals (1989).

6. O. A.Kaibyshev, SuperplasticityofAlloys, Intermetallides, and Ceramics, Springer-Verlag, Berlin (1992).

7. O.A. Kaibyshev, V.V. Astanin, R.Z. Valiev, Dokl.Akad.Nauk, 245, 1979, 1356-1358.

8. R.Z. Valiev, O.A. Kaibyshev, V.V. Astanin, and A.K.Emaletdinov, Phys.Stat.Sol. A78, 1983, 439.

9. O.A. Kaibyshev, R.Z. Valiev, Granitsy Zeren i Svoistva Metallov, Metallurgia, Moscow (1987).

10. R.M. Imayev, O.A. Kaibyshev and G.A. Salishchev, Acta Metall. Mater., 40, 1992, 589-595.

11. R.M. Imayev, O.A. Kaibyshev and G.A. Salishchev, Acta Metall. Mater., 40, 1992, 581-587.

12. O.A. Kaibyshev, R.Z. Valiev, A.K. Emaletdinov, Phys. Stat. Sol. (a), 90, 1985, 197-206.

13. V.N. Perevezentsev, V.V. Rybin, and V.N. Chuvil'deev, Acta Metall.Mater., 40, 1992, 887-894.

14 M.G. Zelin, M.R. Dunlop, R. Rosen, A.K. Mukherjee, J.Appl. Phys., 74, 8 October (1993).

15. V.V. Astanin, K.A. Padmanabhan, and S.S. Bhattacharya, "A Model for Grain Boundary Sliding and its Relevance to Optimal Structural Superplasticity: III The Effects of Localization and Specimen Thickness on Superplasticity in Alloy Supral 100" (to be published).

16. V.V. Astanin, O.A. Kaibyshev, S.N. Faizova, Scr.Met., 25, 1991, 2663-2668.

17. V.V. Astanin, O.A. Kaibyshev, and S.N. Faizova, Acta Metall. Mater., 42, 1994, 2617-2622.

Grain Boundary Structure and Superplastic Behaviour

R. I. Todd

Manchester Materials Science Centre, University of Manchester & UMIST,
Grosvenor Street, Manchester, M1 7HS, UK.

Abstract

The literature concerning the structure of grain and phase boundaries is briefly reviewed, and the important predictions of related theories are stated. The effect of applying a stress to the network of intrinsic secondary grain boundary dislocations (SGBDs) which pervades the grain boundaries of a polycrystal is considered. It is shown that although these dislocations can cross triple points by dissociating into SGBDs in the boundaries beyond, the dissociation products generally contain extrinsic components with a long range stress field. The need to supply the associated strain energy gives rise to a back stress, which can be reduced by various recovery mechanisms, and the strain rate at any time is governed by the climb rate of the SGBDs under the influence of the effective stress. The predictions of this model are compared with the microstructural and mechanical characteristics of superplastic flow, with the conclusion that it is consistent with the whole range of behaviour, including loading transients, internal stress, anelasticity, grain rotation, strain-induced grain growth, grooved and thickened grain boundaries, and intermittent and 'co-operative' grain boundary sliding.

1. Introduction

It is well known that the large tensile elongations which give superplasticity its name stem primarily from the high value of the strain rate sensitivity index, m, in the empirical relationship between the applied stress, σ_a, and the resulting steady state strain rate:

$$\sigma_a \propto \dot{\epsilon}_{ss}^m \qquad (1)$$

The technological importance of this relationship has led to it being seen as the most important test for models of superplasticity, despite the fact that the large number of distinct models which apparently give the correct answer demonstrates that it is not a particularly exacting criterion for success. Fortunately, careful experimental investigations have provided many additional observations, some mechanical (e.g. loading transients, internal stress and anelasticity) and some microstructural (e.g. grain rotation, strain induced grain growth, curved and "thickened" grain boundaries). These are just as important as eq. 1 in testing models in that to be deemed successful, a model must be simultaneously consistent with *all* observations.

Although the importance of lattice dislocation motion in superplasticity has long been a controversial subject, it is almost universally agreed that superplastic flow does not produce many extended configurations of lattice dislocations. When coupled with the observation that a great deal of grain boundary sliding takes place during deformation, it is clear that the key to understanding superplasticity lies in the nature of grain boundaries. A substantial literature on grain boundary structure has built up over the last 25 years or so, yet few of the detailed results have been applied to superplasticity. A summary of the important results of this work is given in the next section. The aim of this paper is to begin to rectify this situation by considering the consequences of applying a stress at high temperature to a well annealed polycrystal whose boundaries have structures in accordance with the predictions of modern grain boundary theory. It is shown that this simple idea leads naturally to many of the mechanical and microstructural attributes associated with superplastic flow.

2. Structure of Grain Boundaries

If the energy of a grain boundary is plotted as a function of the misorientation of the grains which it separates, it is found that some orientations are special in that they allow boundaries of particularly low energy to form. These special orientations give periodic relationships between the lattices of the two grains [1]. This is obvious in retrospect, since for a structure to be special, it must repeat itself, otherwise it cannot be said to possess any particular structure. The low energy structures are so favourable that for grain boundaries separating grains rotated away from the special orientation, most of the boundary retains the low energy structure, and all the extra mismatch becomes localised as intrinsic secondary grain boundary dislocations (SGBD). These are *intrinsic* because they describe the misorientation away from the low energy position in the same way that lattice dislocations accommodate the misorientation across a low angle boundary [2,3]. The low energy structure must be recreated when shifted by the Burgers vectors of the SGBDs, which means that they belong to the lattice made up of vectors joining atoms in one grain to atoms in the other if the grains are assumed to interpenetrate rather than ending at the boundary, and when their relative translation is such that points exist at which an atom from each grain is situated [4]. This lattice is known as the displacement shift complete lattice (DSCL) and generally permits Burgers vectors which are smaller than those of lattice dislocations (hence *secondary*). Because they describe the misorientation of the grains, intrinsic SGBDs do not have long range strain fields. Extrinsic secondary grain boundary dislocations, which do not accommodate the misorientation of the grains also exist. These do have long range strain fields, and must also have DSCL Burgers vectors.

The approach to superplasticity described in this paper is based on this geometrical view of the structure of grain boundaries, and arises from the properties of the DSCL [5]. Subsequent developments exist, notably the structural unit model [6,7], which are more predictive of the details of the grain boundary structure on an atomic scale, and take into account which special structure provides the basis for the boundary. These are nevertheless fully consistent with the ideas of the SGBD model, and can be rationalised in terms of intrinsic SGBDs. The refinements which they introduce do not substantially affect the treatment presented here. Although there is an abundance of experimental verification of the SGBD model for boundaries which can be imaged fully using TEM and HREM (e.g. [8-10]), experimental difficulties have so far precluded its corroboration for the most general grain boundaries of mixed tilt and twist character with high index rotation axes. It is assumed here that general boundaries do conform to these ideas, on the basis that there is no evidence to the contrary. At the very least, a substantial proportion of boundaries certainly do possess this structure. The same theory can be applied to phase boundaries.

3. Consequences of the SGBD Model

1. Because the Burgers vectors of SGBDs are smaller than those of lattice dislocations, SGBDs are confined to the grain boundary. The DSCL contains the lattices of both grains separated by the boundary as superlattices, however, so SGBDs can always undergo reactions which produce lattice dislocations, either by association of the required combination of SGBDs or by dissociation to form a lattice dislocation plus remanent SGBDs. It should be noted that such reactions tend to be unfavourable by the Frank b^2 criterion, and therefore require stress concentrations or thermal activation. Nevertheless, the emission of lattice dislocations by grain boundaries is an experimental fact [11].

2. The DSCL is determined by the relative orientations of the two grains associated with the boundary alone, and not by the orientation of the grain boundary. In general, therefore, the Burgers vectors of SGBDs do not lie in the boundary plane, so the SGBDs are not glissile. They must move by a combination of glide and climb, and in general cause grain boundary sliding and plating or dissolution of material. Further, because many SGBDs are associated with a step in the grain boundary, grain

boundary migration also results from SGBD motion [12].

3. The DSCL is a function of the relative orientations of the grains associated with a boundary, so in general SGBDs cannot simply move unchanged across a triple point. It can be shown, however, that a SGBD in one boundary can always dissociate into integral numbers of SGBDs in the two boundaries beyond the triple point. Such dissociations may be 'easy' or 'difficult' by the Frank b^2 criterion.

4. Deformation by SGBD Movement

Imagine a polycrystal which is well annealed so that its grain boundaries contain a great many intrinsic SGBDs, but few extrinsic SGBDs. When a stress is applied, the SGBDs feel a force and begin to move. If intrinsic SGBDs could be removed and supplied as required at the triple points marking the edges of the grain boundaries, deformation could continue indefinitely, but this is not possible in reality. Although the SGBDs can leave their initial grain boundary by dissociating across a triple point, it is easy to show by counter-example that the dissociation products do not generally represent purely intrinsic SGBDs in the boundaries in which they are situated (fig. 1). The extrinsic components of the products have long range stress fields, and the elastic strain energy required must be supplied by the applied stress, σ_a. This leaves less power to drive the deformation mechanism (viz diffusion controlled climb/glide of the SGBD network) and the deformation process slows down. This can be viewed in conventional terms as the development of an internal back stress, σ_b. Its magnitude can be estimated by assuming that the strain field of the total extrinsic SGBD content when the intrinsic SGBDs in the three boundaries meeting at a triple point have moved a distance equal to nd, is the same as that of a superdislocation with Burgers vector nb_e, where d is an average equilibrium spacing of intrinsic dislocations and b_e is the extrinsic content produced every time the intrinsic network moves up one average spacing. If the long range strain energy is E and the shear modulus is μ then:

$$E \sim \mu (nb_e)^2$$
$$\therefore \frac{dE}{dn} \sim 2\mu n b_e^2$$

The back stress is the stress which supplies the same power as that required to create the extrinsic SGBDs:

$$3\sigma_b L b_i \approx 2\mu n b_e^2$$
$$\therefore \sigma_b \approx \frac{2\mu n b_e^2}{3 L b_i} \qquad (2)$$

in which L is the length of a grain boundary and b_i the average component of the intrinsic dislocations acted on by the applied stress. Equation 2 shows that σ_b increases with n, so that a recovery mechanism is required for sustained flow. Four possible recovery mechanisms are considered here:

A. Grain Boundary Migration. By changing its orientation with respect to the grains which it separates, a boundary can reduce the long range stress field of the extrinsic components which it contains (fig. 2). This process closely resembles polygonisation in conventionally work hardened metals. The mechanism can operate at very low stresses, but in general cannot completely remove the strain energy of the extrinsic SGBDs.

B. Mutual Repulsion of the SGBDs. The extrinsic SGBDs accumulated near a triple point repel one another so that they move off down the boundary (fig. 3). Since they have effectively random Burgers vectors they are eventually annihilated. Modelling of this mechanism suggests that it is too slow to explain superplastic results in region II, and is probably inhibited because the extrinsic SGBDs are not free to move independently because of interactions with the intrinsic grain boundary dislocation

C. An Alternative Dissociation Occurs. When a SGBD crosses a triple point an infinitely large number of dissociations is possible, although most involve unrealistically large numbers of product SGBDs. Nevertheless, it is conceivable that an alternative dissociation to that which is initially preferred may take place, which produces extrinsic components in the opposite sense to those resulting from the initial dissociation (fig. 4). This may be particularly favourable where there are two reactions fulfilling this criterion of closely matched favourability. In this case the extrinsic components of one reaction favour the operation of the other, and the choice of reaction oscillates such that the strain energy of the extrinsic components remains approximately constant. It is unlikely that this is feasible at all triple points, however.

D. Emission of Lattice Dislocations. The tightly packed lattice dislocations either associate, or undergo a dissociation to produce a lattice dislocation which can glide away into the body of the grain (fig. 5). The lattice dislocations produced by such reactions have Burgers vectors of random sense, and so annihilate on average. This mechanism can occur at all boundaries, but is generally unfavourable according to Frank's criterion and so requires high stresses or thermal activation.

Operation of one or more of these mechanisms allows sustained deformation to take place. The strain rate is governed by the rate at which the SGBDs can move by diffusion controlled climb/glide, and so under superplastic conditions is approximately equal to the Coble creep rate as shown by Arzt et al. [13] but with the applied stress replaced by the *effective* stress, $\sigma_e = \sigma_a - \sigma_b$. The recovery processes are difficult to model objectively, so a fully quantitative treatment has not yet been attempted. The fact that the back stress can be modelled and measured allows some quantitative comparisons to be made, however, and the simple basis of the model facilitates microstructural comparisons.

5. Comparison with Experiment

A. Steady State Flow
The steady state applied stress-strain rate relationship cannot be predicted without modelling the recovery process. However, the *effective* stress should be directly proportional to the strain rate. Furushiro et al. [14] measured σ_b as a function of σ_a in the Zn-22%Al eutectoid using stress dip tests, and so were able to deduce σ_e. Their results are shown schematically in fig. 6. The strain-rate is indeed approximately proportional to σ_e, and the *m* value of less than 1 observed is a consequence of the fact that the back stress is approximately equal to the applied stress at low stresses, but falls below it as the stress is raised so that the effective stress increases more rapidly. This behaviour is certainly qualitatively consistent with the current model if the emission of lattice dislocations is considered as the main recovery mechanism, since this unfavourable dislocation reaction should be very difficult at low stresses.

B. Loading Transients
If a superplastic specimen is loaded under constant stress conditions its strain rate decreases continuously over the first 10-40% of deformation [15]. The initial description of the present model described qualitatively how this may occur as a result of increasing back stresses associated with the accumulation of extrinsic dislocations. The approximate form of the predicted loading transient has also been deduced using the following assumptions.
a) Each triple point is associated with a back stress which is initially given by eq. 2, but with $b_e = b_i \sin\theta$. The value of θ is taken to vary randomly between 0 and 90° in recognition of the fact that the amount of extrinsic SGBD produced by each dissociation varies between triple points.
b) No recovery occurs near a triple point until the local value of σ_b reaches a critical level. Recovery subsequently keeps σ_b at this critical level. The critical value of σ_b is the same for all triple points irrespective of their θ value, and is the only adjustable parameter in the treatment. Its value can be

deduced from a single experimental strain rate at a given strain. This assumption is sensible if lattice dislocation emission is the main recovery mechanism, since this is likely to be a very sensitive function of the total extrinsic SGBD content of a boundary.
c) Assuming that the strain rate at any instant is constant throughout the material, this can be calculated as a function of strain by equating the value of the local stress necessary to satisfy this condition to the macroscopically applied stress.
This method has been applied to results from the Sn-38%Pb eutectic and a comparison of the model's predicted loading transient with the experimental results is shown in fig. 8. A value for the diffusion coefficient of a Sn-2%Pb alloy obtained from Coble creep data was used in the model [16]. The circled experimental point was used to calculate the critical back stress at which recovery begins, so that the theoretical line is forced to pass through it. Significantly, the model predicts the general form of the transient, including the following three main features:
(i) The initial very rapid drop in strain rates. It should be noted that in this portion of the curve, no recovery has taken place and so the prediction is absolute (i.e. no adjustables).
(ii) A sharp knee in the curve at a strain of ~0.2%.
(iii) A continued slow fall in the strain rate thereafter. The model predicts that after a strain of 22%, the strain rate has fallen to within 3% of its steady state strain rate, so the range of the transient is of the right order.

C. Anelasticity

If the stress is suddenly removed from an alloy undergoing superplastic deformation, considerable back flow occurs following the initial elastic response. The size of this elastic after-effect, or anelasticity, can be nearly 300 times as big as the elastic response on unloading, and although much of it can be explained in terms of the sintering of deformation-induced cavitation or grain boundary tension driven creep, there remains a contribution up to 30 times as big as the elastic response which cannot be explained in terms of conventional mechanisms [17]. This contribution seems to be driven by elastic strains, and has the same activation energy and apparent threshold stress as superplastic flow [18]. According to the present model, the effect is another manifestation of the internal, or back stress, σ_b. On removing σ_a, the effective stress, σ_e has the value $-\sigma_b$, and this drives the reverse deformation. Thereafter, the developing situation is complicated, as the back stress not only diminishes because the extrinsic dislocations move back across the triple point from which they came, but also the dissociations which they undergo as they return are not necessarily the reverse of those which created them, in general creating new extrinsic SGBDs which oppose further reverse deformation.

There is, nevertheless, one specific prediction which can be tested against experiment: for small σ_a, when $\sigma_a \approx \sigma_b$, the initial anelastic backflow following a constant stress test should closely resemble the initial loading transient when the stress was first applied (but in the reverse direction). This comparison is made for such an experiment in fig. 9. It is clear that at the shortest times studied the elastic after-effect and the loading transient are indeed very similar and have the same form of time dependence. As time goes on the two measurements diverge, as they must since the anelastic strain rate eventually tends to zero whereas the loading transient's strain rate tends to a constant non-zero value. Also plotted is the loading transient predicted using the method of the previous section. This has the correct form and agrees with experiment to within a factor of 2. It should be noted that for much of the range shown the predicted transient is in the regime containing no adjustable parameters.

D. Grain Rotation, Bursts of Grain Boundary Sliding and 'Co-operative' Grain Boundary Sliding

Consideration of specific examples of the crystallography of triple points shows that in many cases the sustained motion of the intrinsic SGBD network is liable to be difficult, for three reasons. (i) At low stresses, where recovery is slow, the accumulation of extrinsic components gives rise to a back stress. (ii) Some of the dissociations required in crossing a triple point are 'difficult' by Frank's criterion, and so are slow, especially at low stresses. (iii) In boundaries where the applied stress produces a force on the intrinsic SGBDs which moves them *away* from the triple point, there is no way of replacing them as they leave this point. The simplest example is shown in fig. 7a: the force on the intrinsic SGBDs in all three boundaries is away from the triple point at which they meet. It is noted that most of the intrinsic SGBDs *can* move in the direction of the force on them without giving rise to long range stresses simply by changing their spacing (fig. 7b). This corresponds to a relative rotation of the grains, which arises despite the fact that the applied stress does not require it.

Continued rotation will eventually rotate the grains to their special, low energy orientations, at which the boundaries are free from intrinsic SGBDs. At this position they provide 'short circuits' for SGBDs down which they can move rapidly without being hindered by interactions with the intrinsic network or causing back stresses to arise. This causes a burst of grain boundary sliding. Such inhomogeneous deformation has been observed experimentally [19]. Since some triple point systems are much more conducive to flow than others, deformation can be expected to be spatially inhomogeneous as well. It may be that this is the explanation for the relatively new observation of 'co-operative' grain boundary sliding [20], these paths of concentrated deformation representing chains of systems in which the crystallography allows sustained flow without the creation of many extrinsic SGBDs.

E. Curved and Grooved Grain Boundaries

Grain boundaries become curved following superplastic flow. As was shown in section 4A, curvature arises according to this model because it can lower the long range strain field of the extrinsic dislocations in the boundary (fig. 2). In reporting this observation in a single phase alloy, Gifkins [21] stated that the curvature was not removed by annealing. In common with many other superplastic phenomena, this is a highly non-classical result, since grain boundary migration is conventionally a fast process. The present model provides an explanation, since the grain boundary has become curved because to do so lowers the energy of the system, i.e. there is no driving force for it to become flat again as long as it contains extrinsic SGBDs. Similar explanations may be envisaged for grooved grain boundaries [22] if the sign of the extrinsic dislocations oscillates along the boundary.

F. Thickened Grain Boundaries

Grain boundaries etch thicker in polished sections following superplastic deformation [23]. The long range strain field of the extrinsic SGBDs in the boundaries increases the ease with which the surrounding material is dissolved by the etchant. The effect may be heightened by the increased segregation of solutes to the strain field.

G. Strain Induced Grain Growth

The movement of SGBDs containing steps causes grain boundary migration. The step height is of the order of the Burgers vector, and since the grain growth rate dL/dT is approximately equal to the boundary migration velocity [24], $dL/dT \sim \dot{\varepsilon} L$. This relationship is observed in nearly all superplastic alloys at intermediate strain rates [25].

6. Discussion

It is evident that even with the simplifications made in considering the general characteristics of a 2-dimension crystal, sustained deformation by the movement of the initially intrinsic network of SGBDs in a well annealed polycrystal is a complicated process. As a result, the model described above is not fully predictive, particularly regarding the relationship between the applied stress and the strain rate at steady state. Clearly there is room for further development, with the priority being to model the recovery processes quantitatively.

Notwithstanding its current shortcomings, it has great philosophical merit, in that it is a unified model, consistent with all the main theoretical and experimental facts concerning both superplasticity and the structure of grain boundaries. This does not arise from a concoction of disparate ideas, but stems naturally from one simple notion, *viz* that the passage of intrinsic SGBDs over triple points leads to the creation of extrinsic dislocations. Nor does its universality merely indicate that it is so nebulous that it could agree with anything: its specific, quantitative predictions concerning loading transients, internal stresses and anelasticity are novel and successful, and it offers new explanations for some of the unusual microstructural aspects of superplasticity.

7. Conclusions

The consequences of deformation using the initially intrinsic network of secondary grain boundary dislocations have been examined. Internal back stresses result from passage of the intrinsic SGBDs across triple points, and the strain rate is approximately equal to the Coble creep rate, but with the applied stress replaced by the effective stress. The model is consistent with the main features of superplasticity and grain boundary structure, and gives good quantitative predictions regarding the loading transient, elastic after-effect and internal stress/strain-rate relationship.

References

1. D. G. Brandon, B. Ralph, S. Ranganathan and M. S. Wald, Acta Metall. **12** (1964) 813
2. D. H. Warrington and W. Bollmann, Phil. Mag. **25** (1972) 1195
3. D. A. Smith and R. C. Pond, Int. Metals Reviews **21** (1976) 61
4. A. P. Sutton, Int. Metals Reviews **29** (1984) 377
5. W. Bollmann, 'Crystalline Defects and Crystalline Interfaces', (1970) Springer Verlag
6. A. P. Sutton, Phil. Mag. **46A** (1982) 171
7. A. P. Sutton, Acta Metall. **36** (1988) 1291
8. T. Schober and R. W. Balluffi, Phil. Mag. **21** (1970) 109
9. F. Cosandey and C. L. Bauer, Phil. Mag. **44A** (1981) 391
10. S. E. Babcock and R. W. Balluffi, Acta Metall. **37** (1989) 2357
11. R. Z. Valiev, V. Y. Gerstman and O. A. Kaibyshev, Phys. Stat. Sol. **97** (1986) 11
12. D. A. Smith, C. M. F. Rae and C. R. M. Grovenor, in 'Grain Boundary Structure and Kinetics', (Ed. R. W. Balluffi), 1979, ASM
13. E. Arzt, M. F. Ashby and R. A. Verrall, Acta Metall. **31** (1983) 1977
14. N. Furushiro, M. Toyoda and S. Hori, Acta Metall. **36** (1988) 523
15. P. K. Chaudhury and F. A. Mohamed, Acta Metall. **36** (1988) 1099
16. J. H. Schneibel and P. M. Hazzledine, J. Mat. Sci. **18** (1983) 562
17. R. I. Todd, Acta Metall Mater. **42** (1994) 2921
18. R. I. Todd, Scripta Metall. Mater. **29** (1993) 407
19. G. Rai and N. J. Grant, Met. Trans. A **14A** (1983) 1451
20. V. V. Astanin, O. A. Kaibyshev and S. N. Faizova, Acta Metall. Mater. **42** (1994) 2617
21. R. C. Gifkins, Bull. Inst. Met. **4** (1958) 117
22. I. I. Novikov, V. K. Portnoy and V. S. Levchenko, Acta Metall. **29** (1981) 1077
23. K. Matsuki, H. Morita, M. Yamada and Y. Murakami, Metal Sci. **10** (1976) 235
24. J. W. Martin and R. D. Doherty, 'Stability of Microstructure in Metallic Systems', 1977, CUP
25. C. H. Cáceres and D. S. Wilkinson, J. Mat. Sci. Lett., **3** (1984) 395

Figure 1 When the applied stress makes intrinsic SGBDs cross triple points, the dissociation products may in general be of intrinsic or extrinsic character. In this and all the diagrams which follow, grain boundaries are represented as tilt boundaries with the rotation axis normal to the page, and all dislocations are of pure edge character. Intrinsic dislocations are represented with Burgers vectors normal to the boundary, and extrinsic dislocations with Burgers vectors parallel to the boundary. **These are schematic devices for the purposes of illustration only.**

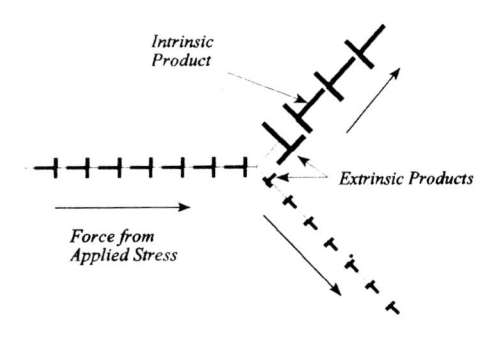

Fig. 2. By changing its orientation, the part of the boundary containing extrinsic SGBDs can lower its long range strain field. This leads to grain boundary curvature.

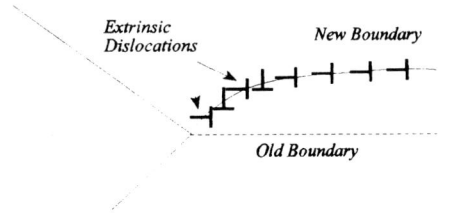

Fig. 3. The extrinsic SGBDs may be able to lower their energy by spreading out owing to their mutual repulsion. This may be hindered by interaction with the intrinsic SGBDs (not shown).

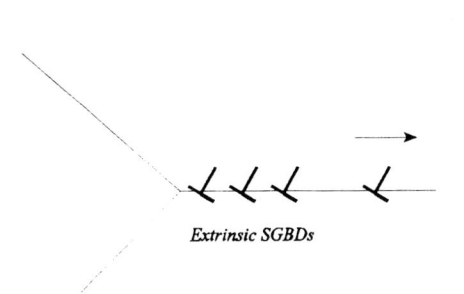

Fig. 4. An alternative dissociation at the triple point may produce extrinsic SGBDs which lower the long range strain field.

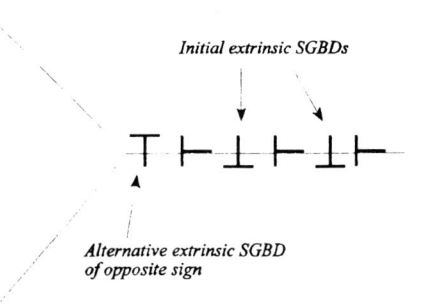

Fig. 5. The strain energy associated with the build up of extrinsic SGBDs is lowered by emission of a lattice dislocation.

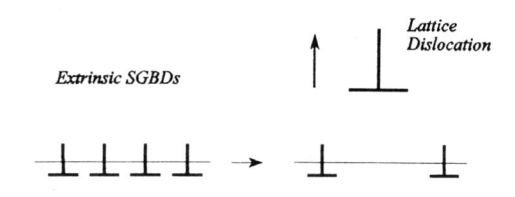

Fig. 6. Schematic representation of experimental results obtained by Furushiro et al using the Zn-22%Al eutectoid [14].

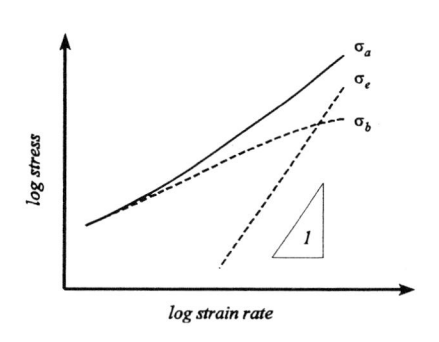

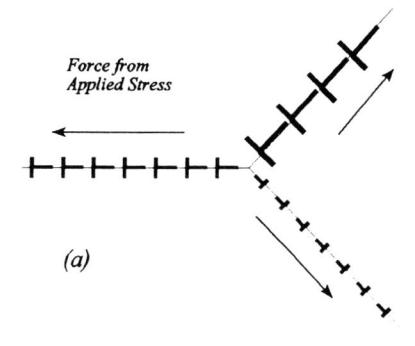

Fig. 7. Even when triple points are difficult to cross, or intrinsic dislocations cannot be supplied at a triple point, most intrinsic SGBDs can move without causing long range stresses to arise if their spacing changes.

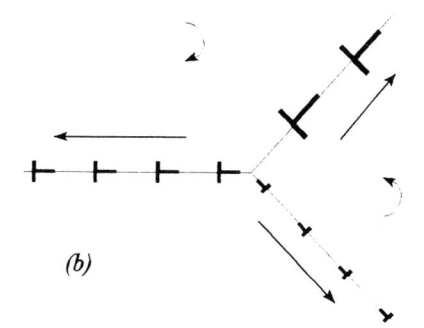

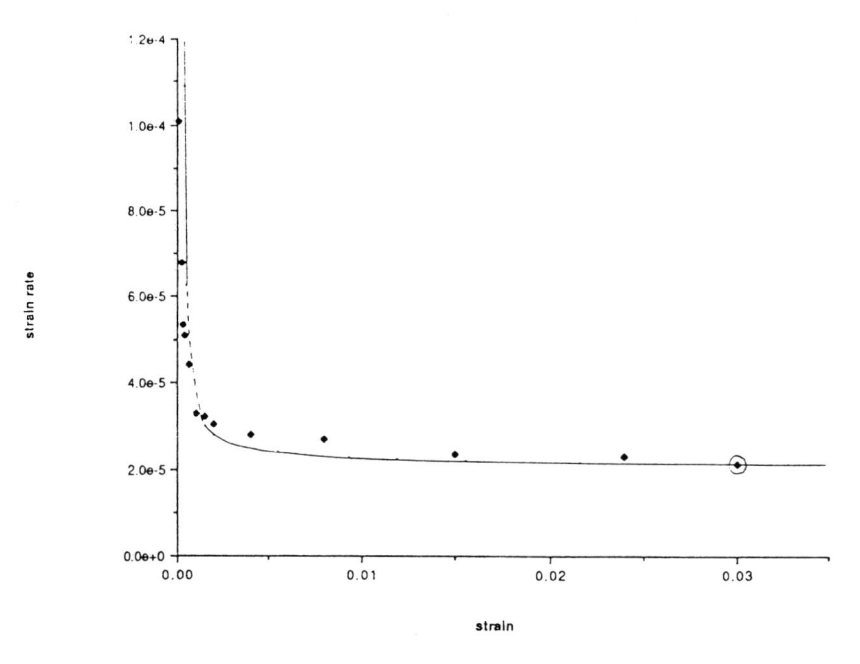

Fig. 8. Comparison of the loading transient, presented as a log-log plot of strain rate against strain, for the model's predictions (line) and a constant stress (1.6MPa) experiment (points) on a specimen of Sn-38%Pb eutectic.

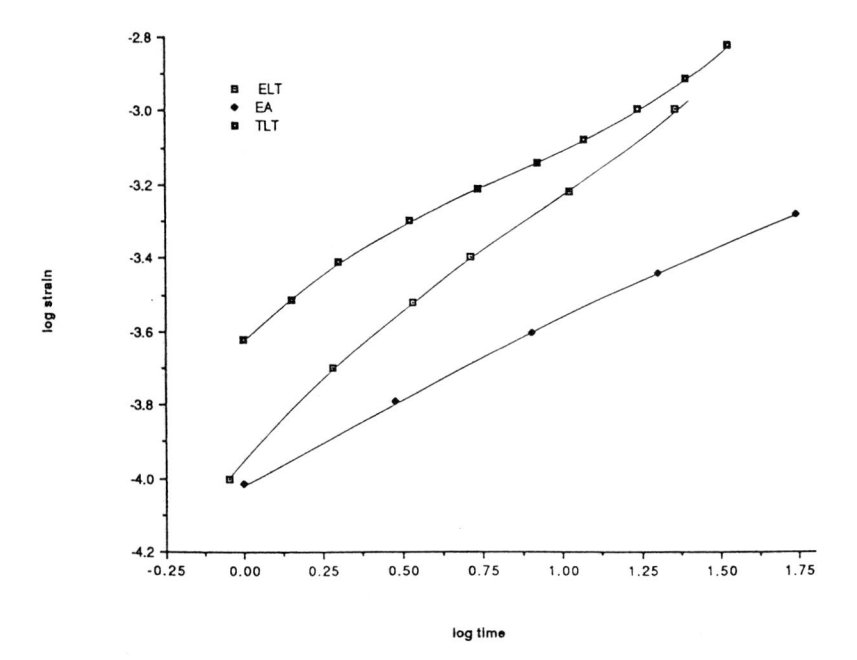

Fig. 9. Plots of log strain against log time for the loading transient (ELT) and the elastic after-effect (EA) of the experiment of fig. 8. The line TLT is the model's predicted loading transient.

Physical Model for Superplastic Flow Rheology in a Wide Strain Rate Range

Sergey A. Larin, Vladimir N. Perevezentsev

Russian Academy of Science, Nizny Novgorod Branch
of Machine Science, Nizny Novgorod 603024, Russia

Abstract

The developed model of superplastic flow of materials is based on the analysis of physical processes taking place in grain boundaries during their interaction with lattice dislocations (LD). It is shown that getting of LD into a boundary and their subsequent spreading give rise to fluid-like boundary parts characterised by low resistance to sliding. The rate of grain boundary sliding (GBS) and intragranular dislocation slip (IDS) depends on the number of fluid-like boundary parts. The object is to get system of equations to describe the regularities of grain boundary sliding and intragranular slip in a wide strain rates range; to simulate the behaviour of curves $\log(\sigma)-\log(\dot{\varepsilon})$ to explain the major features of rheological behaviour of materials under superplasticity.

1. Introduction

According to [1-3], the structural superplasticity (SSP) phenomenon is caused by the transition of grain boundaries to an energetically excited state under the action of bombardment of grain boundaries by LD. The main process responsible for the boundary transition to an excited state is spreading of LD cores. The atomic structure of the boundary exists in a energetically excited ("fluid-like") state for a characteristic time t_d in the region of the spreading dislocation core. The resistance to intergrain shear on the boundary part of length r_d is lowered in the region of the spreading dislocation core, while remaining more high in the rest part of the boundary. The density of excited parts increases with increasing intragranular strain rate $\dot{\varepsilon}_v$ and, hence, increasing LD flow falling on grain boundaries, and the effective resistance to shear in the boundary decreases on the whole. This results in an increase in the rate of the GBS. We show that the processes of IDS and GBS are interrelated under SSP and interdependent. Not only IDS stimulates sliding along grain boundaries but also GBS which is inhomogeneously progressing along the boundaries creates the conditions needed for generation of LD by grain boundaries. The consistent development of these notions permits one to explain the main specialities of IDS and GBS and describe the rheology of superplastic flow.

2. Generation of Lattice Dislocations by Grain Boundaries and Intergranular Deformation

Consider the processes taking place in the grain boundary when LD gets into it. Spreading of the LD core would result at a characteristic time t_d in a decreased resistance to sliding, the local shear U_1 developing in the field of external stresses σ along the fluid-like region of length r_d and in concentration of stresses on the region edges (on the spreading front) Fig. 1. The internal stresses σ^1 occurring at high enough values of σ can relax through plastic shear going from the boundary into the grain interior. As a result of the LD emittance into the grain interior, an orientation misfit dislocations (OMD) inevitably appears in the boundary [4] where its cores is spreading in its turn.

Therefore, even though the excited boundary part appears in a close-to-equilibrium state within the time t_d, two new fluid-like regions are created at its edges. Later on this process is repeated, but in other parts of the boundary. Thus, the process of LD generation by grain boundaries is self-supporting and the intragranular deformation evolves homogeneously.

As a force condition for the LD emittance we may consider the excess of internal stresses $\sigma^{int}(r)$ acting in the boundaries over the value of the resistance to shear σ_0^v in the grain interior $\xi\sigma^{int}(0,5r_d)>\sigma_0^v$ ($\xi=0.5$ is the geometrical multiplier). The fulfilment of this condition is ensured by the presence of stress concentration on the spreading front which is associated with the difference in the material properties in excited and nonexcited boundary parts (note a similarity with the case of the stress concentration in the shear crack edges). Another force condition for IDS realisation is connected with the LD going into the grain boundary. It is necessary that the external stress σ acting on LD should be equal to the mean internal stress $<\sigma^{int}>$ caused by the dislocations distributed in the boundary at a density ρ^s. Since $\sigma=<\sigma^{int}>$, this condition may be written as

$$\sigma = \alpha G \rho^s \Delta b \qquad (1)$$

where G - is the shear modules, Δb - is the Burgers vector of the OMD, $\alpha = 0,1$ - is a geometrical multiplier [9].

As follows from the above, the fluid-like regions continuously increase in number under applied stress even if only one LD is present in the initial state. The kinetics of increase in the number of fluid-like regions per unit boundary length ρ_b is described by the balance equation $\dot{\rho}_b = (\psi \dot{\varepsilon}_v/b)(1-\rho_b r_d) - \rho_b/t_d$ where, the first summand describes the appearance of fluid-like regions due to LD going from the grain interior, the second summand describes disappearance of these regions when the spreading process is over and the third one describes the decrease in ρ_b at a possible coalescence of two adjacent regions. The stationary value of ρ_b^s is obtained from the condition $\dot{\rho}_b=0$:

$$\rho_b^s = \xi \dot{\varepsilon}_v t_d / (1+\xi \dot{\varepsilon}_v t_d r_d/b) \qquad (2)$$

The value of t_d depends on strain rate and the microstructure parameters. In cases where t_d is defined by the time of building of the Burgers vector slip components of OMD into the boundary structure, the expression for t_d has the form $t_d=t_d^\circ \cdot d/(\dot{\varepsilon}_v D_b)^{1/2}$ where t_d° is a dimensionless parameter [5]. The value $\dot{\varepsilon}_v$ in expression (2) seems to depend on the density of grain boundary dislocations ρ and, eventually, is defined by the value of applied stresses σ (note that in the general case, $\rho \neq \rho_b$). In a steady-state condition of deformation for ρ we have

$$\rho^s = \xi \dot{\varepsilon}_v t_r/b \qquad (3)$$

where t_r is the time of internal stress relaxation.
It follows from (1) and (3) that

$$\dot{\varepsilon}_v = k(\sigma/G)/t_r, \quad k = b/(\alpha\xi\Delta b) \qquad (4)$$

Determination of t_r value presents an independent problem and its detailed consideration will be beyond the scope of this work. The relaxation time for a sufficiently wide number of superplastic alloys is defined by the time of the dislocation core spreading up to the moment of coalescence with the spreading cores of the adjacent dislocations. In this case t_r depends on the mean spacing between dislocations $(\rho_s)^{-1}$ and is determined [3]

$$t_r = kT/(G\Omega) \cdot [D_b \delta \cdot (\rho^s)^3]^{-1} \qquad (5)$$

where D_b is the coefficient of grain boundary diffusion, δ - is the diffusion width of boundary, T - is the temperature, k - the Boltzmann's constant, Ω - atomic volume. Using (3), (4), (5) we obtain the dependence of $\dot{\varepsilon}_v$ on σ

$$\dot{\varepsilon}_v = C_v(G\Omega)/(kT) \cdot (D_b \delta / b^3) \cdot (\sigma/G)^4 , \quad C_v = (k/\sigma^4)(b/\Delta b)^3$$

For simulation we can write

$$\dot{\varepsilon}_v = BD_b(\sigma - \sigma^0_v)^4 , \quad B, \sigma^0_v = \text{const} \qquad (6)$$

2. Interrelation of Grain Boundary Sliding and Intragranular Dislocation Slip

In this model, fluid-like regions characterised by a high-level of energetic excitation and increased diffusion permeability appear in polycrystal grain boundaries during deformation. Taking into account the absence of the long-range order in the atomic structure in a normal (not special) boundary and the similarity in its structure to the one of amorphous materials we would simulate the GBS rate in these regions (by analogy with the flow in amorphous materials) by taking the Bingham flow law which, an approximation of low σ, has the form of [6]

$$\dot{U}^l = (\sigma^l - \sigma_0^l)/\mu^l , \quad \dot{U}^p = (\sigma^p - \sigma_0^p)/\mu^p \qquad (7)$$

where $\dot{U}^l$, $\dot{U}^p$ - are the sliding rates, μ^l, μ^p - the viscosity coefficients, σ^l, σ^p - is the mean flow stresses σ_0^l, σ_0^p - the threshold stresses on the excited and nonexcited region respectively. The rate of intrinsic GBS on the boundaries on the whole $\dot{U}^s$ is determined by relation

$$(r_d + r_p)/\dot{U}^s = r_d/\dot{U}^l + r_p/\dot{U}^p \qquad (8)$$

where r_p-is the length of a singe nonexcited region founded from evident relation $(r_d + r_p)^{-1} = \rho_b^s$.

The connection σ^l, σ^p to the applied stress σ is obtained from the equilibrium equation

$$\sigma(r_d + r_p) = \sigma^l r_d + \sigma^l r_p \qquad (9)$$

In order to describe GBS in a wide strain rate interval we should take into account the fact that the rate of intergrain deformation $\dot{\varepsilon}_b$ in a polycrystal is defined by the rates of the two successive processes: plastic shear along the boundary $\dot{\varepsilon}_s$ ($\dot{\varepsilon}_s = \dot{U}^s/d$) and sliding accommodation at boundary triple points $\dot{\varepsilon}_a$ [7]

$$\dot{\varepsilon}_a = A'\sigma^2/d^2 , \quad A' = C_b\Omega/kT \cdot D_b\delta/(Gb^3) = AD_b \qquad (10)$$

Let us analyse the relation of intragranular deformation to the process of GBS. In the steady-state condition of deformation the density of LD in the grain interior ρ^s_v is related to the density of grain boundary sources of LD ρ^s_b (in the absence of LD sources in the grain interior) by relation $\rho^s_v = \rho^s_b \psi(t_e V_d)$ [8], where t_e - the time of the LD loop emittance, the LD life time equal to the time of LD passing through the grain volume with mean rate V_d, ψ -is the coefficient which takes into account the grain shape (ψ -1.5). In the considered model, the value of t_e may be calculated as a time needed for a difference shear equal to the LD Burgers vector be stored on the spreading front during GBS $\zeta(\dot{U}^1 - \dot{U}^p)t_e = b$. Using Orovan relation $\dot{\varepsilon}_v = \rho^s_v b V_d$ we obtain the equation for the rate of intraglanular deformation

$$\dot{\varepsilon}_v = \rho^s_b \psi \zeta (\dot{U}^1 - \dot{U}^p) \qquad (11)$$

This equation essentially represents a condition for conservation of the material continuity at a co-operative going of IDS and GBS. It should be noted that the calculation of superplastic curves $\log(\sigma) - \log(\dot{\varepsilon})$ should take account of the contribution made by diffusion creep (Coble creep) to the total deformation of specimen $\dot{\varepsilon}_{cr} = A_{cr} D_b b^{-3} \sigma$, A_{cr} = const.

Conclusively, the overall strain rate of specimen may be written as

$$\dot{\varepsilon} = \dot{\varepsilon}_b + \dot{\varepsilon}_v + \dot{\varepsilon}_{or} \qquad (12)$$

3. Numerical Results

The obtained system of equations (1-12) makes it possible to describe the SP flow rheology in a wide interval of strain rates. We can solve it minimising the theoretical and experimental $\log(\sigma) - \log(\dot{\varepsilon})$ curves difference of any experimental points. The following values are the parameters of the problem (the constants independent on $\dot{\varepsilon}$, ε and at a fixed temperature):

$$\mu_1, \mu_p, \sigma^0_1, \sigma^0_p, \sigma^0_v, A, B, A_{cr}, D_b, t^0_d \qquad (13)$$

As a rule exact values of this parameters for particular alloys are unknown. However, relying on physical reasons, one can obtain more or less exact ranges of their change. In the absence of complete information on the parameters (13), the problem of constructing a superplastic curve is reduced to the minimisation problem of some specific function under constraints on the parameters variation $par_i^{max} > par_i > par_i^{min}$, i=1,11.

Consider the results of calculation for CDA-638 alloy [10]. The initial data needed for calculation of the SSP curve are given in [10]. Table 1 lists the optimal values of the varied parameters and intervals of their change, which have been obtained from the solution.

Table1. Optimum material parameters and intervals of their change (CDA-638)

	μ_1 MPa/s	μ_1/μ_p	σ^0_v MPa	σ^0_p MPa	σ^0_1 MPa	A	B'	A_{cr}	D_b m/s	t_d s
opt	0.88	0.01	0.	10.	0.	5.93 10^{-6}	0.044	4.7 10^{-12}	2 10^{-13}	0.79
min	0.01	0.001	0.	0.	0.	6 10^{-9}	1 10^{-3}	6 10^{-13}	2 10^{-15}	0.10
max	3000.	0.5	3.	20.	3.	6 10^{-6}	1 10^{2}	6 10^{-9}	2 10^{-9}	1.00

Fig. 2 presents calculated and experimental curves $\log(\sigma)$ -$\log(\dot{\varepsilon})$. It should be emphasised that a good agreement of experimental and theoretical curves is obtained on using reasonable values of parameters. Thus, for example, the calculated optimum grain boundary diffusion coefficient D_b practically coincides with the experimental value [11]. The calculated optimum value of the viscosity coefficient in the excited boundary part is close in its order of magnitude to the values of the viscosity coefficient of liquid Cu [12]. This result is an a close agreement with the present knowledge of the fact that the boundary parts excited by spreading represent quasi-fluid of fluid islands [13].

Fig. 3 shows the curves $\chi_b = \dot{\varepsilon}_b/\dot{\varepsilon}$, $m=d(\log \sigma)/d(\log \dot{\varepsilon})$, $\chi_v = \dot{\varepsilon}_v/\dot{\varepsilon}$, $\chi_{cr} = \dot{\varepsilon}_{cr}/\dot{\varepsilon}$. Here m - is the strain rate sensitive coefficient of flow stress, $\chi_b(\dot{\varepsilon})$, $\chi_v(\dot{\varepsilon})$, $\chi_{cr}(\dot{\varepsilon})$- are the contributions of grain boundary deformation, intragranular deformation, creep Coble deformation respectively, to the total specimen deformation calculated using the optimum values of parameters (see Table 1). The dependence of GBS contribution $\chi_b(\dot{\varepsilon})$ has a characteristic bell-shape similar to the shape of $m(\dot{\varepsilon})$ dependence. The such shape of $m(\dot{\varepsilon})$ and $\chi_b(\dot{\varepsilon})$ dependences is typical for SP alloys and this fact will be found in a number of works [14]. Contribution of the intragranular deformation $\chi_v(\dot{\varepsilon})$ first decreases with increasing $\dot{\varepsilon}$ and then increases and becomes dominants at high strain rates. As it has been expected the minimal value of the contribution $\chi_v(\dot{\varepsilon})$ is obtained at optimal SSP rates corresponding to a maximum $m(\dot{\varepsilon})$. As to the $\chi_{cr}(\dot{\varepsilon})$ diffusion creep contribution , Fig. 3 shows that its value is appreciable only at low $\dot{\varepsilon}$ and the diffusion creep contribution may be neglected at stages II and III of SSP.

Fig. 4 presents the results obtained from calculation of curves for alloy Ti-6%Al-4%V [14] at various values of grain sides.

4. Conclusion

1. A model is developed which describes the main features of superplastic flow and may be assumed as a basis for calculation and analysis of $\log(\sigma)$-$\log(\dot{\varepsilon})$ curves of superplastic alloys.

2. Close interrelation has been found between the processes of intragranular dislocation slip and grain boundary sliding. Interdependence of these processes under superplasticity has been shown.

3. A new mechanism for generation of lattice dislocations by grain boundaries is proposed which explains the experimentally observed explains the experimentally observed specialities of the intragranular dislocation slip is superplastic alloys.

References

[1] V.V.Rybin and V.N.Perevezentsev, Pis'ma v Zh.Tech. Fiz., v.7 (19), 1203-1205, (1981).
[2] V.N.Perevezentsev, V.V.Rybin and A.N.Orlov, Poverkhnost N6, 134-142, (1982).
[3] V.N.Perevezentsev, in Problems of the Theory of Defects in Crystals, 85-100, (Nauka, Leningrad 1987).
[4] A.N.Orlov,V.N.Perevezentsev and V.V.Rybin Grain Boundaries in Metals (Metallurgia, Moscow 1980)
[5] V.N.Perevezentsev,V.V.Rybin and V.N.Chuvil'deev, Poverkhnost N10, 134-142 (1983)
[6] A.S.Argon, Acta Metall. v. 27, 47-58 (1979).
[7] V.N.Perevezentsev,V.V.Rybin and V.N.Chuvil'deev, Poverkhnost N4, 134-142 (1985)
[8] S.A.Larin, V.N.Perevezentsev, V.N.Chuvil'deev, Fisika Metallov i

Metallovedenie N6, 55-61 (1992)
[9] S.A. Shei and T.G.Langdon, Acta Metall. v.26, 639-646 (1978).
[10] V.N.Perevezentsev,V.V.Rybin and V.N.Chuvil'deev, Poverkhnost N11, 101-108 (1985)
[11] A.Needlman and J.C.Rice, Acta Metall. v. 28, 1315-1322 (1980).
[12] L.Battezzati and A.L.Greer, Acta Metall. v.37, 1791-1802 (1989).
[13] Yu.S.Nechaev, Colloque de Physique v.51, 287-292 (1990).
[14] N.Furushiro and H.Shigenori, Trans. of Inst. Metals v.27, 937-941 (1986).
[15] A.K.Ghosh and C.H.Hamilton, Metall. Trans. v. 10, 699-706 (1979).

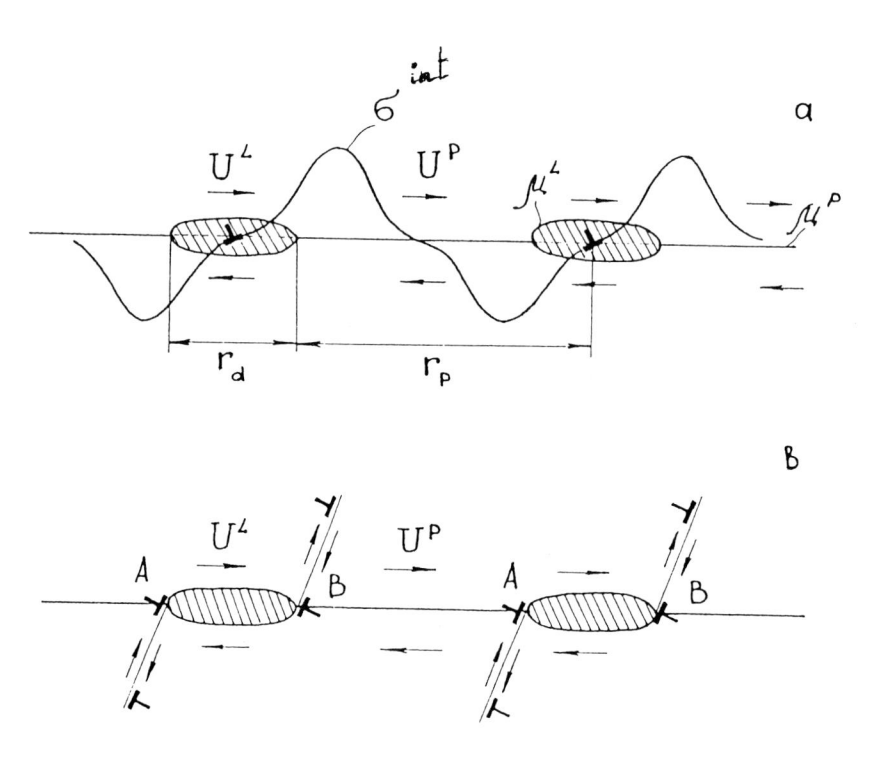

Acknowledgements - The research described in this publication was made possible in part by Grant N R9A000 from the International Science Foundation. The results of this work was obtained in collaboration with Dr. V.Chuvil'deev.

Fig.1 The creation of liquid-like regions of GB during spreading of the lattice dislocation (U^L and U^P - local shear along different parts of the boundary) (a); the emission of the lattice dislocations from the boundary and the appearance of misorientation misfit dislocations (b).

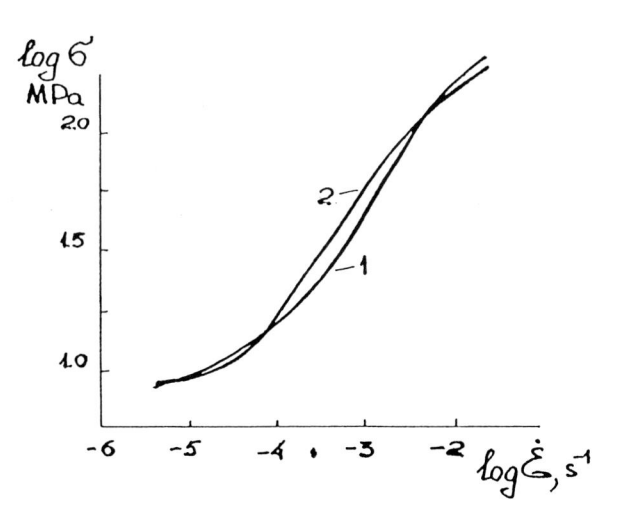

Fig.2 The comparison of the experimental (1) and calculated (2) curves ($\log \sigma - \log \dot{\varepsilon}$) for CDA-638 alloy ($T = 823K, d = 1.3 \mu m$).

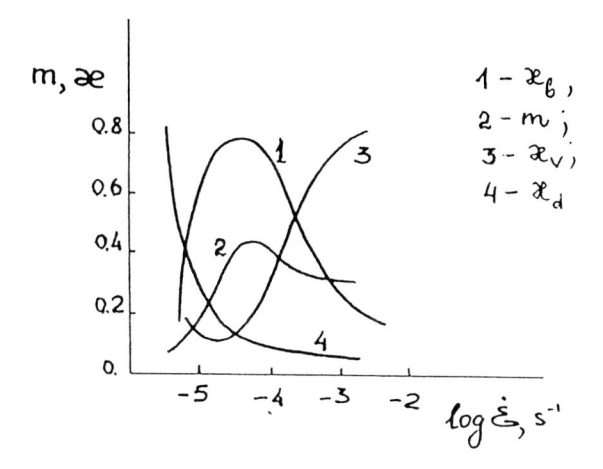

Fig.3 Rate dependences of the coefficient $m = \partial \log \sigma / \partial \log \varepsilon$ and contributions of processes of GB sliding $\aleph_b$, intragranular dislocation sliding $\aleph_v$, and diffusion sliding $\aleph_v$ into the common sample deformation ε of alloy CDA-638.

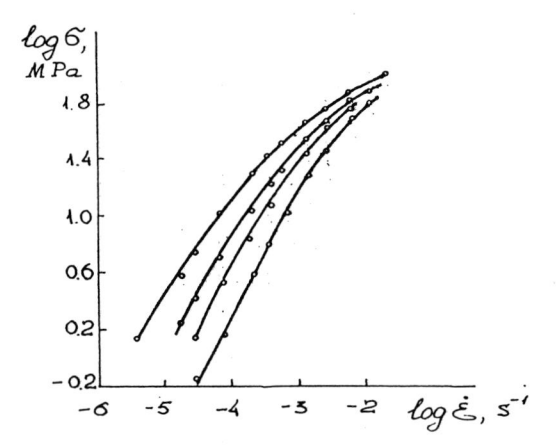

Fig.4 The comparison of the experimental date (o) and calculated curves (solid lines) ($\log \sigma - \log \dot{\varepsilon}$) for $Ti-6Al-4V$ alloy with different initial grain sizes ($1-d = 6.4\mu m; 2 \div 9\mu m; 3 \div 11\mu m; 4 \div 20\mu m$).

The Theory of Evolution of the Microstructure of Superplastic Alloys and Ceramics

Vladimir Perevezentsev
Nizhny Novgorod State University, Russia

Abstracts

The kinetics and regularities of strain induced grain growth and cavitation in superplastic alloys and structural ceramics have been considered. It has been shown that superplastic conditions give rise to additional driving forces for boundary migration and grain growth which are caused by interaction of external and internal stresses with the defects distributed in the boundaries. The mechanisms for nucleation of microcracks and cavities during superplastic deformation have been analyzed. It is shown that microcracks appear at the grain triple junctions under the action of internal stresses caused by disclination-type defects which occur in the process of intragranular deformation. Expressions describing the dependencies of the grain growth rate, mean grain size and the threshold deformation for microcrack nucleation on the strain, strain rate and test temperature were obtained. Peculiarities of realization of the above mechanisms for superplastic microduplex alloys and structural ceramics were analyzed.

1. Introduction

It is well known that superplastic deformation of metallic alloys and structural ceramics is accompanied by intense boundary migration, strain-enhanced grain growth, generation of lattice dislocations by the boundaries and triple grain joints and the nucleation and growth microcavities.It is apparent that all these processes are closely interrelated and exhibit a pronounced strain-rate dependence under definite temperature and strain-rate conditions of superplastic deformation. The purpose of this paper is to give a theoretical description of the above processes. We will try to show that the analysis of the processes of accumulation and accommodation of the defects formed on the boundaries and triple junctions is very fruitful for deciding these problems. The work consists of three parts. The first part presents a brief analysis of the geometrical and kinetics aspects of the accumulation of defects (orientational misfit dislocations and disclinations) induced in the grain boundaries and triple junctions by intragranular slip. The second part gives an analysis of strain-induced boundary migration and grain growth. The peculiarities of its realisation in different materials are discussed. The third part presents an approach to the description of the cavity nucleation process. In contrast to traditional concepts which relate cavity nucleation to grain boundary sliding, we relate it to the intragranular slip.

2. Accumulation of Defects on the Grain Boundaries and Triple Junctions During Intragranular Deformation.

The plastic deformation of polycrystalline solids is essentially nonuniform. The magnitude of strain varies from grain to grain. Such non-uniformity of the plastic deformation gives rise to plastic discontinuities which are localized on the grain (or phase) boundaries. As shown in [1] there are two types of defects induced on the grain boundaries by intragranular deformation. The first type includes defects appearing on the flat boundary and representing orientational misfit dislocations (OMD) in the form of displacements and plastic rotations continually distributed in the boundaries. The second type of defect includes the plastic discontinuities formed at the boundary joints and bends which represent the systems of joint disclinations.

Let us consider the evolution of these defects during high temperature deformation . The lattice dislocations (LD) coming into the boundary out of the interiors of grains can traverse the boundary, leaving in it the OMD (Fig.1a), or they can be fully captured by the boundary, reacting with the LD's, which belong to the adjacent grain, and create OMD.

At a sufficiently high temperature ($T > 0.3 - 0.4 T_m$), dislocations in the general boundary are unstable and spreads at the characteristic time t_d. Of course the discontinuity connected with the dislocation does not disappear but only spreads out along the boundary. Thus the defect contents of the general boundaries under high temperature deformation are an assemblage of orientational misfit dislocations and the products of their spreading (Fig.1b). The kinetics of the accumulation of OMD are determined by the expression:

$$\dot{\rho} = \xi \dot{\varepsilon}_v / b - \rho / t_d \qquad (1)$$

where the first summand describes the appearance of OMD due to the absorption (or emission) of LD by the boundary, and the second summand describes the disappearance of OMD when the spreading process is over ($\dot{\varepsilon}_v$ - intragranular strain-rate; b - Burgers vector; ξ - geometrical coefficient).

In the steady state conditions of deformation the density of these defects is:

$$\rho^s = \xi \dot{\varepsilon}_v t_d / b \qquad (2)$$

The kinetics of accumulation of the Burgers vector density, w_t, of glissile components of OMD which arise in the boundary as a result of spreading of the OMD's cores is determined by two processes - its generation due to the spreading and its disappearance by the movement to the triple junctions, and climb along the adjacent boundary up to the annihilation with similar defects belonging to the other boundary:

$$\dot{w}_t = \rho \Delta b_t / t_d - w_t / t_p \qquad (3)$$

where Δb_t - tangential component of the Burgers vector OMD; t_p - characteristic time of annihilation. As shown in [1], for steady state conditions of deformation $t_d \leq t_p$, therefore:

$$w_t^s \leq \rho^s \Delta b_t \qquad (4)$$

Accumulation of sessile components of OMD on the boundary causes its additional misorientation:

$$\Delta \omega^{s\dot{s}} = -(N^{s\dot{s}} \cdot \Delta \varepsilon^{s\dot{s}}) N^{s\dot{s}} \qquad (5)$$

which depends on the jump of the plastic deformation tensor on the boundary, $\Delta \varepsilon^{s\dot{s}}$, and on the vector of its normal $N^{s\dot{s}}$.

Since the values of $\Delta \varepsilon^{s\dot{s}}$ and $N^{s\dot{s}}$ vary from boundary to boundary, mismatching of additional misorientations occurs at the grain triple junctions which brings into existence wedge disclinations along the lines of the triple junctions (Fig.1c,d). The strength $\Omega^{(k)}$ of the disclination located at an arbitrary k-th junction is defined by the expression $\Omega^{(k)} = \sum_{s\dot{s}} \Delta \omega^{s\dot{s}}$. In polycrystals the disclinations located at the different triple junctions obviously differ in sign and are grouped into dipoles. The strength of dipoles is $\omega = <\Omega>$, and its shoulder, L, is about equal to grain size, d.

The value of ω depends on the rate of accumulation of the defects in grain boundaries $\dot{\omega}^+ \sim \xi \dot{\varepsilon}_v (\Delta b_n / b)$ (b_n - normal component of the Burgers vector OMD) and the rate of its diffusive accommodation $\dot{\omega}^-$. Taking into account possible changes of the dipole shoulder length during grain growth, the expression for $\dot{\omega}^-$ can be written as [2]

$$\dot{\omega}^- = \omega \, [2\dot{d}/d + (G\Omega/kT) \cdot 1/d^2 \cdot (D_v + D_b\delta/d)] \qquad (6)$$

(G - shear modulus; $\Omega \approx b^3$; D_b - GB diffusion coefficient).

So, the final expression for $\dot{\omega} = \dot{\omega}^+ - \dot{\omega}^-$ has the form (at $D_v \ll D_b\delta/d$):

$$\dot{\omega} = \xi\dot{\varepsilon}_v \Delta b_n / b - \omega \, [2\dot{d}/d + c_b (b/d)^3] \qquad (7)$$

(where $c_b = GD_b\delta/kT$).

In the absence of grain growth ($\dot{d}=0$), the dependence of ω on the deformation time t has a simple form:

$$\omega(t) = (\xi\dot{\varepsilon}_v/c_b)(\Delta b_n/b)(d/b)^3 [1 - \exp(-c_b(b/d)^3 t)] \qquad (8)$$

and represented by the curve (1) on Fig.2. The stationary value ω^{st} in this case is determined by the formula:

$$\omega^{st} = \xi\dot{\varepsilon}_v (\Delta b_n/b)(d/b)^3 c_b^{-1} \qquad (9)$$

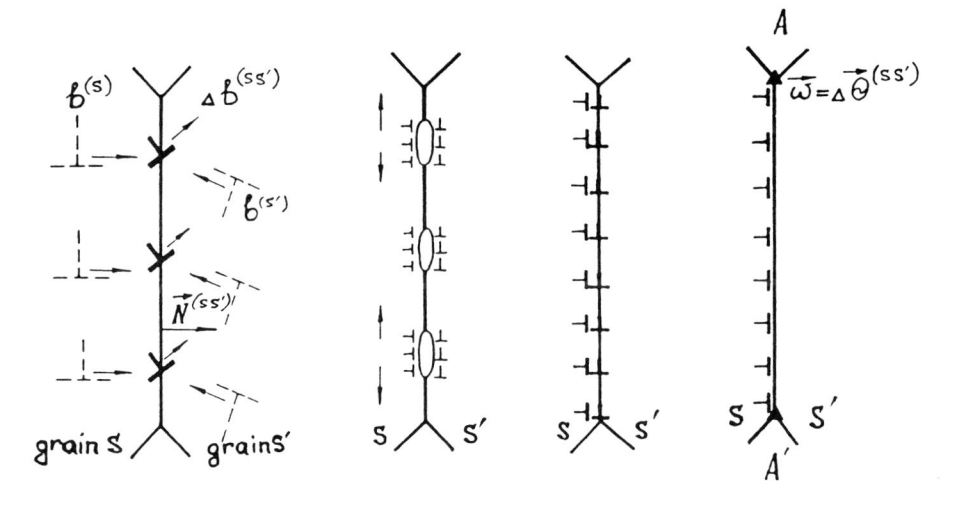

Fig.1 The defect structure of the grain boundary (Gb) as a result of interaction between (Gb) and lattice dislocations: a) the formation of misfit dislocations during plastic deformation of grains S and s'; b) the beginning of the misfit dislocation cores spreading; c) continual distribution of the glissile and sessile components of Burgers vectors of dislocations after their spreading; d) appearance of the wedge disclinations at the triple junctions.

The dependence of the value ω on t for conditions of intensive grain growth was considered in [2], where the grain growth rate was described by the expression $\dot{d}= A(d/b)^{-a}$. Qualitative behaviour of the dependency $\omega(t)$ in this case is represented in Fig.2 (curve 2). As follows from the analysis, in the limiting cases $\beta t \ll 1$ and $\beta t \gg 1$, the value of ω may be approximated by the simple expression:

$$\omega = \xi (\Delta b_n/b) \dot{\varepsilon}_v t = \xi (\Delta b_n/b) \varepsilon_v \qquad (10)$$

Parameter β is determined by the formula:

$$\beta = A(1+\alpha)(b/d)^{(\alpha+1)} \qquad (11)$$

where $A \approx c_b(\sigma/G)$, $\alpha = 1$ in the case of small ω, and $A \approx c_b$, $\alpha = 2$ in the case of large ω. For the intermediate case we have $1 < \alpha < 2$ and $c_b(\sigma/G) < A < c_b$.

3. The Kinetics of Strain-Induced Grain Growth

The interaction of orientational misfit dislocations and junction disclinations with external and internal stresses provide additional driving forces for boundary migration:

$$P = (\sigma + \sigma_i)(\rho \Delta b + \omega) \qquad (12)$$

Here: σ_i - internal stresses from OMD and disclinations. The value of σ_i in the first approximation is:

$$\sigma_i = \alpha_1 G \omega + \alpha_2 G \rho \Delta b \qquad (13)$$

(α_1, α_2 - numerical coefficients). As seen from previous analysis, the value of P essentially depends on the strain rate and σ. Besides, defects located in the boundary and in triple junctions greatly influence the boundary mobility. The effective coefficient of GB-mobility (M) may be represented as:

$$M^{-1} = M_\perp^{-1} + M_\Delta^{-1} + M_b^{-1} \qquad (14)$$

where $M_\perp = c_b(b/d)(1/\rho \Delta b)(b/G)$ - misfit dislocations general mobility [2,3]; $M_\Delta = c_b(b^2/d^2\omega^2)b/G$ - junction disclinations mobility [2,3] and $M_b = D_b \delta b/kT$ - own boundary mobility.
Now we can determine the grain boundary migration rate $V_m = MP$:

$$V_m = [M_b M_\perp M_\Delta / (M_b M_\perp + M_\perp M_\Delta + M_b M_\Delta)] (\sigma + \sigma_i)(\rho \Delta b + \omega) \qquad (15)$$

and its dependence on the strain rate $\dot{\varepsilon}$, and also the dependence of the grain size d on σ, $\dot{\varepsilon}$, ε. The calculation of these dependencies for an arbitrary case is rather complicated and we shall confine ourselves to calculating $V_m (\varepsilon, \dot{\sigma}, d)$ in some limiting cases:

a) Low ω ($\rho \Delta b > \omega$; $M_\Delta > M_b > M_\perp$, $\sigma > \sigma_i$)

In this case we can neglect junction disclinations and the expression for V_m has the simple form:

$$V_m \approx \dot{d} = c_b (\sigma/G)(b/d) \tag{16}$$

Note, that in real conditions the flow stress σ does not remain constant during superplastic deformation. Hence, in order to find the dependence d(t,ἐ), equation (16) should be complemented by the rheological equation relating σ to strain rate, ἐ, and current grain size, d. In the conventional form:

$$\dot{\varepsilon} = Ac_b (\sigma/G)^{1/m} (b/d)^p \tag{17}$$

where A - is the material constant obtained from the analysis of the experimental curves log(σ)-log(ἐ).

So, using (16) and (17) we have:

$$\dot{d} = \dot{\varepsilon}^m / A^m c_b^{m-1}) (b/d)^{1-pm} \tag{18}$$

Note, that in the case of metallic SP alloys, coefficients m and p essentially depend on ἐ and d, which is necessary to take into account in the calculation of curves $\dot{d}(\dot{\varepsilon})$ and $d(\dot{\varepsilon},\varepsilon)$. On the contrary, in superplastic ceramics one can believe that m and p vary weakly in a wide strain-rates range. Besides, for several SP ceramics the relation pm ≅ 1 is fulfilled [4], hence:

$$\dot{d} = \dot{\varepsilon}^m / A^m c_b^{m-1} \tag{19}$$

Fig.2 The qualitative dependence of the disclination strength ω on the deformation time t in the cases when: a) the grain growth is absent (curve 1); b,c) intensive grain growth exists ($\dot{\varepsilon}_3 > \dot{\varepsilon}_2$).

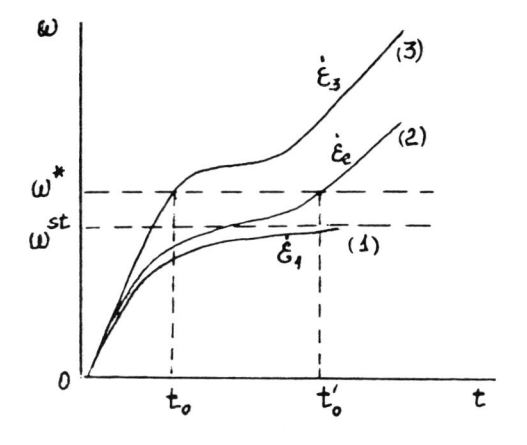

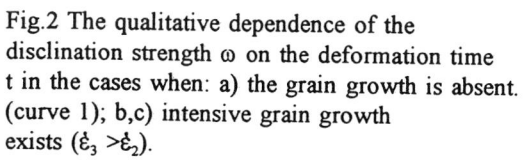

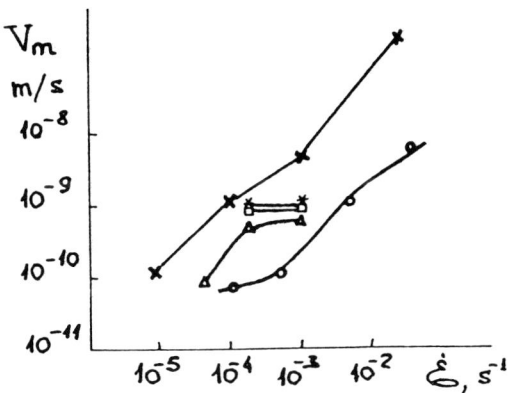

Fig.3 The experimental and theoretical dependencies of grain growth rate, V_m, on the strain rate in superplastic alloys and ceramics: x - (Al-33 %Cu; o - (IN100); Δ,□,★ - (Ti-6Al-4V) (do = 6.48μm, 9μm and 11μm, respectively).

Integrating this expression at a constant $\dot{\varepsilon}$, one can find dependence of grain size d on the strain value $\varepsilon = \dot{\varepsilon} t$:

$$d(\dot{\varepsilon}) - d_o = c_b^{1-m} A^{-m} \dot{\varepsilon}^{(m-1)} \varepsilon \qquad (20)$$

At the experimental value of m = 0.3 [4], typical for composite TZP - ZrO_2, this expression is very similar to the experimentally observed dependence [4]: $d(\varepsilon) - d_o = \lambda \dot{\varepsilon}^{0.6} \varepsilon$.

b) Large ω ($\rho \Delta b < \omega$; $M_\Delta < M_b < M_\perp$, $\sigma < \sigma_i$).

As shown in [2] this case is realized at large initial grain size and/or small diffusivity. The kinetics of the grain growth are determined by interaction between junction disclinations and its mobility. The expression for $\dot{d}$ is:

$$\dot{d}/b = M_\Delta P_\Delta = c_b (b/d)^2 \qquad (21)$$

and the dependence of the grain size on the deformation time is:

$$(d/b)^3 - (d_o/b)^3 = c_b t = D_b \delta G (kT)^{-1} t \qquad (22)$$

Such dependence d (t) has been observed in experimental studies of the SP deformation of several structural ceramics (ZrO_2 [5]; Al_2O_3 [6,7]; Al_2O_3/ZrO_2 composite) [8].

Calculations of the dependencies $\dot{d}(\dot{\varepsilon})$ for metallic SP-alloys [1,2] and structural ceramics [3] in the present model show good coincidence of the theoretical and experimental curves (Table 1, Fig.3). The material constants used in the theoretical calculations are listed in the Table 2.

4. Cavity nucleation

Formation of disclinations in triple junctions leads to the generation of internal stresses near the joints. Calculation of the disclination dipole critical strength ω^*, for which the nucleation of a grain boundary microcrack becomes energetically favourable gives the expression [1]:

$$\omega^* = [2\pi (1-\nu) (2\gamma_s - \gamma_b) / G (d - l_o) \ln(d/b)]^{1/2} \qquad (23)$$

Where γ_s - is the energy of a free surface, γ_b - the surface energy of the boundary; l_o - the microcrack length. At a high temperature, which is typical for superplastic deformation, microcracks are transformed into cavities, the growth of which may occur by well known diffusion or plastic growth mechanisms.

Using the above (part 2) expressions for ω(t) and the criterion $\omega(t) \geq \omega^*$ it is possible to estimate the time t_o, or the threshold deformation $\varepsilon_o = \dot{\varepsilon} t_o$, of the microcrack nucleation. For a large enough strain rate and grain size ($\omega^{st} > \omega^*$) the value ω reaches the critical value ω^* on the linear part of the curve ω(t) (Fig.2). In this case we can easily obtain the expression for t_o and ε_o.

$$t_o = \omega^* / \Psi \dot{\varepsilon}_v; \; \varepsilon_o = (\omega^* / \xi)(\dot{\varepsilon}/\dot{\varepsilon}_v) \qquad (24)$$

Taking into account, that $\dot{\varepsilon}_v = A_1 \sigma_n$ and $\dot{\varepsilon} = A_2 \sigma^{1/m}$, the expression for ε_o may be represented as:

$$\varepsilon_o = const \, \omega^* / \dot{\varepsilon}^{(mn-1)} \qquad (25)$$

For the typical parameters n= 4; 0.3< m <1 the value of ε_o decreases with increasing strain rate $\dot{\varepsilon}$.

Such a dependence of behaviour for ε_o ($\dot{\varepsilon}$) was observed for superplastic alloys [9] and several ceramics [10-12].

At low strain rate and high diffusional accommodation rate (high temperatures and/or very small grain sizes) $\omega^{st}>\omega^*$ and the strength of the disclination $\omega(t)$ can reach the critical value ω^* due to grain growth only. To a rough approximation we believe that when the strength of the disclination $\omega(t)$ reaches ω^{st}, the value $\omega^{st} \sim \dot{\varepsilon}_v d^3$ (see the formula (10)) do not remain constant, but slowly increases during deformation with increasing grain size, $\omega(t) \sim \varepsilon_v d^3(t)$.

Hence, it will reach the critical level ω^* under some critical grain size d^*. Using the expression (10) the value d can be represented as:

$$d^*/b = (\omega^* c_b / \xi \varepsilon_v)^{1/3} \tag{26}$$

If the dependence d(t) has the form $d_{(t)}^k - d_o^k = At = A\varepsilon/\dot{\varepsilon}$, then the critical strain ε_o may be written as:

$$\varepsilon_o = (\dot{\varepsilon}/A) [(\omega^* c_b / \xi \varepsilon_v)^{k/3} - d_o^k] \tag{27}$$

In the case k = 3 the dependence $\varepsilon_o(\dot{\varepsilon})$ similar to the dependence (24). But in the case n=1 the situation changes. Indeed, taking into account relations $\varepsilon_v \sim \sigma^n$ and $\dot{\varepsilon} \sim \sigma^{1/m}$, an expression for ε_o may be written in the form:

$$\varepsilon_o = [(\omega^* c_b / \xi)^{1/3} \cdot \dot{\varepsilon}^{(1-nm/3)} - d_o] / A \tag{28}$$

At the typical values of parameters $n \leq 4$ and $m < 0.75$, the value of the critical strain ε_o increases with increasing of strain rate $\dot{\varepsilon}$. Such an unusual dependence of the threshold deformation on the strain-rate was observed for yttria stabilized tetragonal zirconia [13].

4. Conclusions

1. A united approach for the description of the microstructural evolution in metallic superplastic alloys and structural ceramics has been developed.
2. The expressions are given which describe the grain growth rate and the dependence of the mean grain size on the strain, strain-rate and materials structure parameters.
3. A mechanism of microcrack nucleation in the grain boundary triple junctions is discussed. The strain-rate dependence of the threshold deformation for the microcrack nucleation has been obtained.

Acknowledgements

The research described in this publication was made possible in part by Grant N R9AOOO from the International Science Foundation. The results of this work were obtained in collaboration with Dr. V.Chuvil'deev.

References

1. V.N.Perevezentsev, V.V.Rybin, V.N.Chuvil'deev, Acta Metall. Mater, 40, 1992, 887-924.

2. V.N.Perevezentsev, O.E.Pirozhnikova, V.N.Chuvil'deev, Fiz. Met. i Metalloved. (in Russ.) 4, 1991, 33-41.

3. V.N.Perevezentsev, O.E.Pirozhnikova, V.N.Chuvil'deev, Neorg. Materials (in Russ.) 29, 1993,

421-425.

4. D.J. Schissler, A.H.Chokshi, T.G.Nieh, J.Wadsworth, Acta Metall.Mater. 39, 1991, 3227-3236.

5. T.G. Nieh, J.Wadsworth, J.Amer.Ceram.Soc. 72, 1989, 1469-1472.

6. A.H.Chokshi, J.R.Porter, J.Amer.Ceram.Soc. 69, 1986, 37-39.

7. K.R.Venkatachari, R.Raj, J.Amer.Ceram.Soc. 69, 1986, 135-138.

8. F.Wakai, H.Kato, Adv.Ceram.Mater. 3, 1988, 71-76.

9. I.I.Novikov, V.K.Portnoi, Superplasticity of Alloys with Ultrafine Grains, Metallurgia, Moscow, 1981.

10. Y.Maehara, T.G.Langdon, J.Mat.Sci. 25, 1990, 2275-2286.

11. I-Wei Chen, L.A.Xue, J.Amer.Ceram.Soc., 73, 1990, 2585-2609.

12. A.H.Chokshi, Superplasticity in Adv.Mater., Ed.S.Hori, A.Tokizane, N.Furushiro, Japan Soc. Res. Superplasticity, 1991, 171-180.

13. A.H.Chokshi, T.G.Langdon, Mater.Sci.Technol., 7, 1991, 577-584.

Table 1. The comparison of experimental and theoretical values of the parameter f in the expression: $d = \text{constant} \cdot \dot{\varepsilon}^f$ ($f = \partial(\log d)/\partial(\log \dot{\varepsilon})$)

Parameter / material	$d_o, \mu m$	T(K)	$\dot{\varepsilon}/s^{-1}$	m	f^{th}	f^{exp}
Al – 33%Cu Kashyap (1987)	6.45	813	10^{-5} 10^{-4} 10^{-3}	0.64 0.66 0.67	0.86 0.77 0.92	0.9 0.75 0.9
IN – 100 Menzies (1982)	2.0	1311	$8.6 \cdot 10^{-5}$ $4.3 \cdot 10^{-4}$ $4.3 \cdot 10^{-3}$	0.4 0.5 0.56	0.31 0.96 0.65	0.3 0.9 0.7
Ti – 6%Al – 4%V Ghosh (1979)	6.4	1200	$2 \cdot 10^{-4}$ 10^{-3}	0.66 0.85	1.38 0.22	1.4 0.2
	11.5	1200	$2 \cdot 10^{-4}$	0.66	$9.5 \cdot 10^{-3}$	<<1
ZrO_2 Nieh (1989)	0.3	1820	$1.2 \cdot 10^{-5}$	~0.5	0.61	0.6
Al_2O_3 Venkatachari (1986)	1.6	1690	$2 \cdot 10^{-4}$	~0.5	0.5	0.55
TZP – Al_2O_3 Wakai (1988) Al_2O_3 - phase ZnO_2 - phase	0.5 0.5 0.5	1720	$5.56 \cdot 10^{-5}$ $5.56 \cdot 10^{-4}$	–	0.11 0.12	0.2 0.13

Table 2. The Constants of Materials.

Material	$b \cdot 10^{10}, m$	$\Omega \cdot 10^{29}, m^3$	$G \cdot 10^{-4}, MPa$	$c_b \cdot 10^{-10}$
$c_b \cdot 10^{-10}$	2.5	1.56	5.5	0.164
Al – 33%Cu	2.8	2.19	2.3	2.87
Ti – 6%Al – 4V	2.9	2.44	4.0	15
Al_2O_3	4.75	4.25	15.5	0.129
ZrO_2	2.57		3.7 ÷ 10	1.3 ÷ 3.73

A Composite Model for Superplasticity

Bernard Baudelet and Jianshe Lian*

Génie Physique et Mécanique des Matériaux, Unité Associée au CNRS

ENSPG, Institut National Polytechnique de Grenoble

BP46, 38402, Saint Martin d'Hères, France

*On leave from the Department of Metal Materials Engineering, Jilin University of Technology, Changchun, 130025, P.R. China.

In the last two or three decades, a number of mechanistic models have been proposed to clarify the dominant microstructural features of superplasticity. Both theoretical models and microstructural observations indicate that the most important feature of superplasticity is the role played by grain boundary sliding (GBS). However, dislocation motion or diffusion in grains or near-grain boundary regions must always be invoked to maintain continuous superplastic deformation. Consequently, most models consider superplastic deformation as a GBS process associated with and rate controlled by the accommodation process i.e., the diffusion-accommodation model of Ashby and Verrall, and the dislocation pile-up accommodation models of Ball-Hutchinson and Mukherjee.

A composite model for superplasticity, based on the joint influences of both the behaviour of a composite boundary and creep, is proposed. In this model, superplasticity is considered as a combination of two mechanisms: grain boundary sliding (GBS) and dislocation creep which occur either together or sequentially. Applied to experimental data, it can describe the logarithmic stress vs strain rate curves observed for superplastic materials showing regions I, II and II, by a sequence of controlling mechanisms. Region I is characterised by creep inside the grains which controls the deformation of the composite boundary. The transition from region I to region II reflects the progressive transfer of control to the behaviour of the grain boundary which is accompanied by the development of increasingly wider shear bands in the grains. The composite boundary controlled deformation in region II is progressively replaced by an overall creep in the material to finally reach region III in which the composite boundary completely loses its influence.

Cavitation During Superplastic Flow

Cavitation in Superplastic Materials

N. Ridley and Z.C. Wang
University of Manchester and UMIST, Materials Science Centre
Grosvenor Street, Manchester M1 7HS, UK

Abstract

The effect of microstructure and deformation conditions on aspects of the nucleation, growth and coalescence, of cavities during superplastic flow in a range of materials including metals, intermetallics, MMC's and ceramics, is reviewed. There is evidence that cavities are most likely to nucleate at grain boundary particles or at triple points in quasi-single phase materials, at triple points and grain boundary ledges in microduplex materials free from boundary particles, and at particulate or whisker reinforcement in MMC's. There is strong evidence that cavity growth is usually dominated by matrix plastic flow and that coalescence makes a significant contribution to cavity growth and to the development of large cavities. For fine grained materials which exhibit high strain rate superplasticity at temperatures close to the solidus, the presence of liquid phase appears to inhibit the rate at which cavities grow. In some materials, cavities occur in the form of grain boundary cracks, while in ceramic materials such as aluminas, they propagate as cracks in a direction transverse to the tensile axis. High levels of cavitation and/or the propagation of transverse cracks lead to a pseudo-brittle fracture with little evidence of ductility at the fracture face. Cavitation can be minimised by careful control of processing so as to develop a uniform stable fine-grain microstructure in which dispersoids or reinforcement additions should be small and uniformly distributed. Control of cavitation can be effected by annealing and/or applying a hydrostatic pressure either prior to, during or after deformation, although the most effective procedure is to impose pressure during forming.

1. Introduction

Cavitation can occur in most quasi-single phase and microduplex materials, including metals, intermetallics, metal matrix composites and ceramics, during superplastic (SP) tensile flow. As the literature on cavitation is substantial, particularly for metallic materials, this has led to a reasonable understanding of many aspects of the topic. Cavitation in superplastic materials has also been the subject of a number of major reviews [1-10].

Cavities either nucleate, or develop from pre-existing defects, and their growth, coalescence and interlinkage leads to premature failure. This may manifest itself as a pseudo-brittle fracture. However, because of their high strain rate sensitivities, SP materials are often able to tolerate large volume fractions of cavities prior to failure. Cavities are likely to have a deleterious effect on post-forming properties and to limit the range of applications of commercially formed parts. Most of the current understanding of cavitation is derived from studies on metallic materials. As a consequence, the present paper will review aspects of cavity nucleation, cavity growth and coalescence, and procedures for controlling cavitation in metallic materials, and then examine important features of cavitation in MMC's and ceramics.

2. Cavitation in Metallic Materials.

2.1 Cavity nucleation

It is widely held that during SP flow, strain is accumulated primarily as a result of grain boundary sliding. To maintain structural integrity, it is necessary that boundary displacements are accommodated by the redistribution of matter through diffusion or dislocation processes. When accommodation processes are not rapid enough to meet the requirements imposed by boundary sliding, then stresses which develop at grain boundary features are not relaxed sufficiently rapidly and cavities may nucleate. Experimental observations suggest that cavities are most likely to develop at grain boundary particles in quasi-single phase alloys, and at triple points and boundary ledges in microduplex alloys which are free from coarse particles.

Stowell [1] has proposed a relationship for the critical strain below which cavity nucleation at a grain boundary particle was likely to be inhibited by boundary diffusion stress relaxation. This predicted that cavitation would be minimised by having small grain and particle sizes, and that SP forming should be carried out at as high a temperature and as slow a strain rate as possible, consistent with microstructural stability and sensible commercial practice. Chokshi and Mukherjee [11] developed a similar relationship which included an effective diffusion coefficient, which incorporated all possible diffusion paths. They also modelled cavity nucleation at grain boundary ledges, showing that this was possible for a limited range of conditions.

In contrast to approaches which relate cavity nucleation to incomplete diffusive accommodation of grain boundary sliding, Perevezentsev et al [12] relate nucleation to slip processes. During optimum SP flow, it is proposed that about 25% of the total strain is accumulated by slip. It is the interaction of slip dislocations with various microstructural features such as grain boundary particles which leads to the development of microcrack cavity nuclei when a threshold SP strain is exceeded, e.g. the threshold stress for a particle of radius $r_p = 10^4 b$ would be 0.28. Continuous nucleation during straining is predicted starting with the largest particle when the threshold stress is exceeded. Nucleation in materials undergoing grain growth was also considered, and relationships for the volume fraction of cavities as a function of superplastic strain were developed for quasi-single phase and duplex materials.

Experimental work on a range of materials, mainly alloys, has shown that factors which influence cavity nucleation include those which relate to microstructure such as grain size, the type, size, volume fraction, interfacial energy and distribution of second phase particles, the proportions and properties of the matrix phases, and those which are associated with the deformation conditions such as strain, strain rate, temperature and stress state [9]. There is a strong interaction between microstructural and deformation parameters which is consistent with the relationships developed by Stowell [1], Chokshi and Mukherjee [4], and Perevezentsev et al [12]. Since many of the above factors have been discussed in recent reviews [2-10,13] only selected aspects of cavity nucleation will be considered.

The important role of grain size has been demonstrated for a number of materials, and grain growth has been invoked to explain continuous nucleation reported for Al alloys during SP flow [14-16]. The importance of grain size on cavitation in the highly SP Zn-22%Al alloy has been identified [17], but for the same alloy Park and Mohamed [18] have recently reported a correlation between increasing cavity nucleation at low strain rates, and increasing levels of impurities in the range 6-160ppm. It has been suggested that if impurities segregate to grain boundaries this could decrease the critical nucleus size by either reducing boundary cohesive strength, forming grain boundary precipitates (Fe-rich in the case of SP Zn-22%Al), or reducing the surface energy of vacancy clusters. The role of impurities in cavity nucleation is not clear and could benefit from further investigation.

For duplex materials, the characteristics and properties of the phases present can have a strong influence on the extent of cavitation observed. Microduplex α/β Ti alloys such as Ti-6Al-4V are remarkably resistant to cavitation at their optimum SP deformation temperatures [19]. In SP alloys the relatively soft bcc β-phase is believed to play an important role in the accommodation of grain boundary sliding by volume diffusion and/or plastic flow [20]. At the normal SP deformation temperature of ≈900°C, Ti alloys contain approximately equal volume fractions of the two phases with the β-phase providing accommodation of grain boundary sliding while the α-phase stabilizes the microstructure against substantial grain growth. The Ti alloys also take impurities such as oxides, carbides and nitrides into solution so they are usually free from hard grain boundary particles.

The intermetallic alloy Super Alpha-2 (based on Ti$_3$Al) can be processed to develop a microduplex structure similar to that seen in the α/β Ti alloys, and at its optimum SP deformation temperature (≈960°C), it can sustain considerable tensile elongations (>1000%) and remain completely free from cavitation [21]. However, cavitation attributable to localised grain coarsening has been reported for α$_2$(Ti$_3$Al)+γ alloys [22]. Other intermetallic alloys which may cavitate during SP flow are Ni$_3$Al [23] and duplex alloys based on Ni$_3$Si [24]. For Ni$_3$Al the extent of cavitation increased with increasing grain size and strain rate, as is usually observed for metallic systems, although for the duplex Ni$_3$Si alloy it was reported that cavitation increased with decreasing strain rate for a given strain. It is not clear whether the latter effect is related to grain growth at slower strain rates.

Zelin et al [25,26] have recently reported intersecting rows of cavities in 7475 Al deformed by biaxial bulging and in uniaxial tension, and for γ-TiAl deformed in uniaxial tension. This configuration was attributed to cavity nucleation at intersecting surfaces of co-operative grain boundary sliding (CGBS), and it was suggested that CGBS may be a general phenomena in SP materials.

As was mentioned earlier, it is believed that cavities may nucleate during SP flow or that they may pre-exist. However, it has been proposed recently that small pre-existing cavities are likely to sinter out during heating up and holding at elevated temperature prior to SP deformation [11,16]. This is likely to be true for pre-existing cavities with radii up to ≈1μm, provided they do not contain gases.

2.2 Cavity growth

A cavity located on a grain boundary, whether nucleated or pre-existing, may grow during SP flow by either stress directed vacancy diffusion along grain boundaries which intersect the cavity surface, or by plastic deformation of the surrounding matrix. Relationships which describe the change of cavity radius, r, with strain, ε, have been developed for the different mechanisms. For diffusion controlled growth of small spherical cavities which are equi-sized, it can be shown that $dr/d\varepsilon \propto \sigma_1/r^2$, where σ_1 is the maximum principal stress [27]. Hence, the rate of growth will slow down rapidly as the cavity grows. Provided there is not substantial continuous cavity nucleation during deformation, then the cavity volume fraction $C_v \propto \varepsilon$. When cavities grow to a size greater than the grain size, d, there will be an enhanced growth rate due to additional mass transfer via the boundaries intersecting the cavity. This is termed SP diffusion growth. If vacancy diffusion continues to dominate, then Chokshi and Langdon [28] have shown $dr/d\varepsilon \propto \sigma_1/d^2\varepsilon$, i.e. $dr/d\varepsilon$ is independent of cavity radius, and $C_v^{1/3} \propto \varepsilon$. If growth is plasticity controlled, the model of Hancock [29] gives $dr/d\varepsilon \propto r$; that is, the rate of cavity growth increases linearly with cavity size and is independent of strain rate. Hence, $C_v \propto e^\varepsilon$ or

$$C_v = C_0 \exp(\eta \varepsilon) \qquad (1)$$

$$\text{where} \quad \eta = \frac{3}{2}\left(\frac{m+1}{m}\right)\sinh\left[2\left(\frac{2-m}{2+m}\right)\left(\frac{K}{3} - \frac{P}{\sigma_e}\right)\right] \qquad (2)$$

$$\text{and} \quad \left(\frac{K}{3} - \frac{P}{\sigma_e}\right) = \frac{\sigma_m}{\sigma_e} \quad (3)$$

where η is termed the cavity growth rate parameter, C_o is a constant, m is the strain rate sensitivity, σ_e is equal to the uniaxial flow stress, K is a constant whose value is dependent on test geometry and the extent of grain boundary sliding, P is the imposed pressure and σ_m the mean stress. It has been shown that if 50% of superplastic strain is attributable to grain boundary sliding, then $K=1.5$ and 2.25 for uniaxial and biaxial straining, respectively [9].

Hence, it is seen that the different cavity growth mechanisms have different dependencies of cavity volume fraction on strain. In the case of plasticity control, a linear relationship between $\log C_v$ and ε is predicted (Eqn.1). To reduce the level of cavitation for a given strain it is necessary to reduce the magnitude of η, the cavity growth rate parameter, which is dependent on m and the ratio σ_m/σ_e, (Eqns 2 and 3). An imposed hydrostatic pressure, P will reduce this ratio (Eqn.3), and consequently both η and C_v, to zero when $P=K\sigma_e/3$.

An additional growth process in which cavities grow along grain boundaries to develop a crack-like network has been discussed by Ma and Langdon [30]. This type of behaviour has been observed in several alloys and would be expected when vacancies are being transported by grain boundary diffusion to the edge of cavities more rapidly than surface diffusion can redistribute them within the cavity.

Models of cavity growth can be tested using measurements of cavity growth rates (dr/dε), obtained by quantitative metallography, as a function of superplastic strain. In Fig.1, calculated cavity growth rates as a function of cavity size for the growth mechanisms outlined above are compared with experimental data for a duplex stainless steel. While diffusion control may be important in the early stages of growth, where it is exceedingly difficult to measure growth rates, it is clear that strain controlled growth becomes dominant for void radii above about 1μm. Similar observations have been reported for Al and Cu-based alloys. Only for materials with a fine grain size deformed at a slow strain rate does it appear that superplastic diffusion growth may be significant, as will be discussed later for nano-phase materials.

As discussed above, a plot of cavity volume fraction, which can be obtained using precision density measurements or quantitative metallography, versus strain should also enable the dominant growth mechanism to be identified. This has been done for a range of superplastic alloys and in all cases the data were consistent with plasticity control of growth [6]. However, it should be noted that recently Perevezentsev et al [12] reported measurements of volume fractions of cavities in a superplastically deformed quasi-single phase Mg-base alloy which appear to be consistent with diffusional control of cavity growth.

Fig.2 gives measured cavity volume fraction data for Al-7475 biaxially deformed with and without imposed pressure, which are fully in accord with the predictions of equations (1), (2) and (3) for plasticity dominated growth [31]. That is, a plot of the logarithm of cavity volume fraction versus strain is linear, and the application of imposed pressure has reduced the value of η, the cavity growth rate parameter, and hence the level of cavitation for a given strain. The above equations also predict the effect of stress state and, in particular, that the level of cavitation for a given strain will be higher for equibiaxial tension than for uniaxial tension. This is confirmed by the data in Fig.3, where to avoid the effect of cavity coalescence, continuous nucleation and non-random cavity distributions, the growth of artificially drilled small holes has been examined for both uniaxial and equibiaxial tension [32].

Owen et al [33] have recently modified the single cavity diffusion growth model of Schniebel and Martinez [34] and applied it to very fine grain (nano-phase) superplastic materials, for which the cavities are usually comparable to or larger than then the grain size, and single cavities are intersected by more than one grain boundary. The model, termed the single cavity SP diffusion

Fig.1 Predicted and measured cavity growth rates for a microduplex stainless steel.

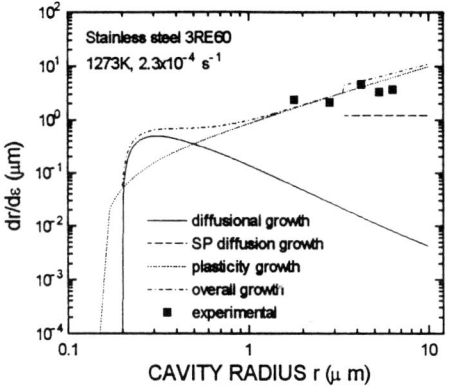

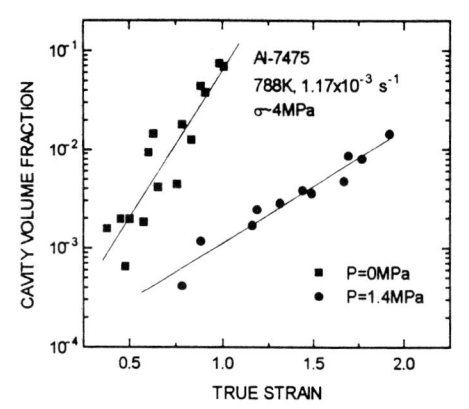

Fig.2 Effect of imposed pressure on cavitation in 7475 Al alloy.

growth model, predicts an enhanced contribution to the cavity growth rate, significantly greater than the earlier superplastic diffusion growth model developed by Chokshi and Langdon [28], although it remains to be tested.

2.3. Cavity coalescence

It has been observed that the predicted sizes to which cavities grow when plastically controlled growth is dominant are often significantly less than is observed experimentally [9]. These differences are attributed to cavity coalescence, for which there is considerable metallographic evidence. The large elongated cavities frequently seen lying parallel to the rolling directions in aluminium and copper-based alloys are almost certainly due to the growth and coalescence of cavities nucleated on closely spaced inclusions or particles. Coalescence extends the size distribution to large radii, and produces the large cavities which have the most deleterious effect on post-forming properties of superplastic alloys. For Al alloys it is likely that in some cases hydrogen outgassing leads to fine localised porosity which provides pre-existing sites for cavity growth and subsequent coalescence [35].

The extent of cavity coalescence during superplastic deformation depends on the size, population density and spatial distribution of cavities. The effect of pair-wise coalescence on the evolution of cavity size distribution in an α/β copper alloy was examined by Stowell et al [36]. Good agreement was obtained between the size distribution measured at a higher strain and that calculated from the distribution measured for a lower strain, except for the largest voids in the distribution. The discrepancies were attributed to a non-random spatial distribution of voids, a feature often seen in cavitated specimens. Pilling [37] has examined the effect of pair-wise coalescence on cavity growth rates in a 7475 Al alloy and obtained good agreement between predicted values and measured growth rates, while Wilkinson and Caceres [38] have proposed that strain hardening due to grain growth can play a significant role in cavity growth and hence in coalescence.

2.4. Control of cavitation

To minimize cavitation during superplastic flow it is necessary to exercise sound microstructural control. The starting materials should be processed to develop a fine uniform grain size and if dispersoids are present, these should be fine and uniformly dispersed. The development of cavitation requires the presence of a local tensile stress. Under conditions of homogeneous compression cavitation is not observed and cavities produced during superplastic tensile flow can be removed by subsequent compressive flow [2].

Cavitation can be controlled by various procedures including annealing and/or the application of hydrostatic pressure prior to, during, or after, superplastic deformation [16]. Annealing before deformation can lead to sintering of pre-existing cavities by vacancy diffusion. For 7475 Al alloys, there is evidence that annealing leads to homogenization and, in particular, to the elimination of incipient melting believed to be associated with metastable constituent particle phases, which causes enhanced cavitation [16,39]. However, as will be seen in the next section, there are circumstances where incipient melting plays a significant role in the minimization of cavitation during SP flow. Annealing in vacuum can be particularly effective for Al alloys leading to hydrogen outgassing, but care must be taken to avoid grain growth and loss of solute by volatilization, e.g. Li.

The application of a hydrostatic pressure during forming is the most effective way of controlling cavitation. Data available for Al alloys deformed in uniaxial, equi-biaxial and plane strain conditions show that increasing pressure reduces the rate at which the overall volume fraction of cavities increases with strain, decreases the level of cavitation for a given strain, displaces to a higher value the strain at which cavitation is first detected, and increases the strain to failure (Fig.2).

Since there is strong evidence that cavity growth is plasticity dominated, then it can be seen from equations (1-3) that to prevent cavity growth during superplastic forming it is necessary for $P \geq K\sigma_e/3$ (or $\sigma_m/\sigma_e \leq 0$) i.e. $P \geq 0.5\sigma_e$ for uniaxial deformation (although $P > \sigma_e/3$ is often quoted in the literature) and $P > 0.75\sigma_e$ for equibiaxial bulging. These predictions are broadly in agreement with experiment. However, this approach assumes that neither nucleation nor diffusion controlled cavity growth will be eliminated as both require $P > \sigma_e$. Diffusional growth is only likely to be important for small cavity sizes and will have little effect on the overall level of cavitation, although Chokshi and Mukherjee [4] have suggested that imposed pressure may exert a significant effect on the inhibition of cavity nucleation. In commercial forming practice, it is unlikely that the pressure would exceed 3.5MPa for safety reasons, so the value of SP flow stress, σ_e, for a given material should not be too high if cavitation is to be avoided. However, even if the condition $P > K\sigma_e/3$ is not met, any level of imposed pressure is beneficial in its effect in reducing cavitation.

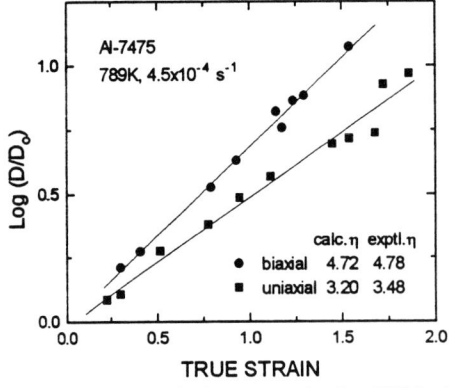

Fig.3 Hole growth in superplastic 7475 Al alloy (Ref.32).

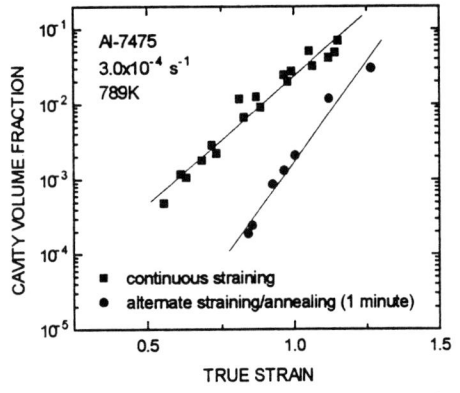

Fig.4 Effect of strain/anneal on cavitation (after Ref.16).

Other methods for the control of cavitation which have been proposed include an alternating strain/anneal technique, in which a short period of straining followed by a period of annealing enables small cavities produced during straining to sinter by diffusion [16]. The significant effect of this procedure is seen in Fig.4, where a cycle of 1 minute straining is followed by annealing for 1 minute. Although annealing following superplastic deformation may remove small cavities, it is less effective with large cavities, which have the most deleterious effect on properties unless long annealing times are used [40]. Post-SPF HIPping can also remove cavities, provided there is no open porosity, but it is likely to be restricted because of cost and limitations on component size. For Al alloys, cavities may also reappear on subsequent heat treatment, probably due to the presence of hydrogen, indicating that cavities are not all completely sintered on HIPping [41].

Recently Conrad and co-workers [42] have shown that the application of an external electrical field during the SPF of 7475 Al alloy reduced the level of cavitation to a significant extent for given strain, and led to the improvement of post SPF properties. The effect on cavitation was due to a reduction in cavity nucleation rate which was attributed to a field-enhanced accommodation of grain boundary sliding involving migration of lattice defects such as dislocations and vacancies, although this was accompanied by an increase in cavity growth rate.

3. Cavitation in Metal Matrix Composites

Mahoney and Ghosh [43] observed enhancement of cavitation in a PM 7000 series Al alloy containing particulate SiC reinforcement. Both SiC particles and oxides from the Al alloy powder starting material provided cavity nucleation sites during SP deformation, additional to those associated with impurity particles normally found in 7000 series alloys. However, flow stresses were relatively low at the optimum strain rate so that cavitation could be inhibited by the application of low hydrostatic pressures, ≤2MPa. Pilling [44] has also demonstrated that hydrostatic pressure applied to a Al-2014/SiC$_p$ and Al-7475/SiC$_p$ composites led to the inhibition of cavitation during superplastic flow and to markedly enhanced strains to failure.

One of the most exciting recent developments which was initially reported for a SiC whisker reinforced Al alloy [45] is that of superplasticity at high strain rates extending up to 10^2 s^{-1} in Al-based composites containing SiC or Si$_3$N$_4$ particles or whiskers [46], and also in metallic alloys, produced by advanced processing methods. These procedures include mechanical alloying, vapour phase deposition and consolidation of amorphous or nanocrystalline powders. The materials are characterised by having nano or near-nano scale grain sizes. Recent work on high strain rate SP is discussed in papers by Higashi and by Iwasaki et al in the current proceedings.

It has become clear that optimum SP strains at high strain rates occur at temperatures close to or just above the incipient melting points of these fine grained materials, such that small amounts of liquid are present at interfaces in the composites and at grain boundaries in the alloys [47]. It was also noted that the actual temperature could be higher than the nominal temperature due to adiabatic heating at rapid strain rates [47]. The corresponding m values are > 0.3 and are consistent with grain boundary sliding, for which there is experimental evidence, as an important deformation mechanism [46]. For MMC's it is proposed that the presence of liquid at the interface of the matrix and the reinforcement results in relief of stress concentrations due to sliding. As a consequence, this limits the development of cavities and gives an enhancement of superplasticity. Higher deformation temperatures give increased amounts of liquid phase such that grain boundaries can no longer sustain normal traction and this leads to failure at low elongations.

While a range of cavitation behaviours has been observed for rapidly deformed MMC's [48,49], it appears that nucleation may occur readily at reinforcement/matrix interfaces at low strains but the cavities grow relatively slowly with strain. Figure 5 shows the distribution of cavity sizes for an Al-Mg-Si composite deformed at 2 s^{-1} and it can be seen that there is little overall growth between strains of 0.2-1.0 when the volume of cavities changes from 0.3 to 0.4%. The

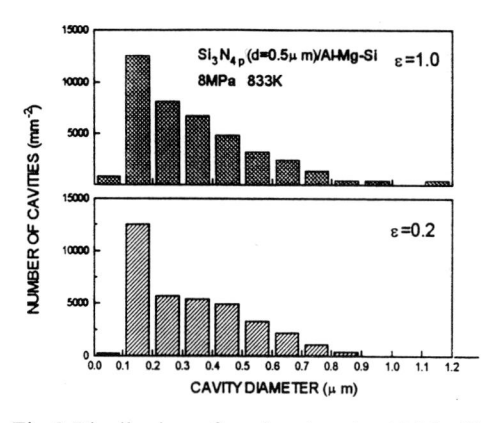

Fig.5 Distribution of cavity sizes in Al-Mg-Si composite at SP strains of 0.2 and 1.0.

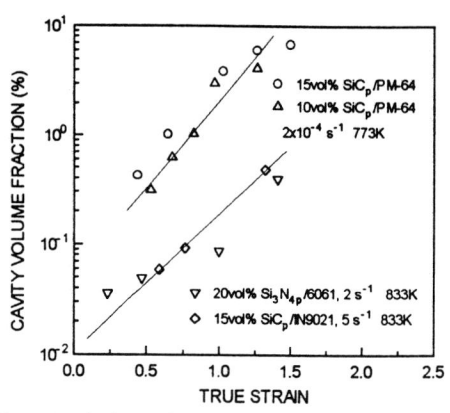

Fig.6 Variation of cavitation with SP strain for conventional and HSRS composites (Ref. 48).

levels of cavitation are lower than those usually obtained in larger grain size composites deformed under conventional conditions. This is illustrated in Fig.6, where the lower curve is for fine grain size composites deformed at a high rate near the solidus compared with slower deformation at lower temperatures for a powder metallurgy composite. It is seen that the curve for high strain deformation shows a lower slope, i.e. a lower cavity growth rate parameter, η. The mechanism whereby the presence of liquid inhibits cavity growth is not clear.

4. Cavitation in Ceramics

Following the observations of large tensile strains in ceramics reported by Wakai and co-workers [50] in 1986 for an yttria stabilized TZP, a range of fine grain monolithic and composite ceramics has been reported to exhibit superplasticity in tension [51]. While many of these materials exhibit cavitation, the extent to which this phenomenon has been systematically studied is relatively limited. As a consequence, attention will, for the most part, be concentrated on 3Y-TZP, alumina (with and without dopants) and composites of these two materials, particularly 20%Al_2O_3/3Y-TZP, which are considered to be free from grain boundary glassy phases. Of particular relevance is the view, supported by experimental observations, that despite the high m values often recorded for fine grain ceramics, the elongation to failure is strongly influenced by the boundary strength [52,53]. When the flow stress reaches the grain boundary cohesive strength, transverse cracks develop and their interlinkage limits the elongation to failure. As would be expected, the existence of impurities or glassy phases at grain boundaries will influence the strain at which failure occurs.

Many feature of cavitation in ceramics are fully consistent with observations made for metallic systems. There is metallographic evidence that cavities can nucleate at triple point junctions for high purity and doped aluminas [54] and 3Y-TZP [55], while for 20%Al_2O_3/3Y-TZP cavities form at both ZrO_2/ZrO_2 and Al_2O_3/ZrO_2 boundaries, possibly nucleating at ledges [56]. The effect of grain size is illustrated by the selective nucleation of cavities at clusters of coarse grains in Y-TZP [55] and Al_2O_3 [52], or at agglomerations of coarse particles in 20%Al_2O_3/3Y-TZP [57]. For Y-TZP there is evidence that the continuous nucleation and coalescence play a role in the evolution of cavitation [58]. The volume of cavities increases as strain rate increases [55,58], consistent with work on 20%Al_2O_3/3Y-TZP [57,59] and as temperature decreases [58]. At tensile elongations of ~200%, Schissler et al [55] reported a volume fraction of cavities of ~30% in the region of fracture. Cavity interlinkage in fine grained ceramics deformed in the SP range leads to a pseudo-brittle fracture, with little signs of necking.

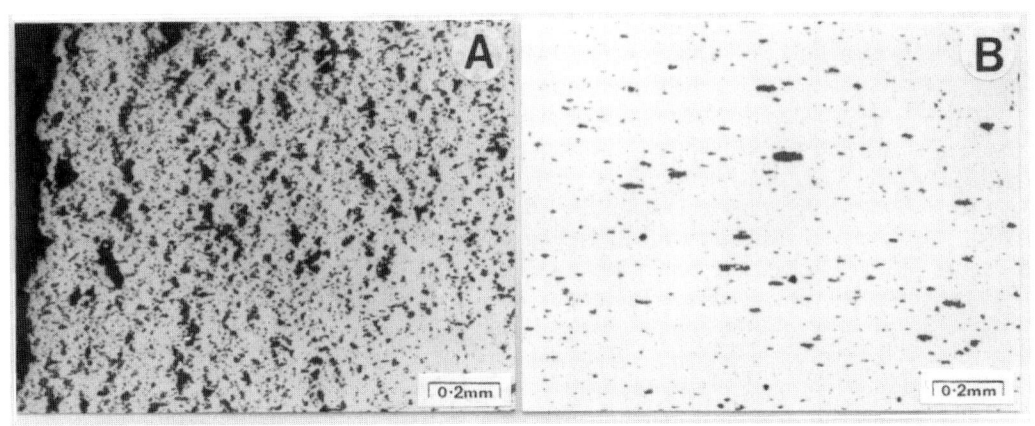

Fig.7 Distribution of cavities in Y-TZP, A - near fracture face; B - in gauge length. 1823K, 10^{-4} s^{-1}, e_f = 480%.

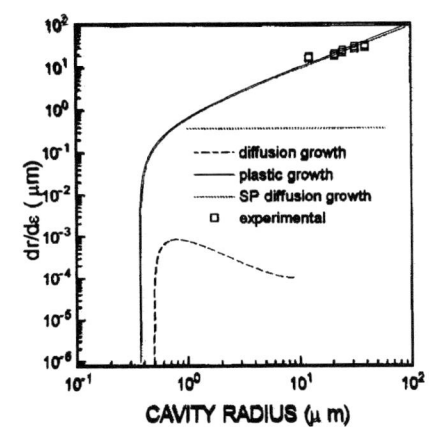

Fig.8 Cavity growth rates for Al_2O_3 doped with CuO, 1673K, 1×10^{-4} s^{-1}.

Two types of cavitation behaviour have been reported for SP ceramics. In most ceramics, including pure and doped Al_2O_3, cavities or cracks grow in a direction perpendicular to the applied tensile stress and are uniformly distributed throughout the gauge section [60]; with increasing strain these grow and interlink to form larger cracks leading to failure at relatively low strains. Pure alumina has a relatively high grain boundary energy. This provides a high driving force for grain growth and may cause grain boundaries to have a low cohesive strength [52]. However, in superplastic 3Y-TZP a cavitation mode similar to that in metallic systems is observed where cavities grow and align parallel to the tensile axis, Fig.7B. Consequently, the elongation to failure is significantly higher than for other ceramic systems. Near the fracture face for Y-TZP, cavities initially parallel to the tensile axis interlink in the transverse direction either by necking as in metallic systems, or more usually by the development and propagation of transverse cracks, Fig.7A. Schissler et al [55] have reported similar observations for Y-TZP, and propose that cavities lying parallel to the tensile axis only develop at high elongations (>400%).

Davies and Ogwu [61] have proposed that the elongated cavities and the relatively high strains to failure (up to 800%) for superplastic Y-TZP may be explained by consideration of the d-state electrons which contribute a 'metallic' character to the bonding.

It has been shown in an earlier section that cavity growth in metals is usually controlled by plastic flow. To assess cavity growth mechanisms in ceramics, artificial cavities were introduced by the authors into fine grain Y-TZP specimens and their shape and size have been investigated as a function of strain. It was shown that the change of mean pore radius, r, with strain, ε, fitted the relationship $dr/d\varepsilon = \eta r$, consistent with plasticity control of growth. Similar data obtained for cavities of 10-40μm for fine grained Al_2O_3 doped with CuO, which is considered to enhance metallic character, are shown in Fig.8, and again are in good agreement with the plasticity controlled growth mechanism. For 20%Al_2O_3/3Y-TZP Wakai and Kato [59] obtained a linear relationship between density loss and true strain, also consistent with plasticity control of growth.

Summary

There is considerable volume of literature relating to experimental aspects of cavitation during superplastic flow in metals and, to a lesser extent, in intermetallics, MMC's and ceramics. Theoretical aspects of cavitation have received much less attention. It is clear that cavities are often associated with grain boundary particles, and with triple point junctions and interphase boundaries, in the apparent absence of boundary particles. The mechanisms and kinetics of cavity nucleation are not clear, nor is the role of pre-existing defects. However, the influence of microstructural parameters and deformation conditions on cavity nucleation is reasonably well understood.

There is considerable evidence that cavity growth is dominated by matrix plastic flow with cavities tending to become aligned parallel to the applied stress. However, for fine grained materials which exhibit high strain rate superplasticity at temperatures close to the solidus, the presence of liquid phase appears to inhibit the rate at which cavities grow, but not the nucleation of cavities. Coalescence makes a significant contribution to cavity growth and to the development of large cavities likely to be most damaging in their effect on post-forming properties. There are few published micrographs showing cavitation in which there is not clear evidence of coalescence.

In several materials, cavities occur in the form of grain boundary cracks, while in ceramic materials such as aluminas, cavities propagate and interlink in a direction transverse to the tensile axis. High levels of cavitation and/or the propagation of transverse cracks leads to a pseudo-brittle fracture showing little macroscopic evidence of ductility at the fracture face. Further studies could be usefully made of the role of continuous cavity nucleation on the relationship of the overall volume of cavities and strain, and on the influence of impurities on cavity nucleation and growth.

Cavities can be minimised by exercising careful control of processing to develop a fine grain structure, with dispersoids or reinforcement particles in a fine and uniformly dispersed form. Control of cavitation can be achieved annealing and/or applying a hydrostatic pressure either prior to, during or after, superplastic deformation. The most effective procedure would be to apply a hydrostatic pressure during forming, although under commercial conditions it is unlikely that the imposed pressure could be much in excess of ~3½MPa (500psi). Hence the elimination of cavitation during superplastic flow could only be feasible for materials with flow stresses of ~5-7MPa, although the application of any level of hydrostatic pressure would be beneficial in reducing the overall level of cavitation.

References

1. M.J. Stowell, in 'Superplastic Forming of Structural Alloys', (Ed. N.E. Paton and C.H. Hamilton), 321-336; 1982. TMS, AIME, Warrendale, PA.
2. N. Ridley and J. Pilling, in Ref.2, 8.1-8.17.
3. M. Suery, in 'Superplasticity', (Ed. B. Baudelet and M. Suery), 9.1-8.19; 1985. CNRS, Paris.

4. A.H. Chokshi and A.K. Mukherjee, in 'Superplasticity and Superplastic Forming', (Ed. C.H. Hamilton and N.E. Paton), 149-159; 1988. TMS, Warrendale, PA.
5. A.H. Chokshi, in 'Superplasticity in Adv. Mater.', ICSAM-91 (Ed. S. Hori, M. Tokizane and N. Furushiro), 171-180. 1991, JSRS, Osaka.
6. J. Pilling, in Ref.5, 181-190.
7. M.J. Stowell, in 'Deformation of Multi-phase and Particle Containing Materials', (4th Riso Int. Symp.) (Ed. J.B. Bilde-Sorensen et al), 119-129; 1983. Riso National Laboratory, Roskilde.
8. B.P. Kashyap and A.K. Mukherjee, Res Mechanica, 17, 1986, 293-355.
9. J. Pilling and N. Ridley, Res Mechanica, 23, 1988, 31-63.
10. N. Ridley and Z.C. Wang, in 'Superplasticity in Adv. Mater., ISCAM-94 (Ed. T.G. Langdon), 177-186; 1994. Trans. Tech. Publications, Switzerland.
11. A.H. Chokshi and A.K. Mukherjee, Acta Metall. 37, 1989, 3007-3017.
12. V.N. Pervezentsev, V.V. Rybin and V.N. Chuvil'deev, Acta Metall. Mater. 40, 1992, 915-924.
13. J. Pilling and N. Ridley, 'Superplasticity in Crystalline Solids', 1989. Institute of Metals, London.
14. J. Pilling, B. Geary and N. Ridley, in' Proc. ICSMA 7', (Ed. H.J. McQueen and J.-P. Bailon), 823-829; 1985. Pergamon Press, Oxford.
15. A.K. Ghosh,in 'Deformation of Polycrystals: Mechanisms and Microstructures', (Ed. M. Hansen et al), 277-283; 1981. Riso National Laboratory, Denmark.
16. A. Varloteaux, J.J. Blandin and M. Suery, Mater. Sci. Technol. 5, 1989, 1109-1117.
17. D.W. Livesey and N. Ridley, J. Mater. Sci. 17, 1982, 2257-66.
18. K.-T. Park and F.A. Mohamed, Metall. Trans. 21A, 1990, 2605-2608.
19. M.T. Cope and N. Ridley, Mater. Sci. Technol. 2, 1985, 140-145.
20. M. Suery and B. Baudelet, in Ref.1, 105-127.
21. N. Ridley, M.F. Islam and J. Pilling, in 'Intl. Symp. on Struct. Intermetallics', (Ed. R. Darolia et al), 63-68; 1993. TMS Warrendale, PA.
22. T. Maeda, M. Okada and Y. Shida, in Ref.5, 311-316.
23. A. Choudhury, A.K. Mukherjee and V.K. Sikka, J. Mater. Sci. 25, 1990, 3142-3148.
24. S.L. Stoner, W.C. Oliver and A.K. Mukherjee, Mater. Sci. Engin., A153, 1992, 465-469.
25. M.G. Zelin, H.S. Yang, R.Z. Valiev and A.K. Mukherjee, Metall. Trans. 24A, 1993, 417-424.
26. W.B. Lee, M.G. Zelin and A.K. Mukherjee, in 'Strength of Materials',ICSMA-10, (Ed. H. Oikawa et al), 823-826, 1994. Jap. Inst. Met., Nagoya.
27. M.V. Speight and W. Beere, Metal. Sci. 12, 1978, 172-176.
28. A.H. Chokshi and T.G. Langdon, Acta Metall. 35, 1987, 1089-1101.
29. J.W. Hancock, Metal Sci. 10, 1976, 319-325.
30. Y. Ma and T.G. Langdon, in Ref.4, 173-178.
31. J. Pilling and N. Ridley, Acta Metall. 34, 1986, 669-679.
32. Z.X. Guo and N. Ridley, Mater. Sci. Technol. 6, 1990, 576-519.
33. D.M. Owen, A.H. Chokshi, Y. Ma and T.G. Langdon, in Nanoceramics (Ed. R. Freer), 61-74; 1993. Inst. Mater., London.
34. J.H. Schneibel and L. Martinez, Scripta Metall., 21, 1987, 495-500.
35. C.C Bampton, M.W. Mahoney, C.H. Hamilton. A.K. Ghosh and R. Raj, Metall. Trans. 14A, 1983, 1583-1591.
36. M.J. Stowell, D.W. Livesey and N. Ridley, Acta Metall. 32, 1984, 35-42.
37. J. Pilling, Mater. Sci. Technol., 1, 1985, 461-465.
38. D.S. Wilkinson and C.H. Caceres, Mater. Sci. Technol. 2, 1986, 1086-1092.
39. C.C. Bampton and J.W. Edington, Metall. Trans. 13A, 1982, 1721-1727.

40. N. Ridley, D.W. Livesey and A.K. Mukherjee, Metall. Trans. 15A, 1984, 1343-50.
41. H. Ahmed and R. Pearce, in 'Superplasticity in Aerospace', (Ed. R. Pearce and L. Kelly), 146-159; 1985. SIS, Cranfield, 1985.
42. H. Conrad, W.D. Cao, X.P. Lu and A.F. Sprecher, Mater. Sci. Eng. A138, 1991, 247-258.
43. M.W. Mahoney and A.K. Ghosh, Metall. Trans. 18A, 1987, 653-671.
44. J. Pilling, Scripta Metall., 23, 1989, 1375-1380.
45. T.G. Nieh, C.A. Henshall and J. Wadsworth, Scripta Metall., 18, 1984, 1405-1408.
46. M. Mabuchi, K. Higashi and T.G. Langdon, Acta Metall. Mater. 42, 1994, 1739-1745.
47. T.G. Nieh, J. Wadsworth and T. Imai, Scripta Metall. Mater. 26, 1992, 703-708.
48. K. Higashi and M. Mabuchi, Mater. Sci. Eng. A176, 1994, 461-470.
49. H.W. Iwasaki, M. Mabuchi and T.G. Langdon, in Ref.10, 537-542.
50. F. Wakai, S. Sakaguchi and Y. Matsuno, Adv. Ceram. Mater. 1, 1986, 259-263.
51. A. Chokshi A.K. Mukherjee and T.G. Langdon, Mater. Sci. Engin. R10, 1993, 237-274.
52. I.W. Chen and L.A. Xue, J. Am. Ceram. Soc., 73, 1990, 2585-2609.
53. W.J. Kim, J. Wolfenstine and O.D. Sherby, Acta Metall. Mater., 39, 1991, 199-
54. C. Carry, in 'Proc. MRS Intl. Meeting on Adv. Mater.', Vol.7, (Ed. M. Kobayashi and F. Wakai), 199-215; 1989. MRS, Pittsburgh, PA.
55. D.J. Schissler, A.H. Chokshi, T.G. Nieh and J. Wadsworth, Acta Metall. Mater. 39, 1991, 3227-3236.
56. T.G. Nieh and J. Wadsworth, Acta Metall. Mater. 39, 1991, 3037-3045.
57. A.H. Chokshi, T.G. Nieh and J. Wadsworth, J. Am. Ceram. Soc., 74, 1991, 869-873.
58. Y. Ma and T.G. Langdon, Acta Metall. Mater, 42, 1994, 2753-2761.
59. F. Wakai and H. Kato, Adv. Ceram. Mater., 3, 1988, 71-76.
60. Y. Yashizawa and T. Sukuma, Acta Metall. Mater. 40, 1992, 2943-2950.
61. T.J. Davies and A.A. Ogwu, Mater. Sci. Technol. 10, 1994, 669-673.

A Comparative Study of Cavitation Characteristics of a Superplastic Aluminium Composite and an Aluminium Alloy

H. Iwasaki *, M. Mabuchi**, K. Higashi*** and T. Mori*

*College of Engineering, Department of Materials Science and Engineering,
Himeji Institute of Technology, Himeji, Hyogo 671-22, Japan
**National Industrial Research Institute of Nagoya, Nagoya 462, Japan
***College of Engineering, Department of Mechanical Systems Engineering,
University of Osaka Prefecture, Gakuen-cho, Sakai, Osaka 593, Japan

Abstract

Cavity formation in a high strain rate superplastic Al-Mg-Si composite, reinforced with 20 vol.% Si_3N_4 particulates, was observed by a high resolution scanning electron microscope and was compared with that in a superplastic Al-Mg alloy. Both the composite and the alloy were deformed in tension under steady state true stress at their optimum superplastic deformation conditions of 8 MPa and 3 MPa at 833 K respectively. Many cavities in the composite were nucleated at/near the interfaces between the matrix and particulates. The number of cavities increased slowly, and the growth was sluggish, with increasing strain. The sizes of most cavities, at least up to a strain of 1.0, were on the order of the particulate sizes, i.e. nearly 0.5 μm, whereas above a strain of 1.4, a few large cavities, almost equal to the grain size, appeared by growth and/or interlinkage of the cavities. On the other hand, in the alloy, a few cavities were formed at the triple points and the second phase/grain boundary interfaces at a strain of 0.2. The size of most cavities was larger than 5μm. Continuous nucleation occurred and cavity growth was associated with the interlinkage of cavities by the development of sharp cracks along the grain boundaries.

1. Introduction

Superplastic materials have the ability to undergo large uniform elongations prior to failure. Large elongations are obtained in the range where high strain rate sensitivity values ($m>0.5$), that is, low stress exponents ($n<2$) are found, since high values of m or low values of n restrict development of necking from the criteria for deformation stability. It is well established [1-7] that cavitation occurs in many kinds of alloys during superplastic flow, and cavitation has an effect on elongations to failure and mechanical properties of the materials after superplastic flow. Therefore, an understanding of cavity formation mechanisms is important not only for its intrinsic scientific value but also for applications for superplastic forming operations.

Recently, many aluminum matrix composites with discontinuous SiC or Si_3N_4 have been reported to show superplastic behavior [8-15]. In particular, it should be noted that some composites exhibited superplastic behavior at high strain rates of more than 10^{-2} s^{-1}. Ghosh and Mahoney [9] reported the effect of the volume fraction of reinforcements and the effect of hydrostatic pressure on the cavity volume fraction for SiC$_p$/PM64 composites. However, a sufficient understanding of the cavitation mechanism has not been developed for superplastic metal matrix composites, and in particular, at high strain rates.

The aim of this paper is to investigate cavity formation for a superplastic Si$_3$N$_{4p}$/Al-Mg-Si composite, which showed superplastic behavior at high strain rates of about 1 s^{-1} and to make comparisons with a superplastic 5083 aluminium alloy. Cavitation behavior is focused on cavity nucleation and cavity growth related to an accommodation process to relax the stress concentration by sliding at grain boundaries and interfaces.

2. Experimental Procedures

The procedure to prepare the Si$_3$N$_{4p}$/Al-Mg-Si composite employed a powder metallurgy method: Powders of both aluminium alloy and Si$_3$N$_4$ particulates (< 0.5 μm in diameter) were mixed by ultrasonic vibrations in an alcohol solvent and were subsequently dried. The mixed powder was then pressed at 873 K for 1.2 ks at a pressure of 390 MPa. Pressed billets were extruded at 773 K with a ratio of 100. The extruded specimen contained homogeneously distributed particulates. The aluminium alloy was a superplastic commercial 5083 Al sheet in thickness of 2.0 mm which was produced by SKY Aluminium Co., Ltd. containing, by wt%, 4.5% Mg, 0.72% Mn, 0.13% Cr, 0.05% Si, 0.08% Fe, 0.01% Ti; balance Al. The grain size, d, at the start of testing (d=1.74L, L is the mean linear intercept grain size) was about 1.9 μm for the composite and 12 μm for the aluminium alloy respectively. Tensile specimens with a gauge length of 5 mm were machined to a gauge diameter of 2.5 mm for the composite, or to a gauge width of 3 mm for the alloy. Tension tests under constant true stress conditions were carried out under optimum superplastic deformation conditions for each material: the temperature was 833 K and the stress was 8 MPa for the composite, and 3MPa for the alloy. The tensile axis was chosen to be parallel to the extrusion or rolling direction for all tests.

The volume of cavities in the gauge length of the deformed specimen was determined by hydrostatic weighing in water with a corresponding gauge head used as a density standard.

Sectioned and mechanically polished specimens after testing to a selected strain were examined using a high resolution scanning electron microscope.

3. Results

3.1 Mechanical properties of composite and alloy

The results of the steady state stress tests for the composite and the alloy were as follows : the strain rates at a fixed strain of 0.1 were 2 s^{-1} and 6 x 10^{-4} s^{-1}, and the elongations to failure were 500%, and 910% respectively, and the strain rate sensitivity m was close to 0.5 for both specimens.

These values in strain rate sensitivity and elongation indicated that these specimens were behaving superplastically.

3.2 Cavity volume fraction

The variation in the cavity volume fraction as a function of true strain is shown in Figure 1 for both the composite and the alloy. It is found that the cavity volume fraction of the composite was larger at $\varepsilon <0.7$, but smaller at $\varepsilon >0.7$ compared with the alloy. It should be noted that the evolution in volume fraction of cavities with strain for the composite was lower than that for the 5083 alloy.

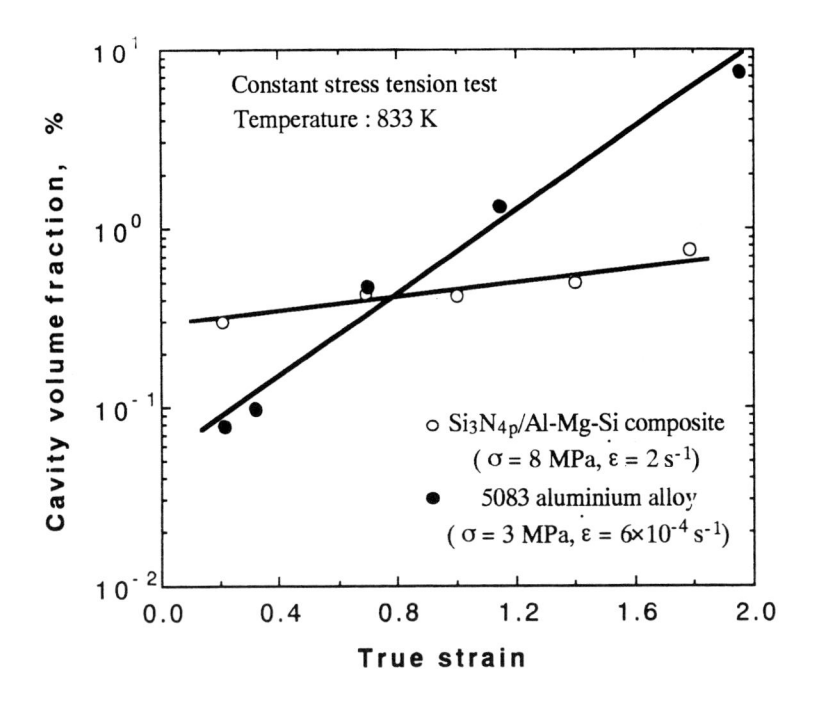

Fig. 1 The variation in the cavity volume fraction as a function of true strain for the Si3N4p/Al-Mg-Si composite and the 5083 alloy

3.3 Cavitation in the early deformation stage

Photomicrographs taken from longitudinal sections of specimens pulled to a strain of 0.2 are shown in Fig.2 for the composite (a) and for the alloy (b), where the tensile axis is horizontal. It is clear that there were no visible cavities in the composite, but some cavities could be seen in the alloy, although the cavity volume fraction of the alloy was smaller than that of the composite at the strain shown in Fig.1. In order to investigate the morphology of the cavities in detail, the specimens were observed by scanning electron microscopy. The results are depicted in Fig.3. Cavities were clearly

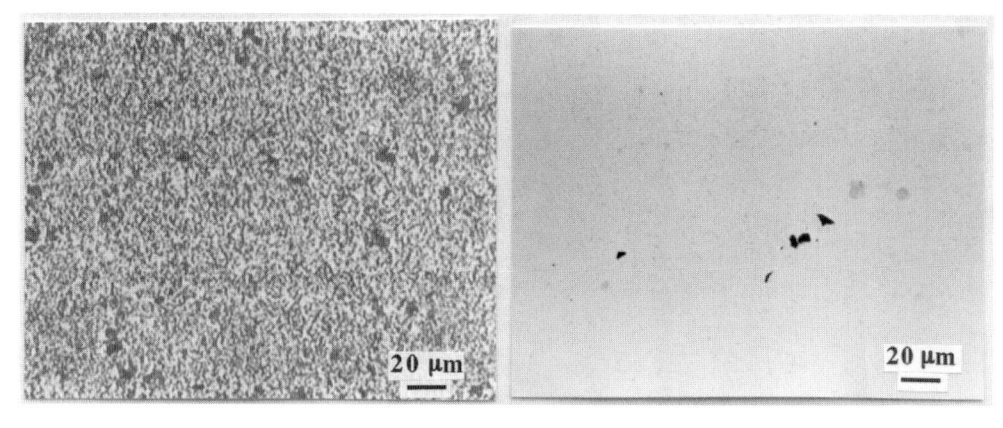

78 *Superplasticity: 60 Years after Pearson*

(a) the Si$_3$N$_{4p}$/Al-Mg-Si composite   (b) the 5083 alloy

Fig. 2 Photomicrographs showing cavitation of the composite (a) and the alloy (b) pulled to a strain of 0.2. The tensile axis is horizontal.

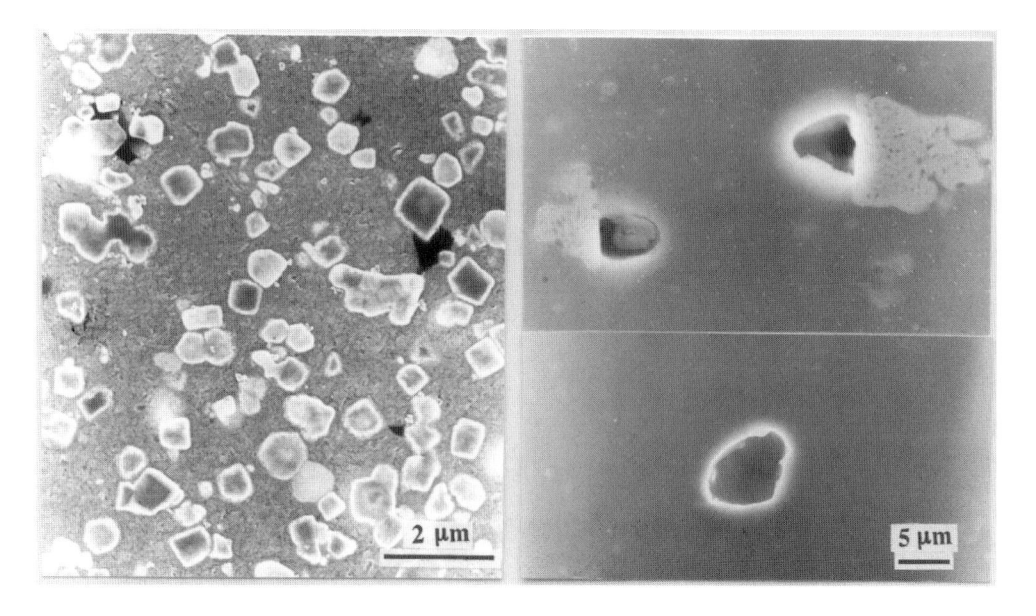

(a) the Si$_3$N$_{4p}$/Al-Mg-Si composite   (b) the 5083 alloy

Fig. 3 SEM micrographs showing cavitation in the composite (a) and the alloy (b) pulled to a strain of 0.2. The tensile axis is horizontal.

observed in both electron micrographs. Most cavities in the composite (Fig.3 (a)) were about the size of reinforcement particulate, i.e. less than 1.0 μm, and they were located at the interfaces between the matrix and reinforcement. On the other hand, cavities in the alloy (Fig.3 (b)) were larger than those of the composite, and they were formed at the triple points of grain boundaries or at the interfaces between the matrix and second phase. The number of cavities in the alloy was much less than that in the composite.

3.4 Cavitation in the middle deformation stage

Figure 4 shows cavities in the specimens pulled to a strain of 1.0 for the composite (a) and the alloy (b). Most cavities in the composite were observed to be less than 1 mm in diameter, being commensurate with the 0.5 μm diameter size of the particulates, and large cavities of the order of a micron were not observed. Essentially, all the cavities were nucleated at the interfaces of the matrix and particulates, and exhibited a variety of shapes. It appeared that there was little variation in both the number and size of cavities throughout the strain range up to 1.0. Cavity growth rates were consequently observed to be correspondingly small. In the alloy, cavities formed at the triple points or at the interface of matrix/second phase were interlinked by a sharp crack developed along a grain boundary as shown in Fig.4 (b). The number of the interlinkages was about 10 / mm^2.

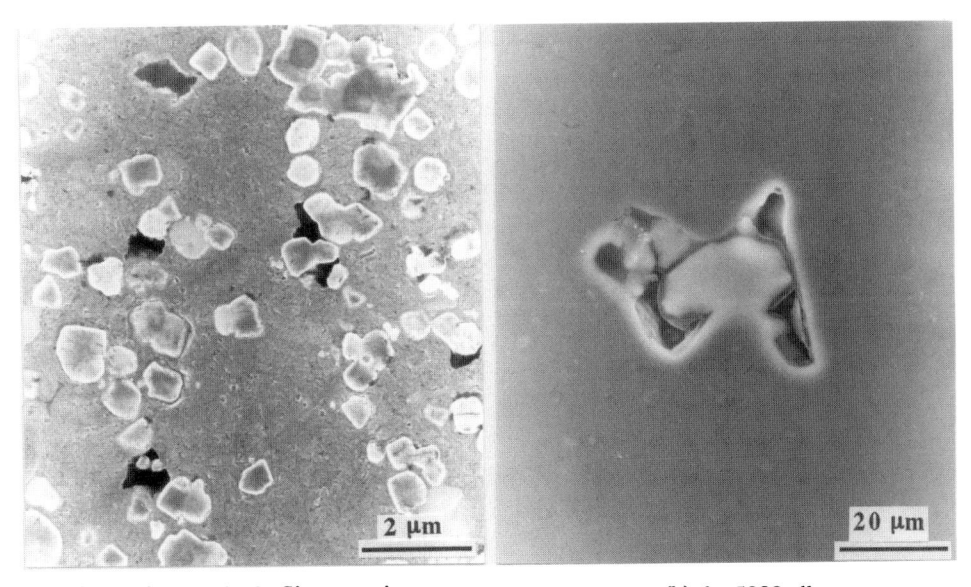

(a) the Si$_3$N$_{4p}$/Al-Mg-Si composite   (b) the 5083 alloy

Fig. 4 SEM micrographs showing cavitation in the composite (a) and the alloy (b) pulled to a strain of 1.0. The tensile axis is horizontal.

3.5 Cavitation in the late deformation stage

In the composite subjected to a strain of 1.0, very few cavities with diameter greater than 1 μm could be observed. However, not only many small cavities, but also a few large cavities, appeared in the specimen deformed to a relatively large strain of 1.4 as shown in Fig.5. The size of the large cavities was greater than the particulate size, i.e. between 2 and 6 μm, which was close to the order of the grain size. The large cavities had angularity in their morphology. It is likely that the large cavities were formed by interlinkage amongst some of the small cavities, not by the coalescence of growing cavities. Figure 6 shows an example of cavity growth by interlinkage with a grain boundary crack in a specimen pulled to a strain of 1.8. On the other hand, cavities in the alloy deformed to a strain of 1.9 were increased and considerably grown by the sharp cracks along grain boundaries as shown in Fig.7. The cracks were mainly developed parallel to the tensile axis and the cavities were also grown parallel to the tensile axis and were as long as 20 to 60 μm.

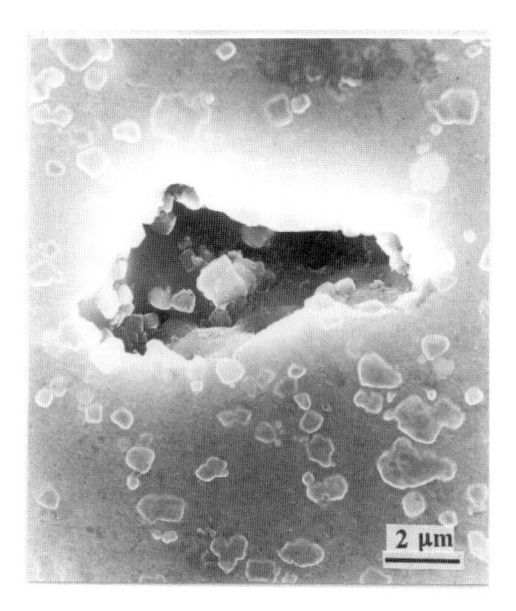

Fig. 5 SEM micrograph showing a large cavity in the composite pulled up to a strain of 1.4. The tensile axis is horizontal.

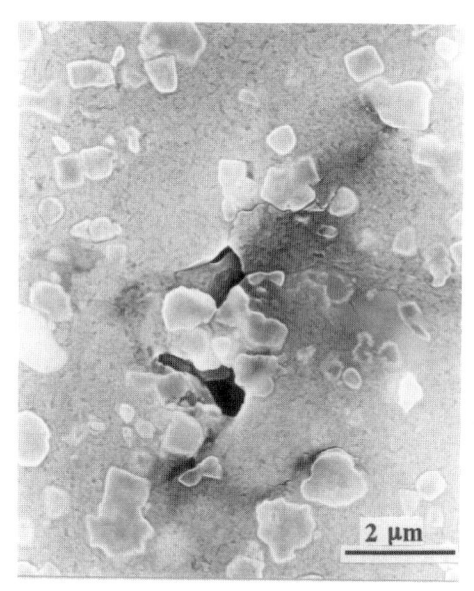

Fig. 6 SEM micrograph showing an interlinkage of cavities by grain boundary cracks in the composite pulled up to a strain of 1.8. The tensile axis is horizontal.

4. Discussion

It is possible that the low rate of cavity formation with strain at high strain rates in the composite is related to the existence of a liquid phase at the interfaces in the composite. The existence of such liquid phases in the composite has been demonstrated by in-situ observations using transmission

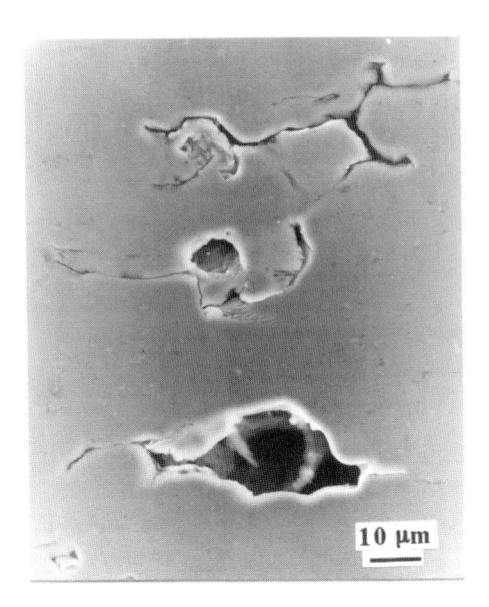

Fig. 7 SEM micrograph showing grain boundary cracks and a large cavity grown by interlinkage in the alloy pulled up to a strain of 1.9 at 833 K. The tensile axis is horizontal.

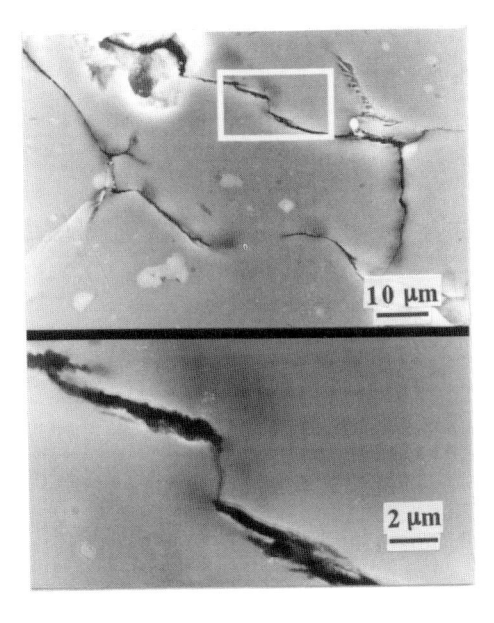

Fig. 8 SEM micrograph showing development of grain boundary cracks at a ledge in the alloy pulled up to a strain of 0.7 at 833 K. The tensile axis is horizontal.

electron microscopy at elevated temperature [18]. The liquid phase was confirmed at grain boundaries with high misfit-angles and at interfaces below the tension test temperature. The stress concentration level at an interface with a liquid phase is much lower than that at an interface without a liquid phase. Therefore, the accommodation required to relax stress concentrations by sliding can be much more easily achieved through rapid liquid phase mass-transfer. It is likely that high strain-rate deformation in the composite can be attributed to the existence of such liquid phases. However, it may also be the case that cavity nucleation occurs much more easily at the interfaces, if a liquid phase is present there, owing to the fact that the bonding force between the liquid and solid is very weak. This might explain why most cavity nucleation occurs at the early stage of deformation or at a strain of approximately 0.2 at 833K; at this point the testing temperature of 833K is slightly above the partial melting temperature of the composite of 829K, as measured by DSC. Liquid phases may form above the partial melting temperature at locations with segregation of solute elements, e.g. grain boundaries and interfaces as confirmed by TEM observation [18]. After nucleation, however, the rate of cavity production may be reduced since the cavities with a liquid phase act as stress relaxation sites, as is indicated by the experimental results.

Regarding the composite specimen deformed to over a strain of 1.4, the formation of large cavities of more than 2 μm in diameter can be observed in Fig.5. Inspection of this figure indicates that the large cavities contain many particulates, and that the inner surfaces are flat like a brittle-fractured surface; the features are a little complicated. Further, it appears that the density of small cavities in proximity to the larger ones is not especially high and consequently their spacing is too large for coalescence to occur. From this fact we deduce that the large cavities are more likely formed by grain boundary fracture due to the existence of stress concentrations rather than by a coalescence of smaller cavities. Figure 6 shows an example of such a grain boundary fracture just prior to the formation of a large cavity.

On the other hand, for the 5083 alloy, cavities were nucleated at the triple points and at the interfaces of matrix/second phase, and continuously nucleated with increasing strain. Furthermore, the cavities grew and interlinked by the development of sharp cracks along the grain boundaries at a large strain of above 1.0. So, the growth rate in cavity volume fraction was larger than that of the composite. These results may be explained by a cavitation mechanism which is not based on the presence of a liquid phase [16, 17]. It seems that grain boundary sliding apparently leads to an increase in stress concentrations at the triple points and at the interfaces between the matrix and second phase, and thereby leads to subsequent high dislocation pile-ups in the matrix because of the absence of a liquid phase. If the accommodation processing by diffusion is incomplete, then, cavities would be expected to nucleate at the triple point or at the interfaces between the matrix and second phase resulting from a release of stress concentration. Consequently, the number and size of the cavities would increase with increasing strain or increasing grain boundary strain.

It is noted that there is a similar phenomenon in cavity growth of both materials, namely, cavity growth by interlinkage with sharp cracks along grain boundaries. Similar crack networks have been reported by Ma and Langdon for two aluminium alloys [19]. It is likely that the crack is a grain boundary fracture, caused when a normal stress on the grain boundary exceeds the boundary bonding strength. Figure 8 shows an example of the initiation of the fracture around a ledge on a sliding boundary. In this case the ledge may generate an excessive normal stress to initiate cracks. It has been described above that the stress concentration can be released easily by liquid phases in the composite. However, if there are some places with stress concentrations because of a lack of liquid phase, it is possible for the grain boundary to fracture locally.

It is likely that the grain boundary fracture is related to cavity growth at a high temperature near to the partial melting point.

5. Summary

The cavitation behavior for the superplastic Si_3N_{4p}/Al-Mg-Si composite and the superplastic 5083 aluminium alloy deformed under a steady state stress of 8 MPa and 3MPa at 833K has been examined by scanning electron microscopy. The results can be summarized as follows;

(1) For the composite, most cavities were formed at interfaces between the matrix and reinforcements, and the cavity size dimensions were almost equal to those of the particulate reinforcements up to a strain of 1.0. Whereas in the aluminium alloy, cavities were nucleated at triple points and interfaces between the matrix and second phase, and cavities were continuously nucleated with strain.

(2) In the composite the existence of a liquid phase at the interface may contribute not only to relax the stress but also to provide sites for cavity nucleation. The reduction of the stress concentration reduces the rate of cavity nucleation and growth.

(3) At large strains of approx. 1.4 for the composite, a few large cavities of more than 2 µm in diameter appear in addition to the smaller ones. The larger cavities appear to be formed by interlinkage, not by coalescence of small cavities. The cavity in the alloy was also grown by interlinkage by grain boundary cracks at a strain of 0.7 to 1.9.

References
1. D.W.Livesey and N.Ridley, Metall. Trans. A, **9A** (1978) 519.
2. T.Chandra, J.J.Jonas and D.M.R.Taplin, J. Mater. Sci., **13** (1978) 2380.
3. C.C.Bampton and J.W.Edington, Metall. Trans. A, **13A** (1982) 1721.
4. D.W.Livesey and N.Ridley, Metall. Trans. A, **13A** (1982) 1619.
5. C.H.Caceres and D.S.Wilkinson, Acta Metall., **32** (1984) 423.
6. A.H.Chokshi and T.G.Langdon, J. Mater. Sci., **24** (1989) 143.
7. A.H.Chokshi and T.G.Langdon, Acta Metall, **38** (1990) 867.
8. T.G.Nieh, C.A.Henshall and J.Wadsworth, Scripta Metall., **18** (1984) 1405.
9. M.W.Mahoney and A.K.Ghosh, Metall. Trans. A, **18A** (1987) 653.
10. J.Pilling, Scripta Metall, **13** (1989) 1375.
11. T.Imai, M.Mabuchi, Y.Tozawa and M.Yamada, J. Mat. Sci. Letter, **9** (1990) 255.
12. H.Xiaoxu, L.Qing, C.K.Yao and Y.Mei, J. Mater. Sci. Lett., **10** (1991) 964.

13. M.Mabuchi, T.Imai, K.Kubo, K.Higashi, Y.Okada and S.Tanimura, Mater. Lett., **11** (1991) 339.
14. M.Mabuchi, K.Higashi, Y.Okada, S.Tanimura, T.Imai and K.Kubo, Scripta Metall., **25** (1991) 2003.
15. M.Mabuchi, K.Higashi, Y.Okada, S.Tanimura, T.Imai and K.Kubo, Scripta Metall., **25** (1991) 2517
16. J.Pilling and N.Ridley, Superplasticity in Crystalline Solids, The Institute of Metals, The Camelot Press, Southampton, England, (1988).
17. A.H.Chokshi and T.G.Langdon, Acta metall. mater., **38** (1990) 867
18. M.Mabuchi and K.Higashi, Philos. Mag. Lett., (1994) in press.
19. Y.Ma and T.G.Langdon, in Proc. Conf. 'Superplasticity', Tokyo, May 1988. Materials Research Society, Vol.7.

Cavitation in Superplastically Deformed Ceramic Materials

Z.C. Wang, T.J. Davies and N. Ridley
Materials Science Centre, University of Manchester/UMIST
Grosvenor Street, Manchester M1 7HS, UK

Abstract

In superplastic metals, cavities usually develop parallel to the tensile axis, while in ceramics they generally propagate in a direction perpendicular to the axis of applied tensile stress and interlink at relatively small strains to cause failure, as in the case of fine grain aluminas. However, in superplastic Y-TZP (yttria stabilized tetragonal zirconia polycrystal), a cavitation mode similar to that in metallic systems exists where cavities grow and align parallel to the tensile stress axis. The elongation to failure in Y-TZP is significantly higher than in other ceramic systems as a consequence.

If domes are bulge formed from Y-TZP and alumina without a suitable imposed hydrostatic pressure, they will contain quite different distributions of cavities. For the former, cavities lie parallel to the axes of tension, whereas the transverse cavities that form in alumina will be linked to the surface.

1. Introduction

Superplastic forming (SPF) is the basis of an important and continually developing technology for the production of near-net shape components in a range of materials. However, many materials undergo cavitation during superplastic flow, and procedures have to be developed to limit its occurrence. While the cavitation behaviour of metallic materials has been intensively investigated, much less is known about cavitation in ceramic materials. The present paper presents some observations of cavitation in two ceramic materials, Al_2O_3 and Y-TZP, following tensile straining.

2. Experimental Observations

The raw materials used were a high purity (>99.99%) fine grained (d<0.4μm) alumina, and 3Y-TZP with average grain size ≈0.25μm. Tensile specimens were slip-cast to net-shape and sintered by HIPing to densities greater than 98%. The details of specimen preparation have been described previously [1]. Tensile tests were carried out using an Instron 4505 tensile machine at constant strain rates in the range 1×10^{-5} to 2×10^{-2} s^{-1} and temperatures in the range 1473K to 1873K. Microstructural observations were carried out on polished and unpolished specimens using scanning electron microscopy (Philips SEM505) and optical microscopy.

In superplastic metals, cavities usually grow and align in a direction parallel to the applied stress [2]; with increasing tensile strain, the aspect ratio (length/width) of individual cavities increases. The volume fraction of cavities is higher near the fracture face and the separation distances between individual cavities are less than in locations away from the fracture. In high purity superplastic aluminas, cavities/cracks grow in a direction perpendicular to the tensile stress, as seen in Figure 1, and are uniformly distributed through the gauge section;

with increased strain these cavities link up to form larger cracks. Yoshizawa and co-workers [3] observed similar phenomena in a MgO-doped alumina which had been strained to 83% elongation at 1706K at a strain rate of 1.2×10^{-4} s^{-1}.

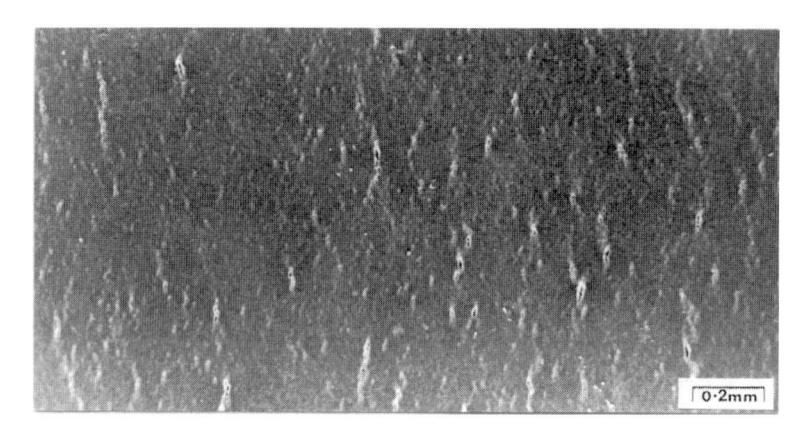

Fig.1 Propagation of cavities/cracks in the direction transverse to the tensile stress in high purity alumina (SEM), d=1.0μm, $\dot{\varepsilon}$=1x10^{-4} s^{-1}, T=1673K, e$_f$=39%; tensile axis horizontal.

Fig.2 Cavities in 3Y-TZP, (A) cavities transverse to stress axis near the fracture face and (B) cavities parallel to stress axis near gauge head (SEM). d<0.33μm, $\dot{\varepsilon}$=5x10^{-5} s^{-1}, T=1723K, e$_f$=340%; tensile axis vertical.

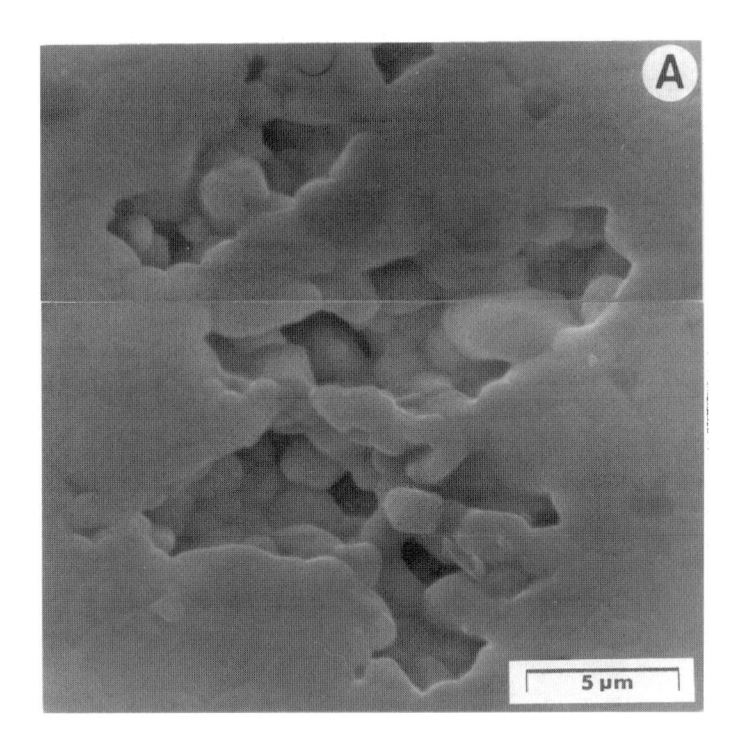

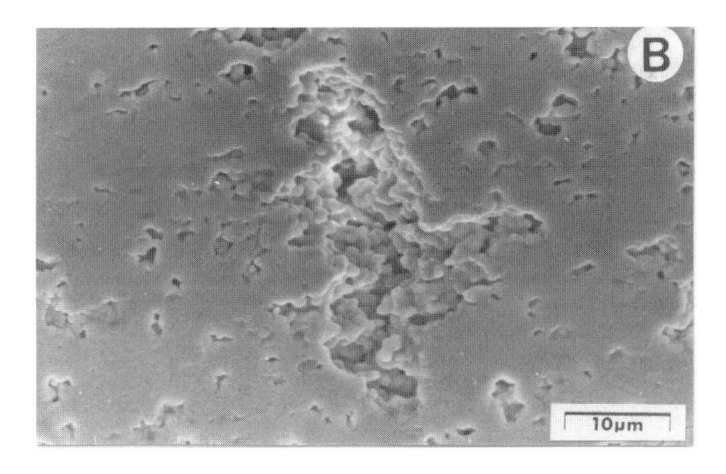

Fig.3 Shows transverse cavities (A) formed from smaller parallel ones by necking and (B) developed from the parallel cavities by cracking; Y-TZP, e_f=480%, 1823K; tensile axis horizontal.

Table 1 Relationships between cavitation modes and maximum elongations of superplastic materials

Materials	Cavitation mode near fracture; away from fracture		Elongation $e_f(\%)$
Metal (an Al bronze)	∥	∥	8000
3Y-TZP-SiO$_2$	⊥	∥	1038
3Y-TZP	⊥	∥	800
Doped Al$_2$O$_3$	⊥	⊥	145
Pure Al$_2$O$_3$	⊥	⊥	40

⊥: Cavities propagate transverse to the tensile stress;
∥: cavities elongate parallel to tensile stress.

By comparison, for yttria-stabilized tetragonal zirconia polycrystal (Y-TZP) specimens, cavities which form near the fracture face during superplastic tensile straining differ from those near the gauge head, as seen in Figure 2. In the region of fracture, cavities that were previously aligned parallel to the tensile axis interlink in the transverse direction and form incipient cracks (Fig.2A), whereas, cavities away from the fracture face elongate parallel to the tensile axis (Fig.2B). Schissler and co-workers [4] made similar observations in their studies on Y-TZP, while Kajihara and co-workers [5] observed cavities aligned parallel to the tensile axis in Y-TZP containing about 10vol% SiO$_2$, when elongated to 1038%. A close scrutiny of the near-fracture face micro-structure in the present work shows that two types of transverse cavities can be identified. The majority are formed by the interlinkage of longitudinal cavities as seen in Figure 3A, and a small proportion, especially the short cavities, are formed by the transverse propagation of previously formed individual longitudinal cavities as seen in Figure 3B. This is in sharp contrast to the behaviour in 'normal' superplastic ceramics [6-7], such as Al$_2$O$_3$, where existing cavities or cracks propagate perpendicular to the tensile axis. These observations are summarized in Table 1.

3. Discussion

The process whereby initially spherical cavities in metals become elongated and align in the direction parallel to the tensile stress can be explained as follows. At an initial stage of superplastic deformation, cavity growth to form near-spherical shapes is controlled by a diffusional mechanism; after this initial stage, a plasticity-controlled mechanism dominates cavity growth and causes the cavity to become elongated parallel to the applied tensile stress. Coalescence of cavities lying along the tensile axis would aid elongation. Final fracture involves a mixture of intergranular and intragranular failure that occurs as a consequence of transverse interlinking of elongated cavities [8].

The reasons for the differences in cavitation behaviour between Al$_2$O$_3$ and Y-TZP are not clear. For Y-TZP, Davies and Ogwu [9] have proposed that the observed plasticity may be explained by consideration of the energies of d-state electrons which contribute a 'metallic' character to the bonding and the development of large extensions and elongated cavities at high temperatures.

When cavities propagate parallel to the tensile axis through the whole specimen gauge length, the recorded elongations to failure are often more than 1000%, as in many metallic superplastic materials. If cavities or cracks propagate transverse to the tensile axis through the whole specimen gauge length, the recorded elongations to failure are often less than 100%, as in aluminas and covalently bonded ceramics. If cavities propagate by a mixture of the above two modes, i.e. parallel to the tensile axis near the specimen gauge head and transversely near the fracture face, the recorded elongations to failure are often more than 100% but less than 1000%, as in some superplastic Y-TZPs. From the viewpoint of engineering applications, it is likely that if a material cavitates, then cavities parallel to the tensile stress would be less deleterious than transverse cavities. For example, when a dome is superplastically formed without a suitable hydrostatic pressure, the surface-linked cavities which develop transverse to the tensile stress (Fig.4a), as in Al_2O_3, are aligned radially, and it will be impossible to remove them by HIPping after forming. However, longitudinal cavities (Fig.4b), as in Y-TZP, will lie in the circumferential direction and can be closed by applying a hydrostatic pressure during or after forming [10].

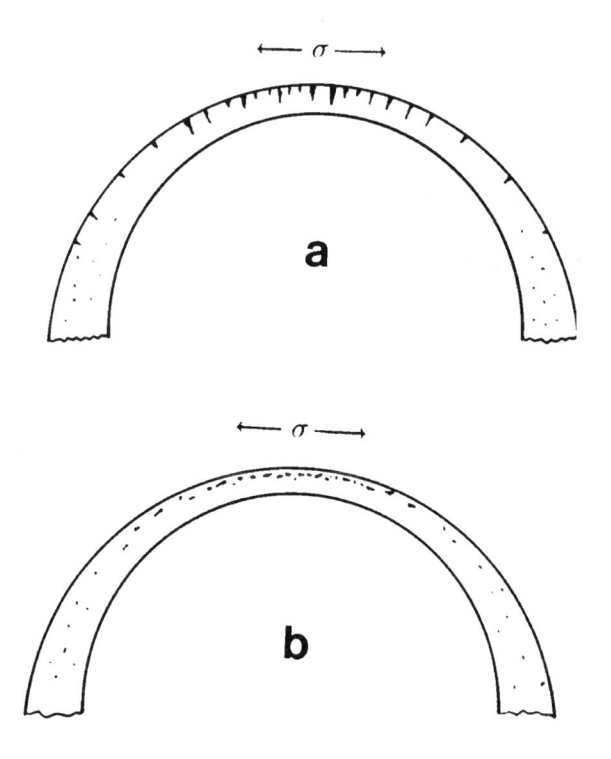

Fig.4 Schematic illustration of cavitation modes. (a) cavities propagate perpendicular to tensile stress and linked to the surface; cannot be removed and (b) cavities elongated parallel to tensile stress; can be eliminated.

4. Summary

Cavities in superplastic metals usually develop parallel to the tensile axis, while in ceramics cavities often propagate in a direction perpendicular to the axis of applied tensile stress and these interlink at relatively small strains to cause failure, as in the case of fine grain aluminas. However, in superplastic Y-TZP (yttria stabilized tetragonal zirconia polycrystal), a cavitation mode similar to that in metallic systems exists where cavities grow and align parallel to the tensile stress axis. The elongation to failure in Y-TZP is significantly higher than in other ceramic systems. If a dome is bulge-formed, the cavitation modes in the two materials are totally different. For Y-TZP, cavities lie parallel to the axes of tension, whereas the transverse cavities that form in alumina are linked to the surface. From the viewpoint of engineering applications, the latter are more deleterious because they cannot be closed once they develop during forming.

5. Acknowledgement

Financial support of this work by the Science and Engineering Research Council is gratefully acknowledged.

References

1. Z.C. Wang, T.J. Davies and N. Ridley, Scripta Metall. Mater. 28, 1993, 301-306

2. M.C. Pandey, J. Wadsworth and A.K. Mukherjee, Mater. Sci. Eng. 78, 1986, 115-125.

3. Y. Yoshizawa and T. Sakuma, Acta Metall. Mater. 40, 1992, 2943-2950.

4. D.J. Schissler, A.H. Chokshi, T.G. Nieh and J. Wadsworth, Act. Metall. Mater. 39, 1991, 3227-3236.

5. K. Kajihara, Y. Yoshizawa and T. Sakuma, Scripta Metall. Mater. 28, 1993, 559-562.

6. T.J. Davies, Z.C. Wang and N. Ridley, "Creep and Fracture of Engineering Materials and Structures", (ed. B. Wilshire and R.W. Evans), p325-334, The Inst. Mater. London, 1993.

7. Z.C. Wang, N. Ridley and Davies, "Creep and Fracture of Engineering Materials and Structures", (ed. B. Wilshire and R.W. Evans), p315-324, The Inst. Mater. London, 1993.

8. W.J Kim, J. Wolfenstine and O.D. Sherby, Acta Metall. Mater. 39, 1991, 199-208.

9. T.J. Davies and A.A. Ogwu, Electronic contribution to superplasticity in ceramics, Mater. Sci. Tech. 10, 1994, 669-673.

10. J. Pilling and N. Ridley, Acta Metall. 34, 1986, 669-679.

Superplastic Flow
at High Strain Rates

Positive Exponent Superplasticity in Metallic Alloys and Composites

Kenji Higashi

*Department of Mechanical Systems Engineering,
College of Engineering, University of Osaka Prefecture,
Gakuen-cho, Sakai, Osaka 593, Japan*

Abstract

High strain rate superplasticity (*i.e.*, superplastic behavior at strain rates over 10^{-2} s^{-1}) has been characterized in very fine grained aluminum alloys and composites, which have been developed by powder metallurgical processing and/or new advanced processing methods to have grain sizes of 50 nm ~ 3 μm. The experimental results on high strain rate superplastic materials are reviewed and related to the temperature and strain rate dependencies of superplastic behavior, *i.e.*, tensile elongation, strain rate sensitivity, activation energy, and cavitation. The optimum superplastic strain rates are found to be strongly dependent upon the refinement of grain structures. Specifically, marked changes in these properties and superplastic flow are closely related to incipient melting points in the materials. A maximum value in elongation, for example, is found at temperatures near or slightly above the incipient melting points. These observations confirm the suggestion that the presence of a small amount of liquid phase at grain boundaries in the alloys and interfaces in the composites not only enhances the strain rate for superplasticity, but also has a strong influence on the deformation mechanisms. A model is proposed in which superplasticity is critically controlled by the accommodation process to relax the stress concentration resulting from the sliding at grain boundaries and/or interfaces, involving an **accommodation helper** such as a liquid phase.

1. Introduction

Research into superplasticity has developed substantially in recent years, to the extent that there are now several important directions for future investigations. One new area is superplastic behavior at relatively high strain rates over 10^{-2} s^{-1} in metallic alloys and metal matrix composites (which is close to commercial hot working rates of 10^{-1} and 10 s^{-1}), when they are produced by a combination of powder metallurgy and/or advanced processing methods. This is clearly noted from **Figure 1**, which shows the relationship between elongation and strain rate for differently processed superplastic aluminum alloys. Especially, *Positive exponent superplasticity* (denoted by this author as superplasticity at very high strain rates over 1 s^{-1}) was observed recently in nano or near-nano scale aluminum alloys, which have been developed by three advanced processing methods, (1) mechanical alloying, (2) consolidation from amorphous powders or nanocrystalline powders and (3) physical vapor deposition [1]. It can be concluded from **Figure 2**, which shows the change of the optimum superplastic strain rate for several aluminum alloys with various grain sizes produced by different processing routes, that the increase in superplastic strain rate strongly depends on the refinement of grain size in materials. However, the precise deformation mechanisms occurring in these high-strain-rate superplastic materials are not understood and deserve further study.

In the present review, the tensile properties of high-strain-rate superplastic materials with fine grained structures (typically, the grain sizes are about 0.05 to 5 μm) are characterized over a wide range of strain rates and temperatures. The purpose of this review is to explore the possible deformation mechanisms of high-strain-rate superplastic materials. On the basis of the new experimental results obtained from the temperature dependence of the flow stress, elongation, strain rate sensitivity exponent, activation energy, and cavitation, it is proposed that the dominant deformation mechanism for all the high-strain-rate superplastic materials is grain boundary sliding, and the presence of a small amount of liquid phase is suggested to be responsible for the observed high strain rate (positive exponent) superplasticity in fine grained aluminum alloys and composites. In addition it is demonstrated that the liquid boundary strongly enhances superplastic elongation in a 7475 alloy with a grain size of about 20 μm, which is a typical conventional aluminum alloy exhibiting superplasticity at low strain rates of about 10^{-4} s^{-1}.

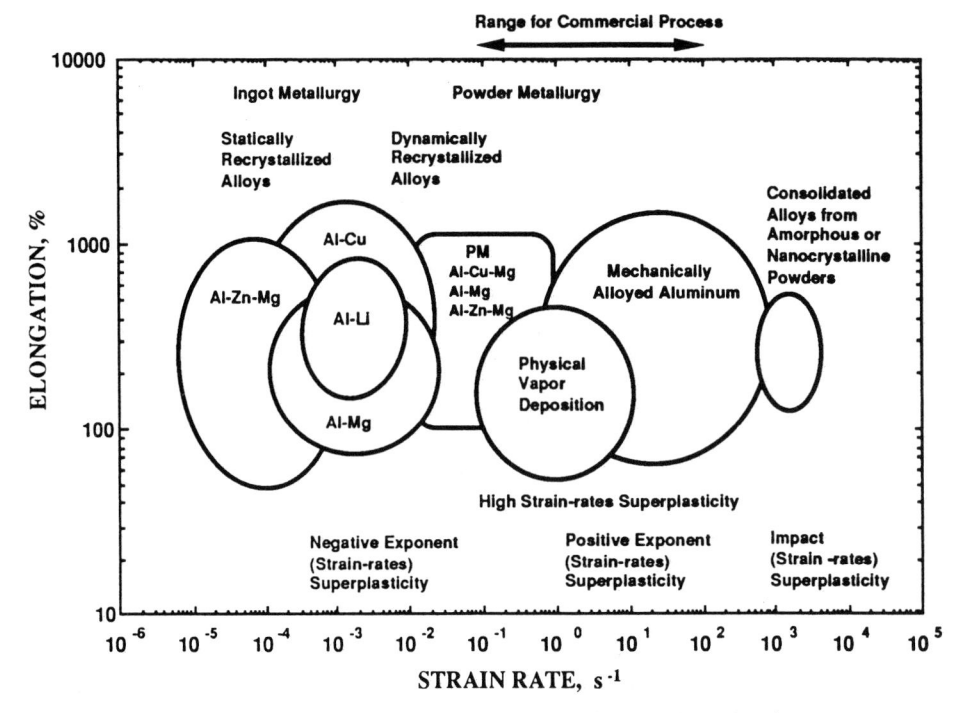

Figure 1 The relationship between superplastic elongation and strain rate for several aluminum alloys produced by different processing routes.

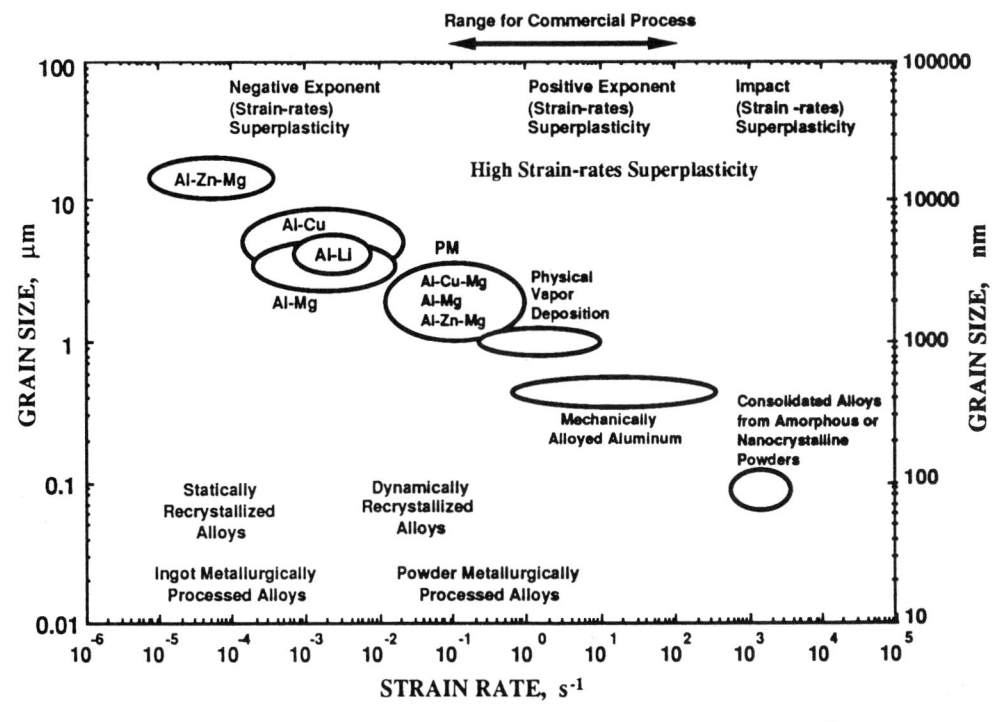

Figure 2 The change of the optimum superplastic strain rate for several aluminum alloys with various grain sizes produced by different processing routes.

2. Materials and Methods

The mechanically alloyed materials, IN9052 (Al-4.0wt%Mg-1.1wt%C-0.8wt%O), IN905XL (Al-4.0wt%Mg-1.5wt%Li-1.2wt%C-0.4wt%O) and IN9021 (Al-4.0wt%Cu-1.5wt%Mg-1.1wt%C-0.8wt%O) were initially obtained as extruded bars, whereas a 15 vol% SiC$_P$/IN9021 composite was obtained in the form of a 1.7 mm thick sheet. All the as–received materials, *i.e.*, both extruded bars and sheet, were subsequently thermomechanically processed followed by warm rolling into thin sheet (1 mm thick). Tensile samples with a 5 mm gauge length of 4 mm width were machined from final rolled sheets. The gauge length of the sample was parallel to the rolling direction. Tensile tests were carried out over a wide strain rate range from 10^{-3} to 1000 s^{-1} in an air at temperatures from 698 to 873 K.

Amorphous powders of Al-14wt%Ni-14wt%Mm alloy and nanocrystalline powders of Al-14wt%Ni-7wt%Mm-1wt%Zr alloy were produced by a high pressure gas atomization technique on an industrial scale at YKK *Ltd*. Bright field electron micrographs taken from the electrolytically thinned powders of amorphous Al-Ni-Mm alloys, and selected-area diffraction, revealed an amorphous structure without any trace of crystallinity [1]. In contrast, a cellular structure with size of about 10 nm and metastable Al$_3$Zr particles of 10 nm in size could be observed in the powders of nanocrystalline Al-Ni-Mm-Zr alloy [1]. Detailed information regarding the processing and microstructural characteristics of each of the materials has been given previously [1]. It is very important to note that both a very fine-grained structure and a uniform distribution of the very fine particulates were obtained by extrusion of both Al-Ni-Mm and Al-Ni-Mm-Zr alloys, without the need for any subsequent thermomechanical treatments. Tensile specimens, which were machined from the final extruded bar, had a gage length of 10 mm and a gage diameter of 5 mm. The constant true strain rate tests were carried out over a high strain rate range from 10^{-3} to 6000 s^{-1} at testing temperatures from 773 to 898 K in air. The tensile axis was selected to be parallel to the extrusion direction for all tests.

The vapor quenched (VQ Al-7wt%Cr-1wt%Fe) alloy was received in thin sheet form (1.6 mm) from RAE (Farnborough). Tensile samples with 5 mm gauge length and 4 mm width were machined from these rolled sheets, with the gauge length parallel to the rolling direction. Tensile tests were carried out in air at temperatures between 873 and 923 K and over a wide range strain rate from 10^{-3} to 100 s^{-1}.

Rapidly solidified aluminum alloy powders of less than 20 μm in diameter used for the composites were supplied by Toyo Aluminum *Ltd*. Si$_3$N$_4$ whiskers of 0.1 ~ 1.5 μm in diameter and 10 ~ 30 μm in length and three types of Si$_3$N$_4$ particulates with different diameters of less than 0.2, 0.5 and 1 μm respectively were supplied by UBE Industries *Ltd*. Both powders of the aluminum alloy and the Si$_3$N$_4$(20 vol.%) reinforcements were mixed in alcoholic solvent under ultrasonic waves prior to drying. The mixed powders were sintered at 873 K for 1.2 ks with a pressure of 390 MPa using a hot press machine. The sintered billets were extruded at a reduction ratio of 100 : 1 at the optimum temperature for each composite. Detailed information regarding the processing and microstructural characteristics of each of the materials have been given previously [2]. Constant strain rate tension tests were carried out over a wide range of temperatures and strain rates. Tensile specimens were machined from the extruded bar and had a gage length of 5 mm and a gage diameter of 2.5 mm. The tensile axis was selected to be parallel to the extrusion direction.

Low strain rate tests (less than 10^{-1} s^{-1}) were performed with an Instron machine, and intermediate strain rate tests (10^{-1} ~ 300 s^{-1}) were done with a hydraulic tensile testing machine, and the dynamic tensile tests (400 ~ 6000 s^{-1}) were done using a split Hopkinson pressure bar system which incorporates a specific attachment. The flow stresses for each sample were determined for a fixed true strain of 0.1. Differential scanning calorimetry (DSC) measurements were performed to determine the incipient melting points. The DSC experiments were carried out from room temperature to about 873 K with a constant heating rate of 10K/min. The thermal analyzer was connected to a computer with a suitable interface. Data from each run were stored in the computer and subsequently analyzed. Disc specimens for TEM examination were cut from a billet by a low speed diamond saw. The discs were mechanically ground to about 20 μm in thickness, and then ion milled with 3 keV argon ions. Specimens were cooled with liquid nitrogen during ion milling to minimize sample heating and irradiation damage. The perforated specimens were then loaded on a single tilt heating stage for TEM examination.

3. Experimental Observation

3.1 Microstructures

Typical superplastic microstructures of the mechanically alloyed materials consist of very fine grains with a 350 ~ 500 nm mean size. Carbon and oxygen are present as fine (<30 nm) carbide (Al_4C_3) and oxide (Al_2O_3, MgO, or LiO_2) particles. The estimated volume fraction of these particles in each material is almost about 5 vol%. The structures of the mechanically alloyed materials are very stable at high temperatures for long times.

The bulk materials consolidated by extrusion only from amorphous or nanocrystalline powders consisted of nano or near-nano scale structures from 50 to 100 nm in grain size. However, these fine structures were unstable above about 773 K, leading to competition between grain growth and superplasticity that exists at high temperatures. Grain growth depends on both the annealing temperature and the holding time, and also heating rate up to the annealing temperature. **Figure 3** shows typical microstructures from the Al-Ni-Mm-Zr alloy heated up to two annealing temperatures of 773 K (left) and 873 K (right) of a very rapid rate of more than 1000 K s^{-1} and then annealed for a constant holding time of 30 s. The microstructure of the Al-Ni-Mm-Zr alloy annealed at 873 K for short times of 30 s consists of grains with a large size of about 1 μm and particulates 0.5 μm in size. Grain growth has already occurred after the short annealing time of 30 s at 873 K. However, nano or near-nano scale grained structures and particulates less than 200 nm in size remain even after annealing for 30 s at 773 K with a rapid heating rate.

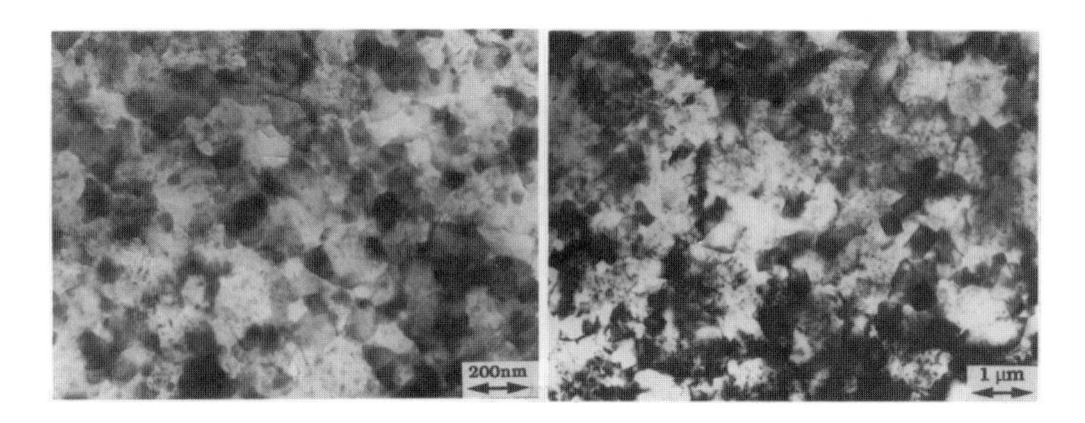

Figure 3 Typical microstructures from the Al-Ni-Mm-Zr alloy heated up to two annealing temperatures of 773 K (left) and 873 K (right) with a very rapid heating up rate and annealed for a constant holding time of 30 s.

A typical microstructure of as-received VQ Al-7wt%Cr-1wt%Fe alloys consists of a very fine grained structure Al-Cr matrix about 100 nm in size and a uniform dispersion of iron-rich precipitates 3~5 nm in diameter. The Al-Cr solid solution is unstable at high temperatures of more than 577 K, resulting in the formation of (Cr,Fe)Al$_7$ particles of sizes more than 100 nm. An operation near high temperatures of 898 K for the optimum superplastic flow produced coarser dispersion of (Cr,Fe)Al$_7$ with 500 nm in size and a solute-depleted matrix. The grain size of the matrix was determined to be approximately 500 nm.

It is found from typical microstructures of the composites after annealing at each optimum superplastic temperature for 1.8 ks that all the grains are equiaxed and the sizes are 1.3 μm for $Si_3N_{4p(0.2\mu m)}$/Al-Mg-Si, 1.9 μm for $Si_3N_{4p(0.5\mu m)}$/Al-Mg-Si, 3.0 μm for $Si_3N_{4p(1\mu m)}$/Al-Mg-Si and 3.3 μm for Si_3N_{4w}/Al-Mg-Si composites, respectively. Scanning electron micrographs reveal that the whiskers in the Si_3N_{4w}/Al-Mg-Si composite are aligned parallel to the extrusion direction and the particulates in the other Si_3N_{4p}/Al-Mg-Si composites are distributed homogeneously. Also no voids are observed at the matrix-reinforcement interfaces in all composites.

3.2 Thermal stability and TEM observation

Typical DSC curves measured from two mechanically alloyed IN9021, IN9052 alloys and four Si_3N_4/Al-Mg-Si composites are shown in **Figure 4** at temperatures between 650 and 873 K. In the case of IN9021, which is essentially a 2xxx-series Al alloy, the first endothermic peak (*i.e.*, incipient melting) occurring at 754 K is caused by the melting of the pseudo–ternary eutectic Al_2CuMg [3]. The largest endothermic peak (860 K) is the solidus temperature which represents the onset of global (or macroscopic) melting of the alloy. In the case of IN9052, however, the DSC curve indicates only one endothermic peak (incipient melting) at 837 K. Since the matrix of IN9052, which is essentially a 5xxx-series Al composition, is a solid solution alloy, this endothermic peak is believed to be a result of Mg segregation to microstructural boundaries including grain boundaries as well as oxide–aluminum alloy matrix and carbide–aluminum alloy matrix interfaces. On the other hand, a sharp endothermic peak and finally a continuous endothermic curve appears in all the Si_3N_{4p}/Al-Mg-Si composites. In a Si_3N_{4w}/Al-Mg-Si composite, however, a weak exothermic peak is found, then followed by a relatively flat area, and finally a continuous endothermic curve appears. These endothermic peaks are likely related to local melting. The melting point of each material by DSC investigations is indicated in Fig. 4. The melting points of the composites are lower than 855 K, which is the published melting point of the Al-Mg-Si (6061) alloy [4]. This suggests that local melting occurs in the composites.

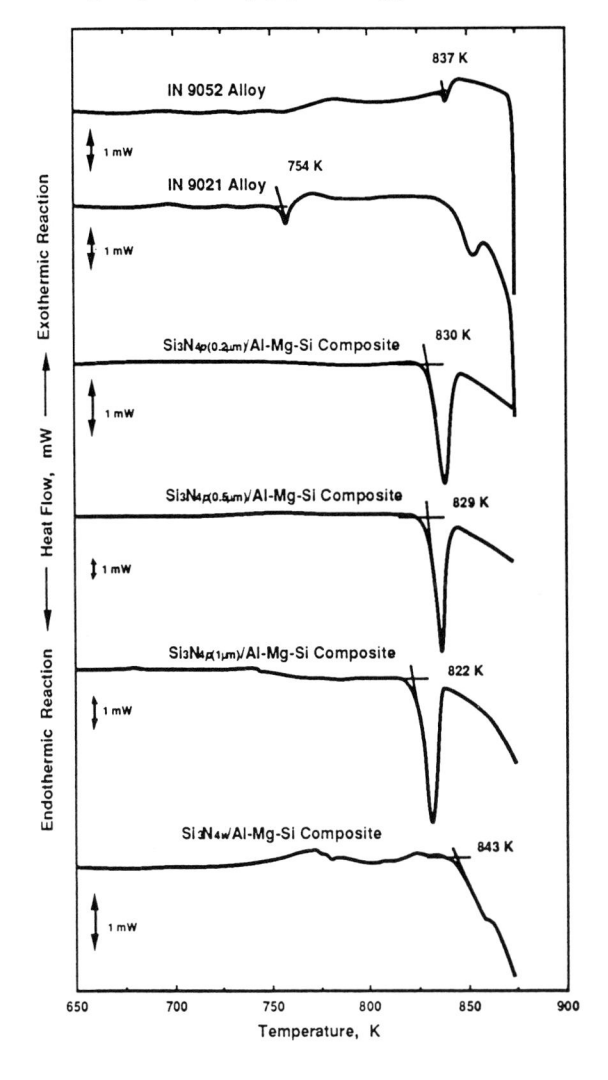

Figure 4 Typical DSC curves measured from two mechanically alloyed IN9021, IN9052 alloys and four Si_3N_4/Al-Mg-Si composites at temperatures between 650 and 873 K.

To further confirm that 754 K is indeed the incipient melting point, an in situ TEM examination of IN9021 was carried out in the temperature range above 754 K. Shown in **Figure 5**(a) is the selected area diffraction (SAD) pattern obtained from IN9021 at 758 K. A weak trace of a diffuse ring begins to be revealed, indicating microstructure disorder; this is most probably caused by the presence of a liquid phase. The amount of the liquid is, however, still limited. As the temperature increases, the amount of liquid phase also increases. At 773 K, the diffuse ring (**Fig. 5**(b)), accompanied by some diffraction spots from the lattice, expands. As shown in **Fig. 5**(c), at yet higher temperature (813 K), the intensity of the diffuse ring further increases. Also, diffraction spots from the lattice dilate, indicating lattice disorder. The pattern in **Fig. 5**(c) is typical of a semi-solid structure. The above results clearly indicate that a partial melting of IN9021 occurs at temperatures above at least 773 K. The sharp endothermic peak, which was observed in the composites, is likely attributed to local melting of matrix occurring at grain boundaries and/or interfaces between the matrix and reinforcements. Direct evidence of the local melting in the composites is also confirmed by the *in situ* TEM investigation [5,6] and local melting results from segregation of Mg or Si at boundaries and/or interfaces between the matrix and reinforcements [7].

Figure 5 Diffraction patterns obtained from in situ TEM observations in IN9021 at (a) 758K, (b) 773K, and (c) 813K, respectively. This result supports the presence of liquid phases.

3.3   Flow stress, elongation and strain rate sensitivity

The variation in elongation-to-failure (top) and flow stress (bottom) of IN9021 tested at optimum superplastic strain rates as a function of temperature is shown in **Figure 6**. The elongation increases with temperature, reaches a maximum value, and then drops rapidly, whereas the flow stress decreases with increasing temperature. It is noted that a discontinuity in flow stress is clearly found, *i.e.*, there is a significant drop in stress from 120 to 60 MPa at temperatures between 730 and 750 K. This strongly suggests that the dominant deformation mechanism in IN9021 changes near the temperatures where the significant drops in stress are found. The changes of the mean values of strain rate sensitivity exponent, m value, in the superplastic strain rate range are shown in **Figure 7** as a function of testing temperature for all mechanically alloyed aluminum alloys. It is evident that the m values of all alloys increase with temperature. The m values at lower testing temperatures are in the range from 0.25 to 0.33 ($3<n<4$) for all alloys, and are therefore obviously lower than those m values of more than 0.5 ($n<2$) obtained at higher testing temperatures. For example, for IN9021 tested at temperatures below 748 K, the m values are about 0.3. However, the m value increases with temperature and becomes greater than 0.5 at temperatures over 773 K. This high value of $m=0.5$, is a typical value for a highly superplastic material and indicates that grain boundary sliding is the dominant deformation mode in mechanically alloyed aluminum alloys. It is noted that values of m increase drastically at the incipient melting points for all the material. A change in m from low– to high–temperature regions indicates a corresponding change in deformation mechanisms.

3.4   Activation energy

The temperature dependence, *i.e.*, activation energy for superplastic flow, in IN9021 is shown in **Figure 8**. The data can be divided into two regions in which the slopes, and thus the activation energies, are different. The activation energy is lower in the low–temperature region than in the high–temperature region. The transition points from one region to another appear to correspond to the incipient melting point of the alloy as marked in the figures. Similar results have been obtained in other alloys and composites exhibiting high-strain-rate-superplasticity.

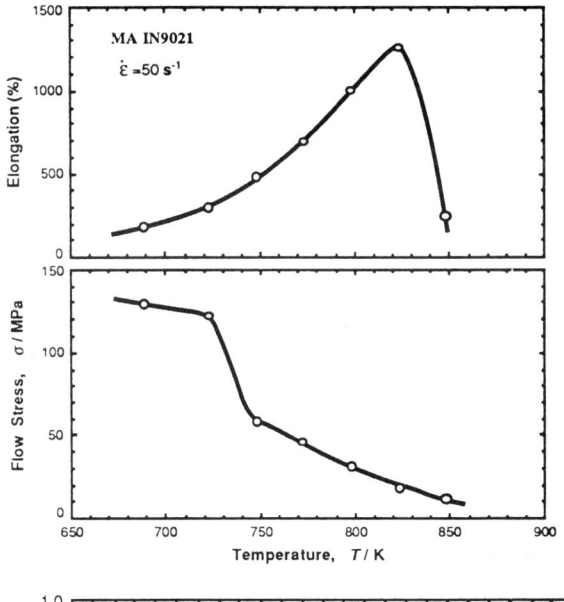

Figure 6 The variation in elongation-to-failure (top) and flow stress (bottom) of IN9021 tested at optimum superplastic strain rates as a function of temperature.

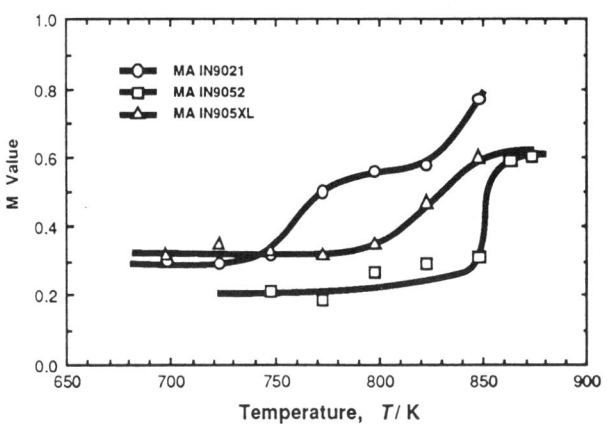

Figure 7 The changes of the mean values of strain rate sensitivity exponent, m value, in the superplastic strain rate range as a function of testing temperature for all mechanically alloyed aluminum alloys.

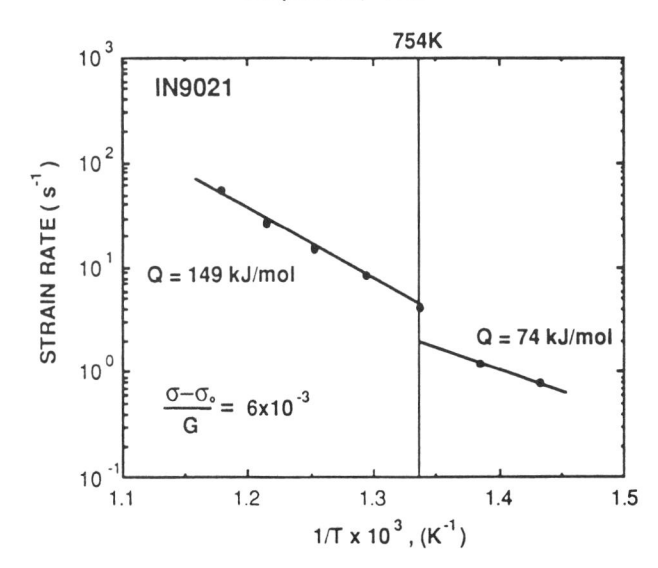

Figure 8 Activation energies for IN9021 deformed at high strain rates at high temperatures. A sudden increase in the activation energy occurs at the temperature near the incipient melting point in the alloy.

It is evident in Fig. 8 that values of m change from low– to high–temperature regions indicating a corresponding change in the controlling deformation mechanism. The measured activation energies for temperatures below (but near to) the incipient melting point, are close to the activation energy for grain boundary diffusion, Q_{gb}, in aluminum which is about 85 kJ mol^{-1} [8]. This result suggests that the sliding mobility of grain boundaries is greatly enhanced in this temperature region and that the deformation strain is accommodated by atoms diffusing along these grain boundaries.

4. Discussion

A summary of the superplastic properties of a number of advanced structural materials is reviewed in Table 1. The materials include the mechanically alloyed materials, the consolidated alloys from amorphous powders or nanocrystalline powders, the vapor quenched alloy, the metal matrix composites, as well as the powder (PM) and ingot (IM) metallurgically processed alloys. T_{op} represents the optimum superplastic temperature, T_i the incipient melting point, T_s the solidus temperature and d the grain size. All the materials are noted to be very fine grained structures from 0.5 to 3 μm in sizes, except IM 7475 which has a coarse grain of about 20 μm. It is interesting to note in Table 1 that larger elongations in all the materials (*i.e.*, relatively coarse grained IM 7475 alloy exhibits a remarkable maximum elongation of 2400 % at 806 K, which is extremely close to the solidus temperature of 808 K) is found near the measured incipient melting point, at which the presence of liquid phases as a result of incipient melting produces dramatic effects upon the superplastic behavior (*e.g.*, elongation, strain rate sensitivity, and activation energy) of the materials. Therefore it is suggested that a liquid phase plays an important role in superplastic deformation for all these aluminum alloys, independent of grain size.

Table 1 Superplastic properties of advanced and conventional materials.

Materials	T_{op} (K)	T_i (K)	T_s (K)	Strain rate (s^{-1})	Stress* (MPa)	m value	Elong. (%)	d (μm)
IN9052	863	837	866	10	15	0.6	330	0.5
IN905XL	848	818	851	20	12	0.6	190	0.4
IN9021	823	754	852	50	18	0.5	1250	0.5
SiCp/IN9021	823	751	866	5	5	0.5	600	0.5
Al-Ni-Mm	885	-	897	1	15	0.5	650	1.0
Al-Ni-Mm-Zr	873	-	898	1	15	0.5	650	0.8
VQ Al-Cr-Fe	898	-	896	1	20	0.5	505	0.5
Si$_3$N$_{4p(1\mu m)}$/Al-Cu-Mg	773	784	853	0.1	5	0.3	640	2.0
Si$_3$N$_{4p(1\mu m)}$/Al-Mg	818	819	866	1	6	0.3	700	1.0
Si$_3$N$_{4p(0.2\mu m)}$/Al-Mg-Si	833	830	855	2	5	0.3	620	1.3
Si$_3$N$_{4p(0.5\mu m)}$/Al-Mg-Si	833	829	858	1	6	0.3	350	1.9
Si$_3$N$_{4p(1\mu m)}$Al-Mg-Si	818	822	853	0.1	5	0.3	450	3.0
Si$_3$N$_{4w}$Al-Mg-Si	833	843	858	0.1	11	0.3	480	3.3
PM Al-Mg-Mn	848	-	845	3x10^{-3}	0.7	0.5	660	3.5
IM 7475	806	-	808	4x10^{-5}	0.3	0.5	2400	19.8

*; Stress at ε=0.1.

However, larger volumes of a liquid, or a continuous liquid layer, can not support normal tractions, and therefore can not contribute to large elongations. Thus, intergranular decohesion at a liquid grain boundary leads to intergranular fractures and very limited elongations. This increase in liquid explains the drop in elongation observed at the highest temperature above the melting point, as shown in Fig 6. There apparently exists a critical amount of liquid phase for the optimization of grain/interface boundary sliding during superplastic deformation. The optimum amount of liquid phase may depend upon the precise material composition and the precise nature of a grain boundary or interface, such as local chemistry (which determines the chemical interactions between atoms in the liquid phase and atoms in its neighboring grains) and misorientation [7]. Only a small amount of liquid phase is present at temperatures close to the incipient melting point, and would be expected to segregate to grain boundaries, and particularly at grain triple junctions. As shown in **Figure 9** when the liquid phase is isolated at only triple junctions (a) or the thickness of the liquid phase along the boundaries is very thin (b), atoms in the solid state across two adjacent grains can still experience a traction force, T_{gb}, from each other. The grain boundary can sustain an applied tensile force. Also, shear stresses τ can be transferred across the boundary. However, at the higher temperatures close to the solidus temperature, macroscopic melting begins to occur, the liquid phase is thick, and atoms across two neighboring grains can no longer experience traction from each other. So the grain boundary can no longer support an applied tensile force and shear stresses cannot be transferred across the boundary. Ductility in this latter case is therefore limited.

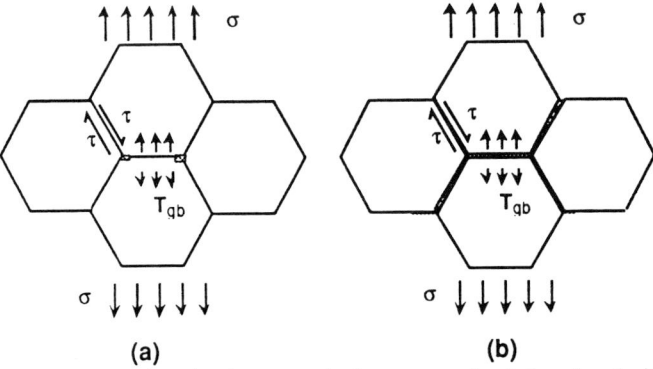

Figure 9 Schematic representation of grain structure in the presence of grain boundary liquid phases.

The superplastic elongation strongly depended on the refinement of grain structures and the accommodation process to relax the stress concentration by the presence of the helpers, such as soft phase, amorphous phase or liquid phase, as summarized in **Figure 10**. It is important to note that the current models have been proposed for superplastic deformation are all considered to be based on solid state of the materials. When if the accommodation process in the solid state by diffusional or plastic flow was not fully adjusted, the generated high concentration in local stress formed cavities around interfaces of the particles or triple junctions of grain boundaries. The results in the present study reveal that the presence of a liquid phase at grain boundaries or liquid boundaries at high temperatures is one of the possible accommodation processes to relieve the stress constraints by grain boundary sliding, and leads to a decrease in the level and the growth rate of cavities. Therefore, a new model of the accommodation process promoted by an isolated liquid phase is proposed as the accommodation processes for the superplastic deformation mechanism in all the advanced and conventional superplastic materials.

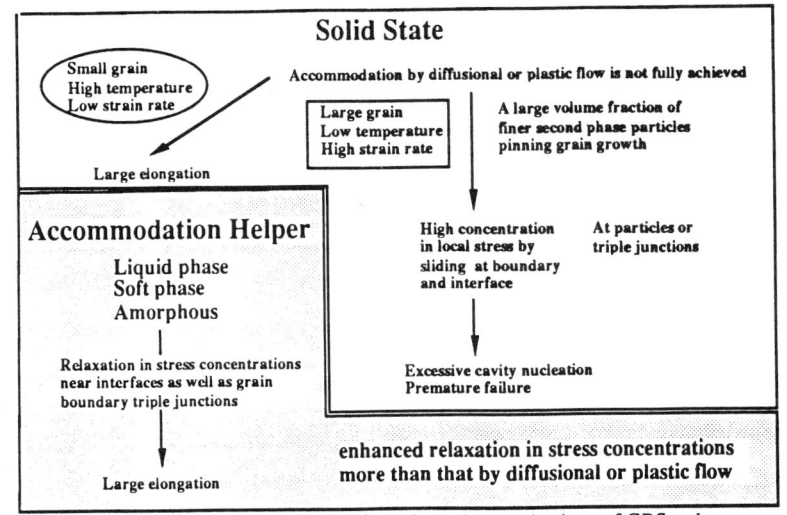

Figure 10 A schematic explanation for superplastic deformation mechanisms of GBS and accommodation.

5. Summary and Suggestion

It has been shown that the superplastic properties of a number of materials are strongly dependent upon the testing temperature. In particular, the temperature dependence of superplastic properties and specifically the tensile elongation, strain rate sensitivity, and activation energy are found to be strongly influence by proximity to the incipient melting points of these materials. For example, the tensile elongation exhibits a maximum at a temperature that is above, but near to, the incipient melting point. The strain rate sensitivity shows a rapid increase at temperatures above the incipient melting point. The activation energy displays a discontinuity at the incipient melting point. These observations confirm the suggestion that the presence of a small amount of liquid phase at interfaces and grain boundaries not only enhances the strain rate but also changes the deformation mechanisms. In addition, the presence of the liquid phase reduces the local stresses, and thus decreases both the cavity nucleation and growth rates. As the amount of liquid phase increases, however, the commonly-held observation that liquid phases degrade material ductility is noted.

The optimum superplastic strain rates, where large elongations were found, strongly depended on the refinement of grain structures. Also the superplastic elongations were critically controlled by the accommodation process to relax the stress concentration resulting from grain boundary sliding by the accommodation helpers such as soft phase, amorphous phase or liquid phase, Table 1 demonstrated that the optimum superplasticity of each alloy is obtained at a temperature close to partial melting point or solidus. It is important for near-nano scale materials, also including conventional superplastic alloys with relatively coarse grain sizes, that the microstructural stability at high temperatures would be achieved with optimum microstructural control, especially interface or boundary design of the accommodation helpers. The new finding of the accommodation helper by a liquid phase in superplastic materials will totally change the traditional understanding of the superplasticity mechanism. The presence of the helpers such as liquid boundaries to relax stress concentrations resulting from the sliding of the grain boundaries and/or interfaces at very high strain rates is very powerful. One of the most important and exciting research areas in the near future is how to design and achieve the optimum distribution of the helpers as well as the refinement of grain size.

Acknowledgment

This work was performed in part under the financial support of the Ministry of Education Science and Culture of Japan as a Grant-in-Aid. The author gratefully acknowledges Drs. T.G. Nieh and J. Wadsworth of the Chemistry and Materials Science Department at Lawrence Livermore National Laboratory, Livermore, CA, U.S.A., Drs. R.W. Gardiner and P.G. Partridge of DRA, RAE Farnborough, U.K. and Drs. A.Inoue and T.Masumoto of the Institute of Materials Research, Tohoku University, Sendai, Japan, for helpful discussions. The author would also like to thank his co-worker of Dr. M.Mabuchi of National Industrial Research Institute of Nagoya, Japan and former student colleagues for their contributions.

References

1. K. Higashi, *Mater. Sci. Eng.*, A166, 1993, 109-118.

2. M.Mabuchi, T.Imai and K.Higashi, *J. Mater. Sci.*, 28, 1993, 6582-6586.

3. J.E. Hatch, "*Aluminum - Properties and Physical Metallurgy*", American Society for Metals, Metals Park, Ohio, 1984, pp. 75.

4. *Aluminum Handbook 4th Edition,* Japan Inst. of Light Metals, Tokyo, 1990, pp. 25.

5. M. Mabuchi and K. Higashi, *Phil. Mag. Lett.*, **40**, 1994, 1-6.

6. J. Koike, M. Mabuchi, and K. Higashi, *Acta metall mater.*, 1994, in press.

7. J. Koike, M. Mabuchi, and K. Higashi, *J. Mater. Res.*, **10**, 1995, 1–6.

8. J.C.M. Hwang and R.W. Balluffi, *Scr. Metall.*, **12**, 1978, 709.

Microstructural Design for Superplastic Metal Matrix Composites

Mamoru Mabuchi[a)] and Kenji Higashi[b)]

a) National Industrial Research Institute of Nagoya, Hirate-cho, Kita-ku, Nagoya, 462, Japan
b) College of Engineering, Department of Mechanical Systems Engineering,
   University of Osaka Prefecture, Gakuen-cho, Sakai, Osaka, 591, Japan

Abstract

Mechanisms of grain refinement by hot extrusion have been investigated in aluminum alloy matrix composites and the concepts of microstructural design for superplasticity are addressed. An interaction of recrystallization and dynamic precipitation, which was enhanced by the presence of reinforcements, was responsible for grain refinement. A reduction in stress concentrations caused by the presence of reinforcements is required to limit development of cavities and to attain large elongations for metal matrix composites. Uniform dispersion of fine reinforcements is effective for a reduction in the stress concentrations. When the sliding process is not accommodated by diffusion and dislocation movement, special accommodation mechanisms are required. Partial melting resulting from segregation plays a vital role in relaxing the stress concentrations. Therefore it is very important to control segregation for processing of superplastic metal matrix composites.

1. Introduction

It is reported [1-9] that many metal matrix composites exhibited superplastic behavior. In particular, it is noted that some composites showed superplasticity at high strain rates (> 10^{-1} s^{-1}). High strain rate superplasticity is very attractive for commercial applications because one of current drawbacks in superplastic forming technology is a slow forming rate which is typically ~ 10^{-3} s^{-1}. The variation in the optimum superplastic strain rate as a function of the grain size is shown in Fig. 1 for aluminum matrix composites, where the optimum superplastic strain rate is the one where a maximum elongation is attained. It is found from Fig. 1 that very small grain sizes (≤ 5 μm) are responsible for high strain rate superplasticity.

Many high strain rate superplastic metals were produced from special methods or special materials such as mechanical alloying (MA) [10-12], physical vapor **deposition** and amorphous powders [12]. On the other hand, it is of interest to note that commercial aluminum alloy matrix composites exhibiting high strain rate superplasticity were processed by thermo-mechanical treatment [2] or hot extrusion [5-8] without special methods such as MA.

Mechanisms of high strain rate superplasticity are in some debate. Nieh *et al.* [13,14] noted that high strain rate superplasticity was found at temperatures close to, or even slightly above, the solidus temperature of the matrix alloy and they noted the importance of the presence of a liquid phase in

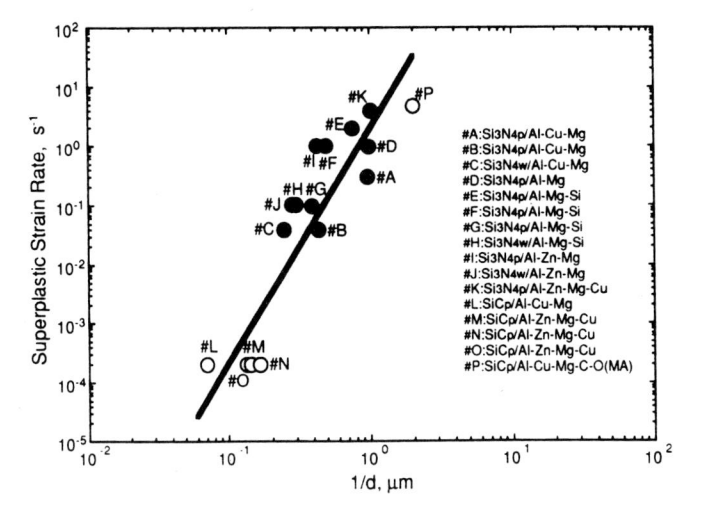

Figure 1 The relationship between superplastic strain rate and the grain size for aluminum matrix composites, where d is the grain size.

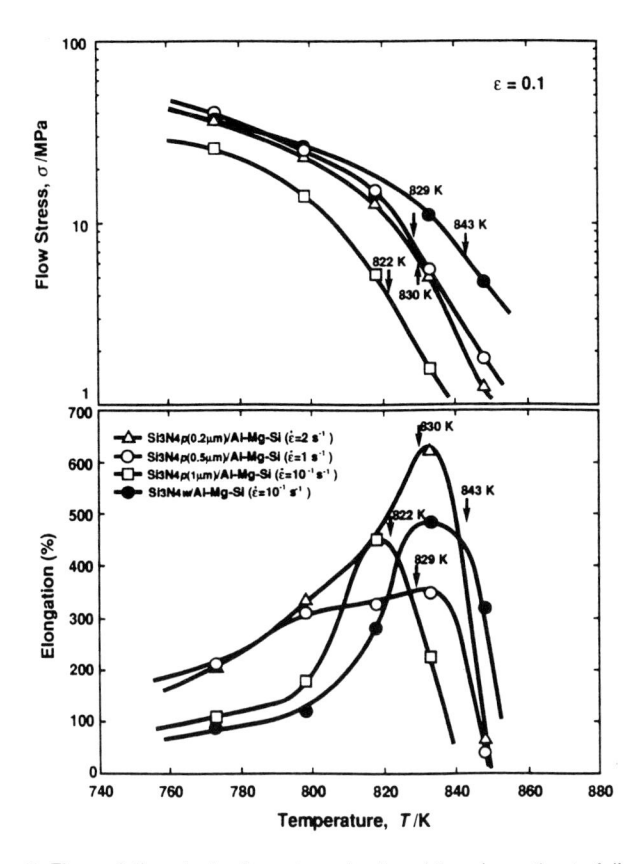

#A:Si3N4p/Al-Cu-Mg
#B:Si3N4p/Al-Cu-Mg
#C:Si3N4w/Al-Cu-Mg
#D:Si3N4p/Al-Mg
#E:Si3N4p/Al-Mg-Si
#F:Si3N4p/Al-Mg-Si
#G:Si3N4p/Al-Mg-Si
#H:Si3N4w/Al-Mg-Si
#I:Si3N4p/Al-Zn-Mg
#J:Si3N4w/Al-Zn-Mg
#K:Si3N4p/Al-Zn-Mg-Cu
#L:SiCp/Al-Cu-Mg
#M:SiCp/Al-Zn-Mg-Cu
#N:SiCp/Al-Zn-Mg-Cu
#O:SiCp/Al-Zn-Mg-Cu
#P:SiCp/Al-Cu-Mg-C-O(MA)

Figure 2 The variations in the flow stress (top) and the elongation to failure (bottom) of four different Si3N4/Al-Mg-Si composites as a function of temperature, where the melting point of each composite is shown by an arrow.

high strain rate superplasticity. The variation in the flow stress and the elongation to failure as a function of testing temperature is shown in Fig. 2 for four Si_3N_4/Al-Mg-Si (6061) composites [15]. It is found that a maximum elongation was attained at the temperature close to the melting point. It was revealed by *in-situ* TEM investigation [16] that partial melting occurred at matrix/reinforcement interfaces and along grain boundaries at superplastic temperatures for superplastic Si_3N_4/Al composites. It is therefore suggested that a liquid phase resulting from partial melting plays an important role in superplastic flow. Recently, an accommodation process by a liquid phase was proposed as a new concept for the accommodation processes for superplastic deformation mechanisms [17]. The new concept is that a liquid phase relaxes stress concentrations caused by the presence of reinforcements and promotes interfacial sliding occurring without excessive cavity formation at the interfaces. However, requirements for superplasticity from the viewpoint of microstructure are not clear in metal matrix composites. The aims of this paper are to investigate mechanisms of grain refinement by hot extrusion and to make clear the concepts of microstructural design for superplasticity in metal matrix composites.

2. Grain Refinement

2.1 Thermo-Mechanical Treatment

Some thermo-mechanical treatments (TMTs) have been developed to produce fine grain sizes for metals. Wert *et al.* [18] developed a TMT for grain refinement of an I/M 7075 aluminum alloy. The TMT consists of solution treatment, overaging, warm rolling and recrystallization treatment. In the TMT, discontinuous recrystallization was used to produce fine-grained microstructures, and consequently a small grain size of about 10 μm was attained. The fine-grained alloy processed from the TMT exhibited superplastic behavior at ~ 10^{-3} s^{-1} [19]. Watts *et al.* [20,21] developed a TMT in an Al-Cu-Zr alloy. In the TMT, fine particles of metastable cubic $ZrAl_3$ inhibited recovery and recrystallization at the warm working and fine subgrains with low angle boundaries were formed by

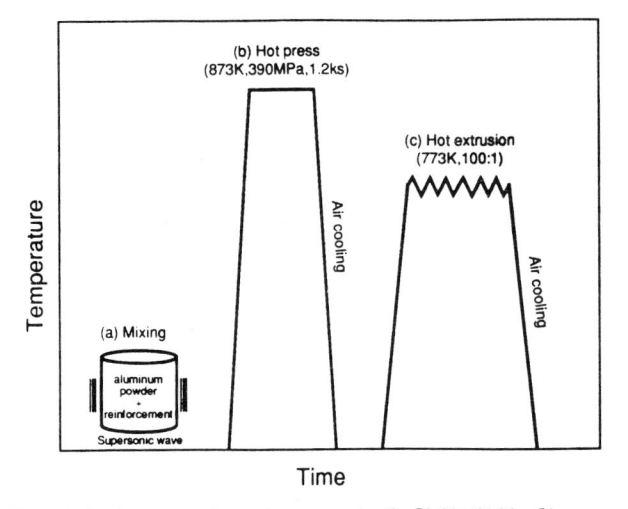

Figure 3 The fabrication procedures for superplastic Si3N4/Al-Mg-Si composites.

heating to the deformation temperature. The low angle grain boundaries continuously became high angle grain boundaries at an initial stage of superplastic flow and the grains were refined. McNelley and co-workers [22-24] developed a TMT in an Al-10wt.%Mg alloy. In the TMT, warm rolling and reheating gave rise to a continued buildup in grain boundary misorientation as dislocations were incorporated into the subboundary walls while precipitation of the β phase stabilized the evolving structure. As a result, grain refinement was attained by continuous recrystallization. TMTs have been applied to make SiC/Al composites superplastic [2-4]. Nieh et al. [2] showed that a SiC$_w$/2124Al composite processed by TMT exhibited superplastic behavior at a high strain rate of 3.3×10^{-1} s^{-1}, which is much higher than the conventional superplastic strain rate range (~ 10^{-3} s^{-1}). Mahoney and Ghosh [3] processed a superplastic SiC$_p$/PM64 composite by powder metallurgy and TMT. Recently it was reported [5-8] that many high strain rate superplastic Si$_3$N$_4$/Al composites were processed by powder metallurgy and hot extrusion. The fabrication procedures for superplastic Si$_3$N$_4$/Al-Mg-Si composites are schematically shown in Fig. 3. It is noted that high strain rate superplastic composites were processed by hot extrusion without special TMTs for grain refinement.

2.2 Effects of Reinforcements

Mechanical properties and microstructures of an extruded 20 vol.% Si$_3$N$_{4w}$/Al-Mg-Si (6061) composite and an extruded Al-Mg-Si (6061) alloy have been examined in order to make clear

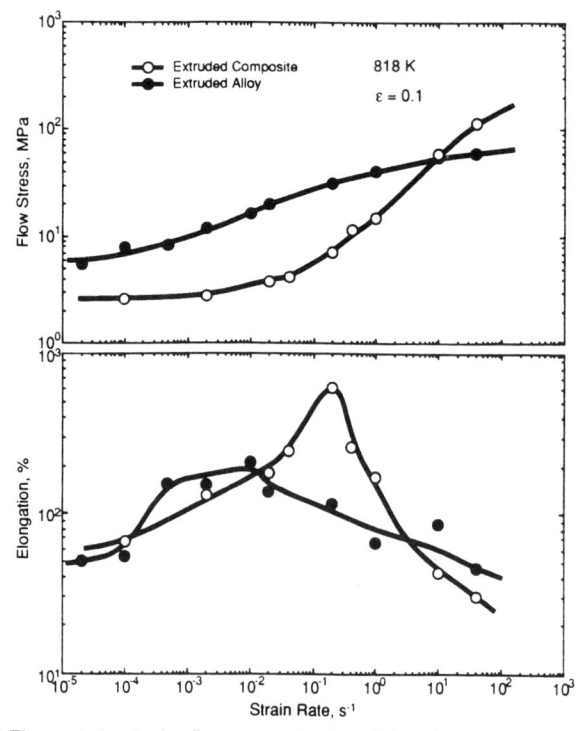

Figure 4 The variation in the flow stress (top) and the elongation to failure (bottom) as a function of the strain rate at 818 K for the extruded Si3N4w/Al-Mg-Si composite and the extruded Al-Mg-Si alloy.

mechanisms of grain refinement by hot extrusion. The variation in the flow stress (top) and the elongation to failure (bottom) as a function of the strain rate at 818 K is shown in Fig. 4 for the extruded composite and the extruded alloy, where extrusion is carried out at 773 K with a reduction ratio of 100 : 1. The relation between the flow stress and the strain rate at elevated temperatures for crystalline materials may be given by

$$\sigma = K\dot{\varepsilon}^m \qquad (1)$$

where σ the flow stress, K a constant incorporating structure and temperature dependencies, $\dot{\varepsilon}$ the strain rate and m the strain rate sensitivity. It is accepted that one of mechanical characteristics of superplastic flow is high m, which is commonly associated with grain boundary sliding as the predominant deformation mechanism, and m would be greater than 0.3 for superplastic flow. It is well known that for superplastic metals, there is a sigmoidal relationship between the logarithmic flow stress and the logarithmic strain rate, dividing the behavior into three regions. High m (> 0.3) is found in an intermediate strain rate range, which is the superplastic region. In this region, the elongations to failure are large. Both the high and low strain rate ranges exhibit values of $m < 0.3$ and there are decreases in the elongation to failure in these regions. Another important characteristic of superplastic flow is grain size dependence of mechanical properties [25-29]. The flow stress decreases and a superplastic region shifts to a higher strain rate range with decreasing grain size. The extruded composite showed high m (> 0.3) at $4 \times 10^{-2} \sim 10$ s^{-1}. It is noted that a large elongation of about 600 % was attained at a high strain rate of 2×10^{-1} s^{-1}. However, the extruded alloy showed high m at lower strain rates of $5 \times 10^{-4} \sim 2 \times 10^{-2}$ s^{-1} and a maximum elongation of about 200 % was attained at 10^{-2} s^{-1}. It follows that the extruded composite had higher superplastic potential than the extruded alloy.

Microstructures of the extruded composite and the extruded alloy are shown in Fig. 5. The extruded composite had equiaxed and small grains of about 3 μm. It is noted that the very small grain size was attained by hot extrusion without special TMT for the composite. Pinned dislocations, dislocation networks and precipitates of 0.1 ~ 1 μm in diameter were observed for the extruded composite. It is recognized that when recrystallization occurs continuously, pinned dislocations and dislocation networks remain, on the other hand, when recrystallization occurs discontinuously, sweeping of a high-angle boundary removes pinned dislocations and dislocation networks. The fact that pinned dislocations and dislocation networks were observed in the grains of the extruded composite suggests that the operative mechanism of grain refinement by hot extrusion is continuous recrystallization occurring during hot extrusion. For the extruded alloy, low angle grain boundaries were observed, and grains with high angle grain boundaries were elongated to the extrusion direction and were coarser than those of the extruded composite. Clearly the reinforcements played a vital role in grain refinement. In general, particles and grain boundaries are responsible for an increase in dislocation density occurring during deformation because dislocation generation is enhanced to accommodate strain gradients caused around particles and grain boundaries. In particular, large (> 1

Figure 5 Microstructures of (a) the extruded Si3N4w/Al-Mg-Si composite and (b) the extruded Al-Mg-Si alloy.

μm) and non-deformed particles give rise to intense strain gradients around the particles [30] and stimulate recrystallization in deformation zones adjacent to the particles [30,31]. Therefore, reinforcements cause very intense strain gradients, and so the driving force for recrystallization is increased and recrystallization is accelerated for composites, compared with alloys. It is concluded that acceleration of recrystallization due to the presence of reinforcements plays a vital role in grain refinement.

2.3. Mechanisms of Grain Refinement

The presence of reinforcements gives rise to enhancement of recrystallization. However, very small grain sizes can not be attained only by enhancement of recrystallization because limitation of grain growth is required for control of the grain size. A TEM micrograph of microstructure during hot extrusion for the Si_3N_{4w}/Al-Mg-Si composite is shown Fig. 6. It is found that a number of fine particles were precipitated. The mean size of the precipitates was about 20 nm. The fine precipitates were not observed prior to hot extrusion. It is suggested that precipitation occurred dynamically during hot extrusion. Christman and Suresh [32] showed that nucleation of precipitation was enhanced in a composite, compared with a matrix alloy, because of preferential precipitation of solute atoms at dislocations increased by the presence of reinforcements. Therefore dynamic precipitation of the fine particles is probably enhanced by the presence of reinforcements. The fine precipitates likely play a vital role in limitation of grain growth because of high pinning potential. It is therefore suggested that an interaction of recrystallization and dynamic precipitation results in the very small grain size.

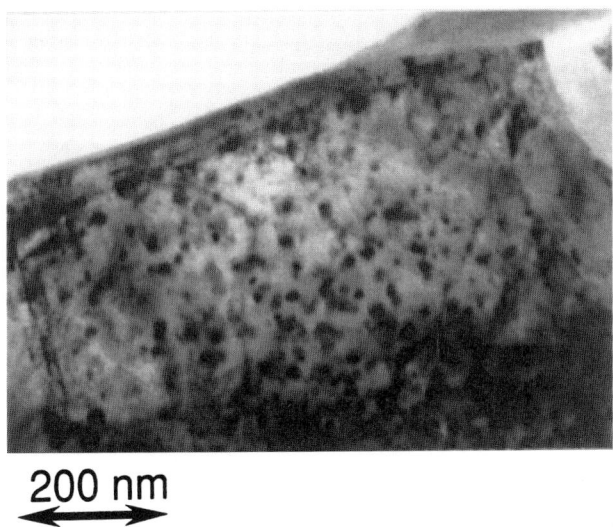

Figure 6 A TEM micrograph of microstructure during hot extrusion for the Si3N4w/Al-Mg-Si composite.

3. Microstructural Design for Superplasticity in MMC

A small grain size is required for superplasticity in composites as well as metals. However, only this requirement is not sufficient to attain large elongations for composites because reinforcements provide the sites for excessive cavity formation during superplastic flow because stress concentrations are caused by the presence of reinforcements [17]. Raj and Ashby [33] showed that the nucleation rate of cavities increases exponentially with increasing stress. A reduction in the stress concentrations at the matrix/reinforcement interfaces is required to attain large elongations. The local normal stress at the interface [34] may be given by

$$\tau = \frac{kTa^3\dot{\overline{U}}}{1.6\Omega\lambda^2 D_L(1+5\frac{\delta D_{GB}}{aD_L})} \quad (2)$$

where τ the stress at the interface, k the Boltzmann's constant, T the absolute temperature, a the particle size, $\overline{\dot{U}}$ the average sliding rate of a grain boundary, Ω the atomic volume, λ the space between the particles, D_L the lattice self-diffusion coefficient, δ the grain boundary thickness and D_{GB} the grain boundary diffusion coefficient. It follows that the local stress increases exponentially with increasing particle size. Uniform dispersion of fine reinforcements is effective for a reduction in the stress concentrations.

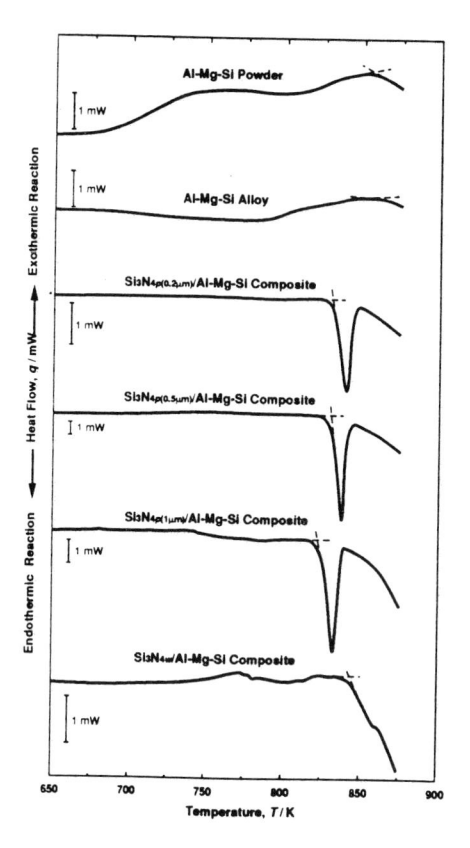

Figure 7 The DSC experimental data from 650 to 873 K for Si3N4/Al-Mg-Si composites, Al-Mg-Si powder and an extruded Al-Mg-Si alloy.

Even if reinforcements are dispersed finely and uniformly, sliding processes are not always accommodated by diffusion and dislocation movement. In such cases, special accommodation mechanisms are required for superplastic flow. Recently, a new accommodation mechanism by a liquid phase was proposed [17]. It should be noted from Fig. 2 that the optimum superplastic temperature where the maximum elongation was attained depended on materials though the chemical composition of the matrix is the same for any composites. The DSC experimental data from 650 to 873 K are shown in Fig. 7 for the Al-Mg-Si (6061) alloy and the Si_3N_4/Al-Mg-Si (6061) composites. The melting point of each material from the DSC results is listed in Table 1, where the onset temperature for the melting point is determined from an intercept of two dotted lines indicated in Fig. 7. The melting points of the rapidly solidified powder of the matrix alloy and the extruded matrix alloy were 855 K, which is the same as the published data of the Al-Mg-Si alloy (6061) [35].

It is found from Table 1 that the melting points of the Si_3N_4/Al-Mg-Si composites were lower than 855 K. Partial melting is responsible for melting at temperatures lower than 855 K [16]. It is reported from microchemical analysis by a field-emission TEM [36] that the boundaries are segregated mostly by Si, O and N and that the interfaces are segregated mostly by Mg and O. Such segregation gives rise to partial melting. The difference in the partial melting temperatures of the composites is probably resulting from the difference in concentration of solute segregation.

Table 1 Summary of melting points by the DSC investigations.

Al-Mg-Si Powder	855 K
Al-Mg-Si Alloy	855 K
$Si_3N_{4p(0.2\mu m)}$/Al-Mg-Si Composite	830 K
$Si_3N_{4p(0.5\mu m)}$/Al-Mg-Si Composite	829 K
$Si_3N_{4p(1\mu m)}$/Al-Mg-Si Composite	822 K
Si_3N_{4w}/Al-Mg-Si Composite	843 K

Large elongations were not obtained when an operating temperature was much higher than the melting point, as shown in Fig. 2. This indicates that too larger volume fraction of a liquid phase does not contribute to large elongations. It is associated with intergranular decohesion at the liquid grain boundaries leading to intergranular fractures. The presence of an isolated or discontinuous liquid phase is required to limit intergranular decohesion. Therefore it is very important to control of segregation.

4. Summary

Very small grain sizes (≤ 5 μm) and a reduction in stress concentrations at interfaces are required to attain high strain rate superplasticity in metal matrix composites.

Many high strain rate superplastic aluminum matrix composites have been processed by hot extrusion. Grain refinement by hot extrusion was attributed to an interaction of recrystallization and dynamic precipitation, which was enhanced by the presence of reinforcements.

Uniform dispersion of fine reinforcements is effective for a reduction in the stress concentrations. When sliding processes are not accommodated by diffusion and dislocation movement, special accommodation mechanisms are required for superplastic flow. Partial melting resulting from segregation plays an important role in relaxing the stress concentrations. Therefore it is very important to control segregation for processing of superplastic metal matrix composites.

References

1. M.Y.Wu and O.D.Sherby, *Scripta Metall.*, (18), 1984, pp. 773-6.
2. T.G.Nieh, C.A.Henshall and J.Wadsworth, *Scripta Metall.*, (18), 1984, pp. 1405-08.
3. M.W.Mahoney and A.K.Ghosh, *Metall. Trans. A*, (18A),1987, pp. 653-61.
4. J.Pilling, *Scripta Metall.*, (23), 1989, pp. 1375-80.
5. T.Imai, M.Mabuchi, Y.Tozawa and M.Yamada, *J. Mater. Sci. Lett.*, (9), 1990, pp. 255-57.

6. M.Mabuchi, T.Imai, K.Kubo, K.Higashi, Y.Okada and S.Tanimura, *Mater. Lett.*, (11), 1991, pp. 339-42.
7. M.Mabuchi, K.Higashi, Y.Okada, S.Tanimura, T.Imai and K.Kubo, *Scripta Metall. Mater.*, (25), 1991, pp. 2003-6.
8. M.Mabuchi, K.Higashi and T.G.Langdon, *Acta Metall. Mater.*, (42), 1994, pp. 1739-45.
9. K.Higashi, T.Okada, T.Mukai, S.Tanimura, T.G.Nieh and J.Wadsworth, *Scripta Metall. Mater.*, (26), 1992, pp. 185-90.
10. T.G.Nieh, P.S.Gilman and J.Wadsworth, *Scripta Metall.*, (19), 1985, pp. 1375.
11. T.R.Bieler, T.G.Nieh, J.Wadsworth and A.K.Mukherjee, *Scripta Metall.*, (22), 1988, pp. 81-86.
12. K.Higashi, *Mater. Sci. Eng. A*, (A166), 1993, pp. 109-18.
13. T.G.Nieh and J.Wadsworth, *Mater. Sci. Eng. A*, (A147), 1991, pp. 129-42.
14. T.G.Nieh, J.Wadsworth and T.Imai, *Scripta Metall. Mater.*, (26), 1992, pp. 703-8.
15. M.Mabuchi and K.Higashi, *Mater. Trans. JIM*, (35), 1994, 399-405.
16. J.Koike, M.Mabuchi and K.Higashi, *Acta Metall. Mater.*, 1994, in press.
17. M.Mabuchi and K.Higashi, *Phi. Mag. Lett.*, (70), 1994, 1-6.
18. J.A.Wert, N.E.Paton, C.H.Hamilton and M.W.Mahoney, *Metall. Trans. A*, (12A), 1981, pp. 1267-76.
19. C.C.Bampton, J.A.Wert and M.W.Mahoney, *Metall. Trans. A*, (13A), 1982, pp. 193-8.
20. B.M.Watts, M.J.Stowell, B.L.Baikie and D.G.E.Owen, *Metal Sci.*, (10), 1976, pp. 189-97.
21. B.M.Watts, M.J.Stowell, B.L.Baikie and D.G.E.Owen, *Metal Sci.*, (10), 1976, pp. 198-206.
22. T.R.McNelley, E.-W.Lee and M.E.Mills, *Metall. Trans. A*, (17A), 1986, pp. 1035-41.
23. E.-W.Lee, T.R.McNelley and A.F.Stengle, *Metall. Trans. A*, (17A), 1986, pp. 1043-50.
24. S.J.Hales, T.R.McNelley and H.J.McQueen, *Metall. Trans. A*, (22A), 1991, pp. 1037-47.
25. F.A.Mohamed and T.G.Langdon, *Acta Metall.*, (23), 1975, pp. 117-24.
26. 5. R.C.Gifkins, *Metall. Trans. A*, (7A), 1976, pp. 1225-33.
27. S.-A.Shei and T.G.Langdon, *Acta Metall.*, (26), 1978, pp. 639-46.
28. A.Arieli, A.K.S.Yu and A.K.Mukherjee, *Metall. Trans. A*, (11A), 1980, pp. 181-91.
29. C.H.Hamilton, C.C.Bampton and N.E.Paton, in *Superplastic Forming of Structural Alloys*, N.E.Paton and C.H.Hamilton, eds., The Metallurgical Society of AIME, San Diego, 1982, p. 173-89.
30. F.J.Humphreys, *Metal Sci.*, (13), 1979, pp. 136-45.
31. F.J.Humphreys, *Mater. Sci. Eng. A*, (A135), 1991, pp. 267-73.
32. T.Christman and S.Suresh, *Acta Metall.*, (36), 1988, pp. 1691-704.
33. R.Raj and M.F.Ashby, *Acta Metall.*, (23), 1975, pp. 653-66.
34. R.Raj and M.F.Ashby, *Metall. Trans.*, (2), 1971, pp. 1113-27.
35. *Aluminum Handbook 4th Edition*, Japan Inst. of Light Metals, Tokyo, 1990, p. 25.
36. J.Koike, M.Mabuchi and K.Higashi, *J. Mater. Res.*, (10), 1995, 1-6.

Superplasticity in Mg-Al-Ga alloys produced by rapid solidification processing

T.Shibata and A.Uoya
YKK SIMST, Tomiya Kurokawa Miyagi, 981-33, Japan

K.Higashi
Dep. Mech. Eng., Univ. Osaka Pref., Sakai, Osaka, 593, Japan

Y.Yamaguchi, A.Inoue and T. Masumoto
IMR, Tohoku Univ., Sendai 980, Japan

Abstract

The extruded bulks of rapidly solidified Mg-Al-Ga alloys show the extremely high specific strength (3×10^5 N.m.kg^{-1}) superior to a one of the Ti-6Al-4V T6 alloy. As-extruded Mg-Al-Ga alloys are composed of a fine grained or sub-grained texture of nearly 600nm and the intermetallic compounds from 100nm to 300nm in diameter. The matrix refinement and dispersed compounds were expected to result in high strain rate superplasticity. The superplasticity was first obtained in the strain rate range between 10^{-3} and 10^{-1}s^{-1}, and the value of fracture elongation reached 1080% at a strain rate of 10^{-2}s^{-1} and at a temperature of 573K.

1. Introduction

Rapid solidification processing has been used to obtain new metastable crystalline phase with extended solid solubility and amorphous phase. The present authors have prepared several Mg-based amorphous or metastable crystalline alloys with high strength and good ductility. Rapidly solidified Mg-Al-Ga alloys are strengthened by the homogeneous dispersion of intermetallic compounds into the h.c.p. Mg supersaturated solid solution and its tensile strength reaches 600MPa. The specific strength (σ_B/ρ) of these alloys exceeds significantly the value of Ti-6Al-4V T6. The grain size of the matrix and the particle one of the intermetallic compounds in $Mg_{89}Al_8Ga_3$ alloy are nearly 600nm and from 100 to 300nm, respectively. The most straightforward microstructural feature that can be modified to achieve high strain rate superplasticity is to decrease the grain size. The strain rate range of superplasticity, for example, that of conventional commercial aluminum alloys, is relatively slow compared to many manufacturing processes [1, 2]. So many methods have been developed for attainment of fine grain structure in metal-based materials.

Present authors have reported that high strain rate superplasticity occurs in RS+PM $Al_{88.5}Ni_8Mm_{3.5}$ [3] and $Al_{89.7}Ni_8Mm_{1.5}Zr_{0.8}$ alloys [4]. The superplastic strain rate range of 0.1 to 10 s^{-1} for Al-Ni-Mm and Al-Ni-Mm-Zr alloys is many orders of magnitude higher than that for typical commercial superplastic alloys [1]. it would be expected that high strain rate superplasticity occurs in the Mg-Al-Ga alloys.

In this paper, it is demonstrated to identify the structure's change of the Mg-Al-Ga alloy[5] through the processing and examine superplastic behavior in the deformation range of strain rates from 10^{-4} to 1 s^{-1} and temperatures between 523 and 623 K. Based on the mechanical data, superplastic deformation mechanisms are considered and established a constitutive equation for superplastic deformation of the Mg-Al-Ga alloy.

2. Experimental Procedure

The ingots were prepared by the induction melting a mixture of pure Mg(99.9 mass%), Al(99.9%) and Ga(99.99%) metals in a purified argon atmosphere. The chemical composition ratios are expressed in atomic percentage. From the master alloy ingot, a rapidly solidified ribbon with a cross section of 12x0.1mm^2 was produced by a single roller melt-spinning technique. The rotation speed of the copper roller is 31m s^{-1}. The ribbons were slightly pulverized for easy compaction to a container with 24mm in diameter. The powders were densificated at temperatures between 443 and 463K and under applied loads between 620 and 930MPa. The Hot-pressed billet was extruded into a bar with a diameter of 7mm in the temperature range of 443 to 463K and at applied pressures range of 770 to 1040MPa. The resulting extrusion ratio was 10:1. The structure of the melt-spun ribbons and extruded bulks was examined by X-ray diffractometry and transmission electron microscopy(TEM). Vickers hardness and tensile strength were measured by a Vickers hardness tester with a load of 0.25N and an Instron-type tensile testing machine. In order to determine the superplastic behavior, constant true strain rate tensile tests were carried out over a strain rate range from 10^{-4} to $1\ s^{-1}$ at a testing temperature between 523 to 623 K in air. The tensile specimens had a gage length of 5 mm and gage diameter of 2.5 mm.

3. Result and Discussion

Transmission electron micrographs of as-quenched ribbon and as-extruded bulk of the Mg-Al-Ga alloy are shown in Figure 1 (a) and (b), respectively.

As-quenched ribbons have a network structure on the grain boundaries and very fine precipitates inside of the grains. The network structure grows as a granular shape at the first stage and precipitates as an aggregate on the grain boundaries of the hcp Mg matrix through the extrusion. As the result of the warm extrusion, the grain size of the matrix turns from 300 to 600nm. The granular and the globular-like compounds are identified to be $Al_{12}Mg_{17}$ and Mg_5Ga_2, respectively, from results of the compositional analyses by EDX and X-ray diffraction analyses. In addition, the diffraction pattern differed between the longitudinal section and the cross section of the extruded bulk and the structure seemed to have the preferred orientation. This result suggests that the solidification as melt-spinning has occurred from the contact side with roller, leading to the unidirectional growth of h.c.p. Mg phase. Furthermore, the extrusion pressure seems to rearrange the texture. In order to examine the texture of the extruded bulk, neutron diffraction analyses were conduced for Mg-Al-Ga alloy as shown in Figure 2. Neutron diffraction analyses is enable to examine the texture quantitatively because of the Neutron have a good penetration through the metals. The diffraction pattern obtained from the cross-section to the extrusion direction indicates that c-axis of the

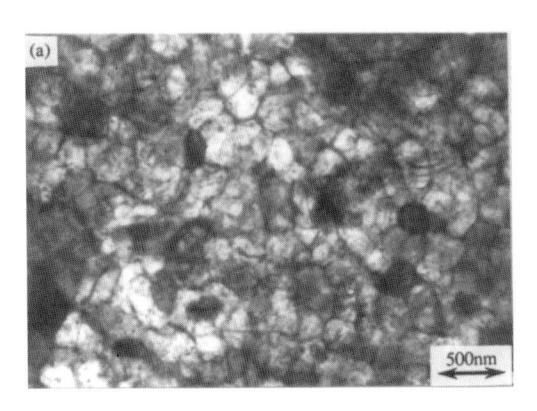

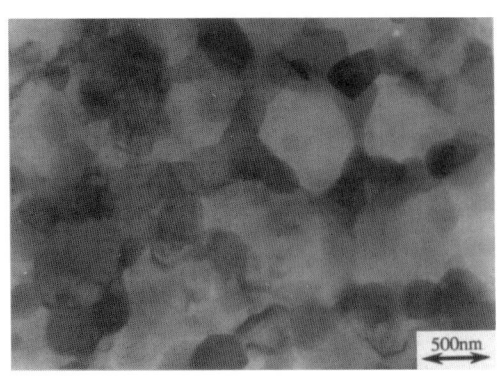

Fig.1 Transmission electron micrographs of (a) as-quenched ribbon and (b) as-extruded bulk of the Mg-Al-Ga alloy.

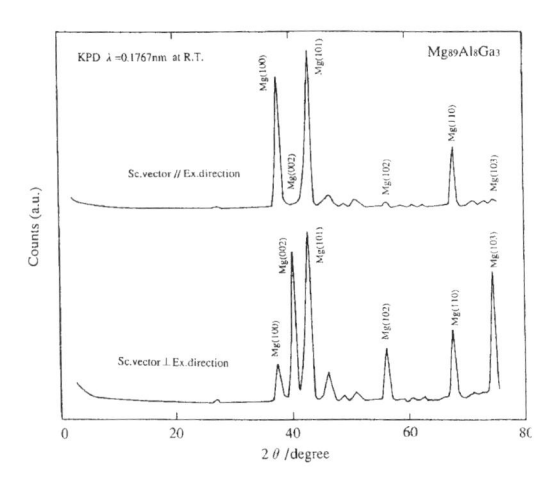

Fig.2 Neutron diffraction patterns of as-extruded bulk of the Mg-Al-Ga alloy.

h.c.p. Mg orientate perpendicularly to the extrusion direction because of peak intensity of (002) being zero. Furthermore, if we suppose that the extrusion direction is a*-axis[100], the calculation was doing as shown in following function.

$$I_{obs.} = kf(\Phi) I_{calc.}$$

scanning vector ⊥ extrusion direction

$$I_{obs.} = k' f(\frac{\pi}{2} - \Phi) I_{calc.}$$

scanning vector // extrusion direction

angles at a diffraction beam with a*-axis, Φ

When the values of Φ were x-axis and the values of $I_{obs.}/I_{calc.}$ plotted a diagram, the function of $f(\Phi)$ can be drawn by a curve as shown in Figure 3. This fact indicates the supposition that a*-axis [100] of the h.c.p. Mg be toward the extrusion direction have been correct. During the extrusion, intermetallic compounds of Al12Mg17 and Mg5Ga2 precipitate still more and disperse homogeneously from the supersaturated h.c.p. Mg solid solution. The Vickers hardness number (Hv) of rapidly solidified ribbon is 118. On the other hand, the Hv of extruded bar is 163.

Dispersion strengthening by intermetallic compounds are interpreted to contribute to the achievement of the high Hv and strength of the Mg-Al-Ga alloy. Typical mechanical properties of the Mg-Al-Ga bulk alloy at room temperature are shown in Table 1 with those of IM Mg-Al-Zn alloy, high strength aluminum 7075 alloy and titanium alloy. It is particularly noticed that the specific strength of the Mg-Al-Ga alloy is about 1.25 times as high as that for the Ti-6Al-4V alloy. A typical microstructure after annealing at 573 K for 1.2 ks of the Mg-Al-Ga alloy consists of fine grains of 2 μm in size and intermetallics of Al12Mg17 of 1 μm in diameter. The grain size, that of 0.6 μm in an as-extruded condition, increases with increasing temperature to 3 μm at 623 K.

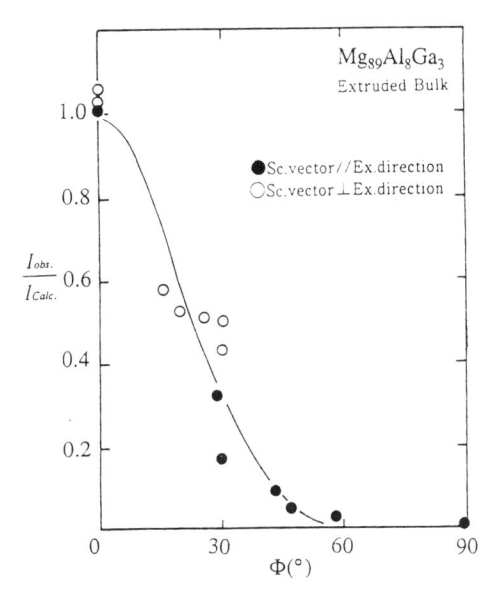

Fig.3 Angles Φ at a diffraction beam with a* - axis vs intensity ratio between calculation and observation.

TABLE 1. Typical mechanical properties of the Mg-Al-Ga bulk alloys and the conventional alloys at room temperature

alloy	Hv (ribbon)	ρ /Mg·m^{-3}	σ_B /MPa	$\sigma_{0.2}$ /MPa	ε_P /%	E /GPa	Hv	$\sigma_B \cdot \rho^{-1}$ /10^5 N·m·kg^{-1}	$E \cdot \rho^{-1}$ /10^7 N·m·kg^{-1}
Mg90Al8Ga2	111	1.87	595	499	3.5	48	152	3.18	2.6
Mg89Al9Ga2	112	1.88	613	474	1.3	48	154	3.26	2.6
Mg89Al8Ga3	118	1.91	630	498	2.6	48	163	3.30	2.5
AZ91D		1.81	234	152	3	45	78	1.30	2.5
7075 T6		2.80	573	503	11	72	173	2.05	2.6
Ti-6Al-4V T6		4.43	1167	1030	7	113	365	2.63	2.6

Typical true stress - true strain curves for the Mg-Al-Ga alloy are shown in Figure 4 at strain rates from 10^{-3} to 1 s^{-1} at temperature of 573 K. The curves of which strain rates are below 10^{-1} s^{-1} exhibit strain hardening which may be cause by dynamic grain growth.

The flow stress and elongation of the Mg-Al-Ga alloy as a function of strain rate are shown in Figure 5 for test temperatures between 523 to 623 K.

The flow stress increased with strain rate in a typical sigmoidal curve, as has been previously observed for superplastic materials. In the equation

$$\sigma = k\dot{\varepsilon}^m$$

(σ is the true stress, $\dot{\varepsilon}$ is the true strain rate, k is a constant incorporating structure and temperature dependencies), the strain rate sensitivity, m is the slope of this curve ($d\ln\sigma/d\ln\dot{\varepsilon}$). In general, large elongations are obtained in the range of both temperatures and strain rates where high m values are found. For the Mg-Al-Ga alloy deformed at a temperature of 573 K, the strain rate sensitivity of less than 0.2 was obtained in the strain rate range below 10^{-3} s^{-1}

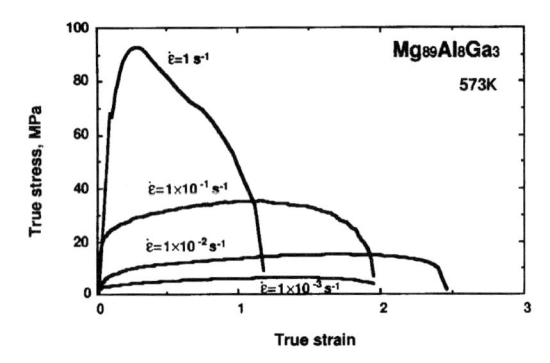

Fig.4 Typical true stress-true strain curves for the Mg-Al-Ga alloy at 573 K and at strain rate from 10^{-3} to 1 s^{-1}.

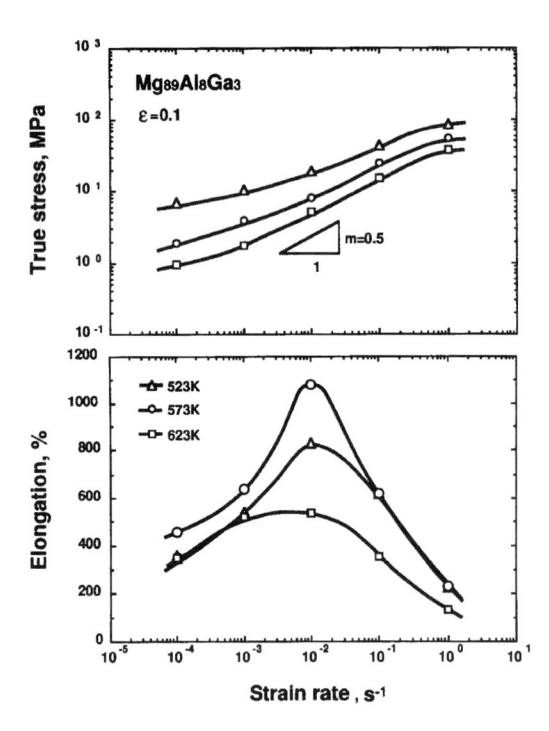

Fig.5 The data for flow stress and elongation of the Mg-Al-Ga alloy at testing temperature between 523 and 623 K as a function of strain rate.

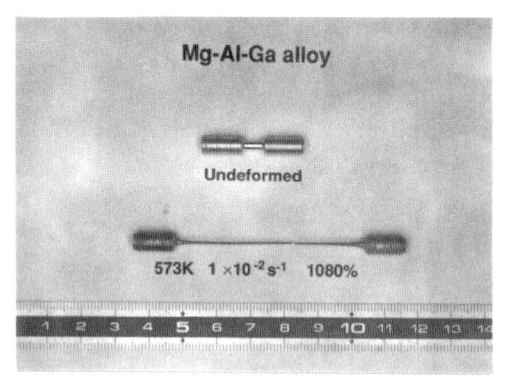

Fig.6 Fractured specimen of the Mg-Al-Ga alloy deformed at 573 K at a strain rate of 1×10^{-2} s^{-1}.

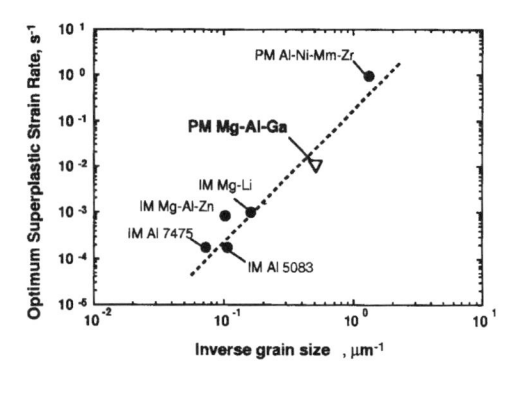

Fig.7 Relationships between inverse grain size and optimum superplastic strain rate for several superplastic light metals produced by RS+PM methods or IM methods.

and above 10^{-1} s^{-1}. On the other hand, at the strain rate range from 10^{-3} to 10^{-1} s^{-1}, a high m value of about 0.5 was obtained. This high m value of 0.5 suggests that grain boundary sliding could be a primary deformation mechanism in the Mg-Al-Ga alloy at high temperatures. Furthermore, at high strain rates above 10^{-1} s^{-1}, where dislocation processes dominate the deformation, the m value decreased again to 0.2-0.3.

For the Mg-Al-Ga alloy, the peaks of elongation were found at the temperature range of 523 to 623 K and at the strain rate of nearly 10^{-2} s^{-1}. Especially, a maximum elongation of 1080 % is obtained at the strain rate of 10^{-2} s^{-1} at 573 K. The elongations decrease at the strain rates below 10^{-3} s^{-1} and above 10^{-1} s^{-1}. The strain rate range from 10^{-3} to 10^{-1} s^{-1} for many elongations corresponds to that for high m value of 0.5. It is noticed that large elongations over 100 % were obtained at all deformation condition in this work, compared with the poor ductility at room temperature.

The undeformed and the fractured specimens are shown in Figure 6, where uniform elongation of 1080 % was obtained at 573 K and 10^{-2} s^{-1}.

Figure 7 shows the relationship between inverse grain size and optimum superplastic strain rate for several superplastic alloys. It is well established that fine grain size is an essential prerequisite and fine grain size needs to remain stable throughout the deformation condition for superplasticity. And it is well-known that superplastic strain rate increases with decrease of grain size. The grain size of the Mg-Al-Ga alloy at 573 K, for optimum superplastic condition, was about 2 μm in diameter. That was comparatively fine with the grain size of conventional IM alloys. So optimum superplastic strain rate for the Mg-Al-Ga alloy was a few orders of magnitude higher than those for typical commercial superplastic aluminum alloys such as 5083 [6] and 7475 [1] and other magnesium alloys [7,8].

axis[100] was toward the parallel to the extrusion direction. The grain size of magnesium matrix at 573 K, which was optimum temperature for superplastic deformation of the Mg-Al-Ga alloy, was about 2 μm in diameter. As a result, high strain rate superplasticity was observed. Maximum elongation of about 1080 % was obtained at 573 K and strain rate of 10^{-2} s^{-1}, which strain rate in the Mg-Al-Ga alloy is higher than those in other magnesium alloy and conventional superplastic aluminum alloys previously reported.

References

[1] John Pilling and Norman Ridley, SUPERPLASTICITY IN CRYSTALLINE SOLIDS. The Institute of Metals, London, 1989
[2] J. Wadworth, T. G. Nieh and O. D. Sherby, Superplasticity in Advanced Materials, The Japan Society for Research on Superplasticity, Osaka, Japan, 1991, 13-22
[3] K. Higashi, T. Mukai, S. Tanimura, A. Inoue, T. Masumoto, K. Kita K Ohtera and J. Nagahora, Scripta.Met., 26, 1992, 191-196
[4] K. Higashi, A. Uoya, T. Mukai, S.Tanimura, A. Inoue T. Masumoto and K. Ohtera, Mater. Sci. Eng., A181/182, 1994, 1068-1071
[5] T. Shibata, M. Kawanishi, J. Nagahora, A. Inoue and T. Masumoto, Mater. Sci. Eng., A179/180, 1994, 632-636
[6] H. Iwasaki, K. Higashi, S. Tanimura, T. Komatubara and S. Hayami, Superplasticity in Advanced Materials, The Japan Society for Research on Superplasticity, Osaka, Japan, 1991, 447-452
[7] P. Metenier, G. Gonzalez-Doncel, O. A. Rauno, J Wolfenstine and O. D,. Sherby, Mater. Sci. Eng., A125, 1990),195-202
[8] H. Takuda, S. Kikuchi and N. Hatta, J. Mat. Sci. 27, 1992, 937-940

4. Conclusion

By rapidly solidified technique and powder metallurgy method, Mg-Al-Ga alloys were produced with fine microstructure. The present alloys exhibited high mechanical strength exceeding 600MPa with its tensile strength because of the matrix refinement and compounds' dispersion. The h.c.p. Mg matrix had preferred orientation whose c-axis was toward the perpendicular to extrusion direction and a*-

Superplasticity in Ceramics and Intermetallics

SUPERPLASTIC CERAMICS AND INTERMETALLICS
AND THEIR POTENTIAL APPLICATIONS

J. Wadsworth and T.G. Nieh

Lawrence Livermore National Laboratory
P.O. Box 808, L-353, Livermore, CA 94550

Abstract

Recent advances in the basic understanding of superplasticity and superplastic forming of ceramics and intermetallics are reviewed. Fine-grained superplastic ceramics, including yttria-stabilized tetragonal zirconia polycrystal, Y- or MgO–doped Al_2O_3, Hydroxyapatite, β-spodumene glass ceramics, Al_2O_3–YTZP two-phase composites, SiC–Si_3N_4, and Fe–Fe_3C composites, are discussed. Superplasticity in the nickel-base (e.g., Ni_3Al and Ni_3Si) and titanium-base intermetallics (TiAl and Ti_3Al), is described. Deformation mechanisms as well as microstructural requirements and effects such as grain size, grain growth, and grain-boundary phases, on the superplastic deformation behavior are addressed. Factors that control the superplastic tensile elongation of ceramics are discussed. Superplastic forming, and particularly biaxial gas-pressure forming, of several ceramics and intermetallics are presented with comments on the likelihood of commercial application.

Superplastic Ceramics

The forming and shaping of ceramics are difficult because the melting points of ceramics are relatively high and, consequently, the temperatures required to thermally activate plastic deformation in ceramics are also high. In addition, the propensity for grain boundary separation in ceramics is well known. In the 1950s, extensive efforts were made in the western world to hot fabricate ceramics using conventional metallurgical processes such as extrusion, rolling, and forging [1,2]. The goal was to produce near-net-shape parts in order to avoid the expensive machining of ceramics. A number of structural oxides, including CaO, MgO, SiO_2, ZrO_2, BeO, ThO_2, and Al_2O_3, were studied. As a result of this work, an improved understanding of ceramic deformation was developed but certain problems, and in particular the requirement for relatively high forming temperatures, still existed. For example, the temperature required for hot forging Al_2O_3 was found to be about 1900°C which is extremely high from a practical standpoint. Subsequently, the concept of thermomechanical processing of ceramics was more or less abandoned.

The first observation of fine structure superplasticity in ceramics, in a yttria-stabilized tetragonal zirconia, is generally attributed to Wakai in 1986 [3], although an elongation to failure of ~100% in polycrystalline MgO was reported in 1965 by Day and Stokes [4]. Many other claims have been made of superplastic behavior in ceramics, but nearly all are based upon tests carried out only in compression [5-7]. It is important to point out that superplasticity refers to high ductility in <u>tension</u> and so enhanced plasticity data from compression tests cannot necessarily be considered to be convincing evidence of superplasticity. This is because ceramics often exhibit high values of strain rate sensitivity exponent (m is often equal to 1) in compression tests, but nonetheless show very limited tensile ductility. Tensile ductility at elevated temperatures is limited primarily by grain boundary cavitation which is initiated by tensile stresses but usually suppressed by compressive stresses. Since 1986, a number of fine-grained polycrystalline ceramics have been demonstrated to be superplastic in tension. These include yttria-stabilized tetragonal zirconia polycrystal (YTZP) [3,8], Y_2O_3- or MgO-doped Al_2O_3 [9], Hydroxyapatite [10], β-spodumene glass ceramics [11], Al_2O_3-reinforced YTZP (Al_2O_3-YTZP) [12,13], SiC-reinforced Si_3N_4 (SiC-Si_3N_4) [14,15], and iron-iron carbide (Fe-Fe_3C) [16] composites. The area of superplastic ceramics has been the subject of considerable interest and some review papers are now available [17,18].

Mechanical Properties

A summary of the microstructure and properties of some of the superplastic ceramics and ceramic composites are listed in Table 1. Generally, superplastic flow in ceramics is a diffusion-controlled process and the strain rate, $\dot{\varepsilon}$, can be expressed as

$$\dot{\varepsilon} = A \cdot d^{-p} \cdot \sigma^n \cdot exp(-\frac{Q}{RT}) \qquad \text{Equ. 1}$$

where d is the grain size, σ is the flow stress, Q is the activation energy for flow, R is the gas constant, T is the absolute temperature and A, p, and n are constants. The values for p, n, and Q vary according to the microstructure, the specific flow/diffusion law, and sometimes the impurity content in a material.

Table 1 Data for Superplastic Ceramics and Ceramic Composites in Tension

Material	Microstructure	Testing Parameters	Max. Elong., %	Material Variables*	Ref.
Monolithics					
3YTZP	$d = 0.3$ μm	450°C 5×10^{-4} s^{-1}	120	$n = 2$, $p = 2$, $Q = 590$ kJ/mol	[3]
3YTZP	$d = 0.30$ μm no glassy phase	1550°C 8.3×10^{-5} s^{-1}	800	$n = 1.5$, $p = 3$, $Q = 510$ kJ/mol	[19]
3YTZP	$d = 0.3$ μm	1450°C 4.8×10^{-5} s^{-1}	246	$n = 2$, $p = $ NA, $Q = 580$ kJ/mol	[20]
Al$_2$O$_3$-500 ppm Y$_2$O$_3$	$d = 0.66$ μm	1450°C ~10^{-4} s^{-1}	> 65	$n = $ NA, $p/n = 1.5$, $Q = $ NA	[9]
Hydroxyapatite	$d = 0.64$ μm	1050°C 1.4×10^{-4} s^{-1}	> 150	$n > 3$, $p/n = 1$, $Q = $ NA	[10]
β-spodumene glass	$d = 0.91$-2.0 μm >4 vol% glassy phase	1200°C 10^{-4} s^{-1}	> 400	$n = 1$, $p = 3.1$, $Q = 707$ kJ/mol	[11]
Composites					
20wt% Al$_2$O$_3$/YTZP	$d = 0.50$ μm no glassy phase	1650°C 4×10^{-4} s^{-1}	620	$n = 2$, $p = 1.5$, $Q = 380$ kJ/mol	[13]
20wt% Al$_2$O$_3$/YTZP	$d = 0.50$ μm	1450°C 10^{-4} s^{-1}	200	$n = 2$, $p = $ NA, $Q = 600$ kJ/mol	[12]
10 vol% ZrO$_2$/Al$_2$O$_3$	$d = 0.5$ μm	1400°C 10^{-4} s^{-1}	>100	$n = ?$, $p = ?$, $Q = ?$	[21]
20wt% YTZP/Al$_2$O$_3$	d(ZrO$_2$) $= 0.47$ μm d(Al$_2$O$_3$) $= 1.0$ μm	1550°C 2.8×10^{-4} s^{-1}	110	$n = 2$, $p = $ NA, $Q = 700$ kJ/mol	[22]
30wt% TiC/Al$_2$O$_3$	$d = 1.2$ μm	1550°C 1.2×10^{-4} s^{-1}	66	$n = 4$, $p = $ NA, $Q = 853$ kJ/mol	[23]
β'-SiAlON	$d = 0.4$ μm with glassy phase	1550°C 10^{-4} s^{-1}	230	$n = 1.5$, $p = $ NA, $Q = $ NA	[24]
20 wt% SiC/Si$_3$N$_4$	$d = 0.2$-0.5 μm with glassy phase	1600°C 4×10^{-5} s^{-1}	>150	$n = 2$, $p = $ NA, $Q = 649$-698 kJ/mol	[25]
20 vol% Fe/Fe$_3$C	$d = 3.4$ μm	1000°C 10^{-4} s^{-1}	600	$n = 1.6$, $p = 2.9$, $Q = 200$-240 kJ/mol	[16]

The specific functional relationship among the material variables are dependent upon the material, and strain rate and temperature ranges. The n value is noted in Table 1 to be less than 2 for all materials, except for hydroxyapatite. Also, superplastic flow in these materials is noted to be dependent upon grain size. The grain size exponent, p, ranges from 1 to 3, which is within the range of various diffusion controlled flow models. Grain sizes in all the superplastic ceramics with tensile elongation greater than 100% except for Fe/Fe$_3$C, are noted to be less than 1 μm, which is much smaller than the grain size found in superplastic metals (typically, ~10 μm). Because of the fineness of the microstructure, grain growth, and in particular dynamic grain growth, usually occurs during superplastic deformation of ceramics [26-28]. Grain growth normally follows a mathematical expression of the form [26,29]

$$D^\ell - D_o^\ell = kt \qquad \text{Equ. 2}$$

in which $\ell \sim 3$, and D and D_o represent the instantaneous and initial grain sizes, respectively, k is a kinetic constant which depends primarily on temperature and grain boundary energy, and t is the time. Since the tensile ductility of superplastic ceramics decreases with increasing grain size, dynamic grain growth deteriorates the properties of superplastic ceramics. In addition, dynamic grain growth can affect the accurate value of strain rate sensitivity. For example, the reported strain rate sensitivity exponent values for superplastic YTZP vary from 0.3 [8,30,31] to 0.5 [3,20,27,32], in part, resulting from microstructural evolution during deformation [33]. For example, Nieh and Wadsworth [33] have determined the true m value to be about 0.67 (i.e., one measured under constant structure conditions), by normalizing the stress with the square of the final grain size; this result is illustrated in Fig. 1.

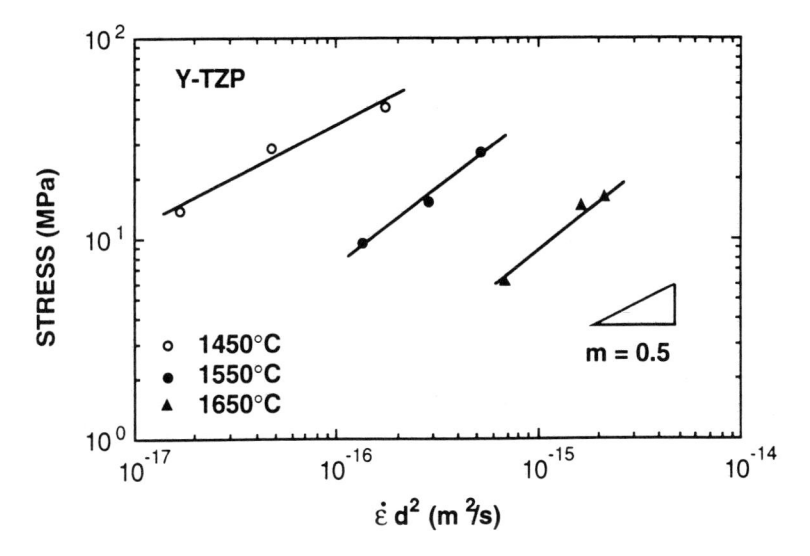

Fig. 1 Flow stress as a function of grain size normalized strain rate. The strain rate sensitivity exponents are less than 0.5 at all temperatures. (from Ref. [33])

In contrast to metals, ceramics generally exhibit high values of m, the tensile ductility of ceramics is nonetheless limited, indicating that necking stability does not govern the tensile ductility of ceramics. Kim et al. [34] and Chen and Xue [17] independently found that flow stress plays a dominant role in determining the tensile elongation of superplastic ceramics. Plotted in Fig. 2 is the true fracture strain of many superplastic ceramics as a function of flow stress; the true fracture strain is a linearly decreasing function of the logarithm of the flow stress. This result is directly related to the fact that when the flow stress is lower than the grain boundary strength of the material, intergranular failures do not occur and the material deforms plastically. As the flow stress is increased, so is the likelihood that the cohesive strength of grain boundaries will be reached. Once this level of stress is attained, intergranular cavitation and cracking occur, and the elongation to failure is decreased.

To illustrate the reverse effect of strengthening on ductility, Dougherty et al. [35] recently performed experiments with Al_2O_3 particle and SiC whisker reinforced fine-grained YTZP. Experimental results showed that the addition of SiC whiskers to YTZP causes a significant strengthening effect, but not for the Al_2O_3 particle reinforced composite. As shown in Fig. 3, at 1550°C and a strain rate of 10^{-3} s^{-1}, the flow stresses of both YTZP and a 20 wt% (28 vol%) Al_2O_3 particle-reinforced YTZP are both less than 30 MPa, whereas the flow stress of 20 vol% SiC whisker reinforced YTZP is almost 200 MPa. All of the three materials have a fine-grained matrix (~0.5 μm). As a result of the strengthening effect, the maximum elongation value of the SiC/YTZP composite was only 50%. In contrast, both the YTZP and Al$_2$O$_3$ particle reinforced YTZP composite behave superplastically.

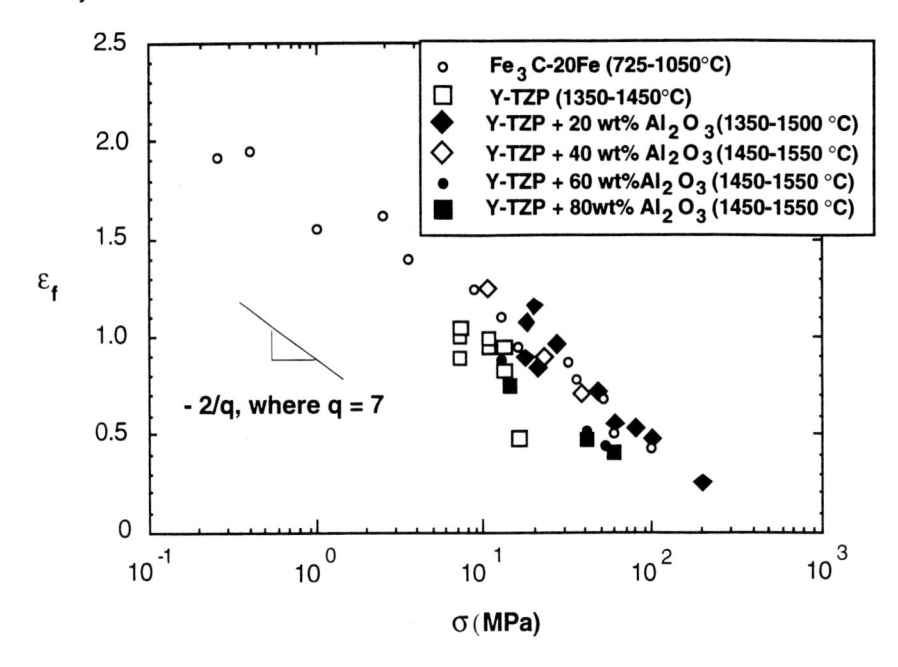

Fig. 2 Fracture strain as a function of the logarithm of the flow stress for superplastic iron carbide and YTZP based ceramics. (from Ref. [34])

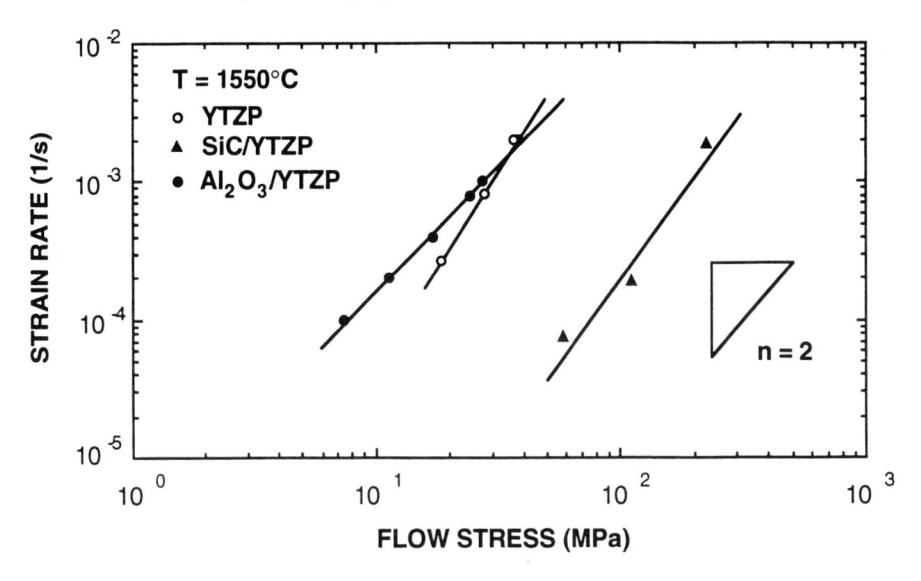

Fig. 3 Direct comparison of the strengths of YTZP and SiC/YTZP and Al$_2$O$_3$/YTZP composites at 1550°C. (from Ref. [35])

Grain-Boundary Structure

A general question concerning the microstructure of superplastic ceramics is: "Is the presence of a grain-boundary glassy phase necessary in order to produce superplasticity?" Nieh et al. [36] have presented several pieces of experimental evidence to show that there is no grain boundary glassy phase in their superplastic YTZP and 20 wt%Al_2O_3/YTZP samples. These experimental results include high resolution lattice images of grain boundary triple junctions in YTZP, shown in Fig. 4(left), in which lattice fringes from adjoining grains can be followed to their intersections at both the grain boundary interface and the triple junction. In addition, X-ray photoelectron Spectroscopy (XPS) from the intergranular fracture surfaces of superplastically deformed specimens, shown in Fig. 4(right), indicate the absence of low melting point glassy phases in YTZP and 20 wt%Al_2O_3/YTZP. These results suggest that the presence of a grain boundary glassy phase is unnecessary for superplasticity in fine-grained ceramics. Although the presence of a liquid phase at grain boundaries may not be a necessary prerequisite for superplasticity in ceramics, its presence can definitely affect some of the kinetic processes, such as grain boundary sliding. For example, Wakai et al. [37] and Chen and co-workers [17,38] have both demonstrated that the presence of grain-boundary glassy phases can, in general, reduce the superplastic forming temperatures for various ceramics, including YTZP and Si_3N_4.

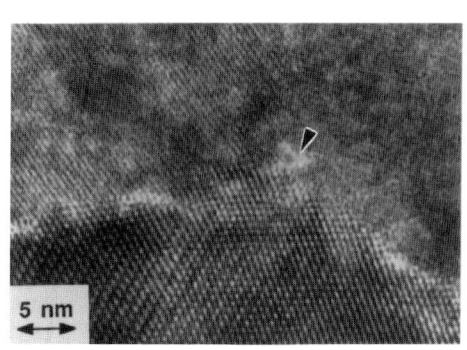

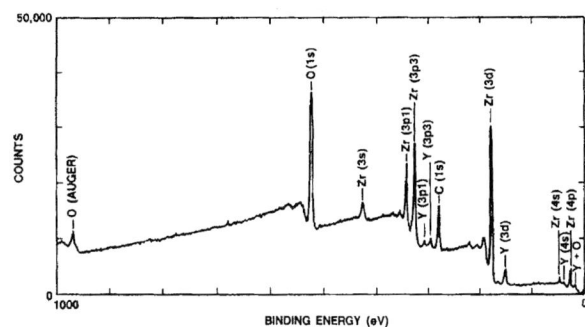

Fig. 4 (left) High resolution lattice image of a grain boundary triple junction in fine-grained superplastic YTZP, and (right) XPS spectrum from the fracture surface of a superplastic YTZP sample, indicating the absence of any second phase. (from Ref. [36])

Superplastic Intermetallics

Although large tensile elongations (~100%) for an intermetallic (Sendust, Fe-9.6wt%Si-5.4wt%Al) were indicated as early as 1981 [39], the observation of true superplastic intermetallics was not reported until 1987 [40]. Up to the present time, several intermetallics of the $L1_2$ structure (e.g., Ni_3Al [40-42] and Ni_3Si [43,44]), Fe_3Al [45], titanium aluminide (TiAl [46-48]), and trititanium aluminides (Ti_3Al [49,50]), have been demonstrated to be superplastic. Similar to the case for superplastic metals, superplasticity in intermetallics has been recorded in both quasi-single-phase and two-phase materials. The strain rate sensitivity ranges typically from 0.3 to 1. One interesting observation is that some intermetallics exhibit superplastic properties, even in the relatively coarse-grained (>10 μm) conditions [39,43,45].

Nickel-base

Two Ni-base intermetallics, Ni_3Si and Ni_3Al, have been demonstrated to be superplastic. A summary of the superplastic properties of Ni-base intermetallics is listed in Table 2. It is noted that the superplastic deformation of Ni_3Si and Ni_3Al can be well described by a classical equation (Equ. 1). The exact values for n, p, Q, and the material constant A in Equ. 1 for each material, of course, are determined by the temperature and strain rate regions in which superplastic properties of the material are characterized.

Table 2 Property data of superplastic Ni-base intermetallics

composition (wt%)	d, (μm)	T, (°C)	$\dot{\varepsilon}$, (s^{-1})	m	el. (%)	Ref.
Ni-9%Si-3.1%V-2%Mo (duplex)	15	1080	8 x 10^{-3}	0.5	710	[43]
Ni-9%Si-3.1%V-4%Mo (duplex)	>10	1070	10^{-3}	0.43	560	[43,51]
Ni$_3$(Si,Ti) (single-phase)	4	850	6 x 10^{-5}	0.43	180	[44]
0.24 at% B-doped Ni$_3$Al (single-phase)	1.6	700	5 x 10^{-5}	0.42	160	[41]
Ni-8.5%Al-7.8%Cr-0.8%Zr-0.02%B (duplex)	6	1100	8.3 x 10^{-4}	0.75	640	[42]

In the case of duplex Ni$_3$Si alloy, superplasticity was observed over a limited temperature range from 1000 to 1100°C, but over a wide strain rate range from 6 x 10^{-4} to 1 s^{-1}. A tensile elongation of over 200% can be generally obtained under all the above test conditions, and a maximum elongation to failure of 710% has been recorded at 1080°C at a strain rate of 8.0 x 10^{-3} s^{-1}. Strain rate as a function of flow stress is plotted in Fig. 5. The strain rate sensitivity, m, is determined to be about 0.5, which is a typical value in many superplastic metals, at a strain rate > 4 x 10^{-3} s^{-1}. In the low strain rate region, the m value increases and approaches one, probably resulting from the fact that the testing temperatures are over 0.9 T$_m$. This near-Newtonian-viscous behavior indicates that the deformation mechanism either changes from grain boundary sliding to a diffusion-type mechanism, such as Coble Creep or Nabarro-Herring Creep, or it may be a result of slip accommodation by a dislocation-glide mechanism [52]. The activation energy for the m = 0.5 region is calculated to be about 555 kJ/mol which is relatively high in comparison to the value of activation energy for superplastic deformation measured in a fine-grained nickel base superalloy MA754 of 267 kJ/mol [53]. In another study, Takasugi et al. [54] processed single-phase Ni$_3$(Si,Ti) into a fine-grained condition (~4 μm) and characterized the superplastic behavior of the alloys. The fine-grained material exhibits superplasticity (elongation = 180%) at temperatures between 800 and 900°C and at strain rates between 6.0 x 10^{-5} and 10^{-3} s^{-1}. These deformation temperatures are only about 0.8 T$_m$, where T$_m$ is the absolute melting point of Ni$_3$(Si,Ti), whereas superplasticity is found at about 0.85-0.94 T$_m$ for duplex Ni$_3$Si alloys. This is primarily attributed to the fact that grains in the single-phase Ni$_3$Si coarsen quickly at T > 850°C, while the microstructures of the duplex alloys are more thermally stable.

In the case of Ni$_3$Al, superplasticity has been produced in both single-phase [41] and duplex [42] alloys, and in both powder-metallurgy [40,42] and ingot-metallurgy [41] products. In the case of PM alloys, the superplastic IC-218 (composition of Ni-18 at%Al-8 at%Cr-1 at%Zr-0.15 at%B) had a duplex microstructure, containing about 10 to 15% of disordered γ phase in an ordered γ' phase matrix, and a grain size of about 6 μm. The alloy exhibited superplasticity (maximum elongation = 640%) at temperatures from 950 to 1100°C and strain rates from 10^{-5} to 10^{-2} s^{-1}. At strain rates > 7 x 10^{-2} s^{-1}, the strain rate sensitivity value is about 0.32. It increases to 0.75 over the strain rate range from 7 x 10^{-2} to 4 x 10^{-4} s^{-1} below which the m value appears to increase to an even higher value (i.e., it approaches 1). Single-phase, boron-doped (0.24 at.%) Ni$_3$Al has been reported to be superplastic [41]. The material is of ultrafine grain size (~1.6 μm). A maximum elongation of 160% was recorded at 700°C and at a strain rate of less than 10^{-4} s^{-1}. Superplasticity in this single-phase Ni$_3$Al takes place via dynamic recrystallization. Although the single-phase Ni$_3$Al has a finer grain size than the duplex Ni$_3$Al (1.6 μm vs 6 μm), superplastic elongation of the single-phase alloy is much inferior to that of the duplex alloy. A lower superplastic elongation in the single-phase Ni$_3$Al is believed, in part, to be related to the less stable grain structure under high temperature and dynamic conditions.

One of the interesting characteristics of the superplastic duplex Ni$_3$Si is that the temperature range is quite limited, ranging only from about 1000 to 1100°C; Ni$_3$Si is single-phase below 1000°C. However, superplasticity can exist over a relatively wide strain rate range. For example, at 1080°C, a change in strain rate over three orders of magnitude, from 10^{-3} to 1 s^{-1}, only results in a decrease in tensile elongation from 650 to 300%. An elongation value of 300% is still considered to be superplastic, and a strain rate of 1 s^{-1} is considered to be very high; in fact, such a strain rate is within the range for conventional forging. This offers a technological benefit of superplastic forming of Ni_3Si at high strain rates. In contrast to duplex Ni_3Si, the superplastic strain of duplex Ni_3Al depends strongly on the strain rate. For example, the tensile elongation of Ni_3Al reduces from 440% to only 100% when the strain rate increases by one order of magnitude (from 8×10^{-4} to 8×10^{-3} s^{-1}) at 1050°C [42].

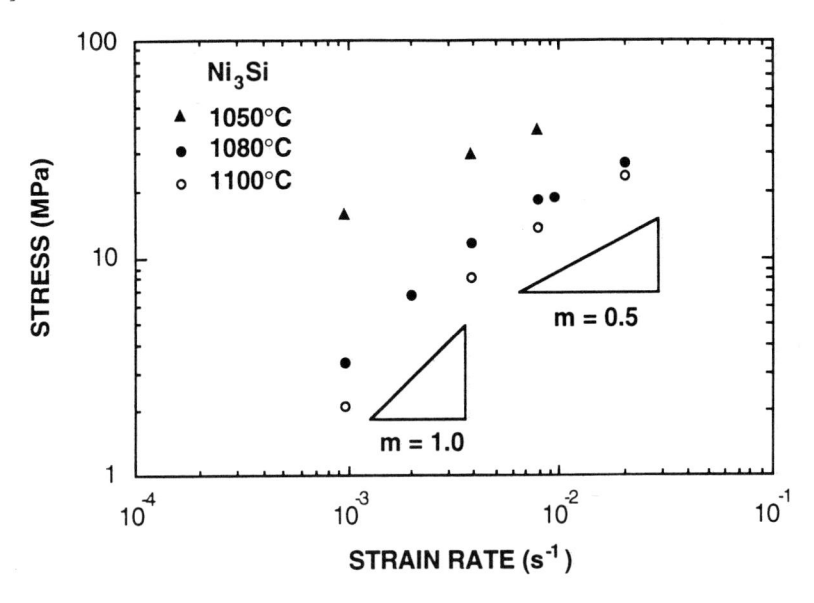

Fig. 5 Strain rate as a function of flow stress for superplastic duplex Ni_3Si. The stress exponent, m, is approximately 0.5 and approaches one in the high strain rate region. (from Ref. [43])

Titanium-base

Superplasticity has been observed in both TiAl (γ) and Ti_3Al (α_2). In the case of Ti_3Al, only duplex alloys, Ti-25Al-10Nb-3V-1Mo (super α_2) [50] and Ti-24Al-11Nb (α_2) [50,55] have been shown to be superplastic. These alloys were thermomechanically processed to fine-grained (about 5 μm) conditions. Superplasticity exists in a narrow temperature range (950 to 1020°C); within this temperature range the volume fraction of β phase is about 0.4-0.6 [55]. Maximum tensile elongation values of about 500% and 1350% were recorded in the α_2 and super α_2 alloys, respectively, at a strain rate of approximately 10^{-5} s^{-1}. The measured strain rate sensitivity values of both alloys are above 0.5, and it is believed that a grain boundary sliding mechanism is responsible for the observed superplasticity. It was particularly pointed out by Ridley et al. [55] that super α_2 remained free from cavitation after large superplastic tensile elongation (>1000%). This suggests the alloy is suitable for stretch forming without the need to impose back pressure to inhibit cavitation.

Studies of superplastic TiAl are of the most recent origin. A summary of published work on superplastic TiAl is listed in Table 3. All materials, except PM Ti-47Al, were produced by ingot casting techniques. As shown in Table 3, superplasticity is only observed in fine-grained, two-phase ($\gamma + \alpha_2$) TiAl alloys, with the exception of the Ti-43Al-13V alloy which is ($\gamma + \beta$). The fine microstructure is achieved if a composition is selected in the two-phase region where approximately 50 vol% of each of the two aluminides, TiAl and Ti_3Al, coexist [46].

Table 3 Property data of superplastic TiAl

composition (at%)	d, (μm)	T, (°C)	$\dot{\varepsilon}$, (s^{-1})	m	el. (%)	Ref.
Ti-43Al	5	1000–1100	10^{-5}–2×10^{-2}	0.5	275	[46]
Ti-50Al	<5	900–1050	2×10^{-4}–8.3×10^{-3}	~0.4	250	[56]
Ti-47Al-2Nb-1.6Cr-0.5Si-0.4Mn	20	1180–1310	2×10^{-5}–2×10^{-3}	0.65	470	[48]
Ti-47.4Al	8	927	10^{-4}	NA	~400	[47]
Ti-43Al-13V	NA	800–1143	3×10^{-4}–10^{-1}	NA	580	[57]
PM Ti-47Al	2	950	10^{-4}–10^{-3}	0.3	NA	[58]

A typical strain rate-flow stress curve for the fine-grained, equiaxed TiAl is presented in Fig. 6. The curves divide into two different stress exponent regions as a function of the flow stress. At high values of the flow stress, m = 0.25 is found. In this region, the rate-controlling deformation mechanism was suggested to be a diffusion-controlled dislocation process. At low values of the flow stress, m = 0.5 is observed at all temperatures. At the highest temperature of testing, 1100°C, the m = 0.5 region extends to a strain rate of as high as 10^{-3} s^{-1}. In this region, the material is superplastic and the rate-controlling deformation process is attributed to a grain boundary sliding mechanism. An elongation-to-failure value of only 275% was obtained at a strain rate of 2.5 x 10^{-4} s^{-1} at 1050°C, suffering the sample from oxidation during testing in air. (A higher maximum elongation of 470% has been recorded from a two-phase, relatively coarse-grained (~20 μm) TiAl tested in vacuum (3 x 10^{-3} Pa) [48]). The apparent activation energy, in the m = 0.5 region, was determined to be equal to 390 kJ/mole which is close to that measured from the creep of a fine-grained Ti(53 at.%)-Al(47 at.%) [59].

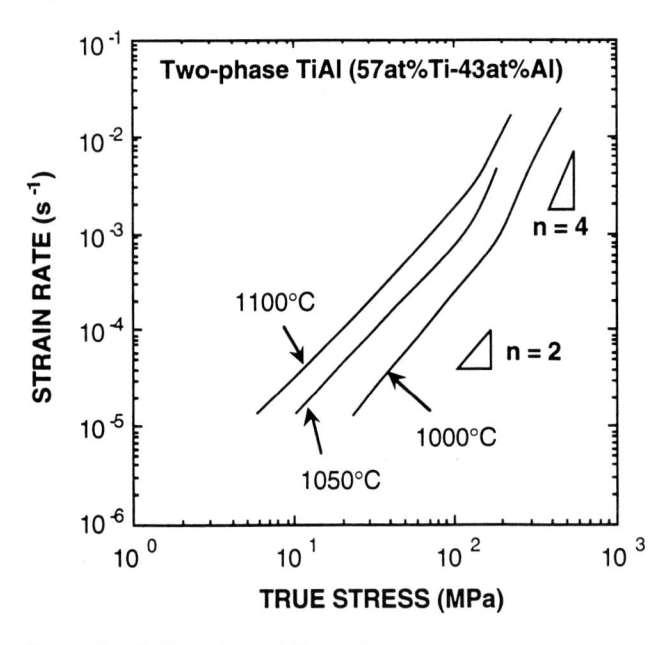

Fig. 6 Strain rate as function of flow stress for fine-grained, two-phase equiaxed TiAl. (data from Ref. [46])

Superplastic Forming of Ceramics

Improved understanding of superplastic ceramics has now advanced to the stage that technological application of ceramic superplastic deformation is receiving increasing attention. Examples of superplastic forming include the extrusion of YTZP powders [29], closed die deformation of YTZP [60], punch forming of YTZP sheet [61] and biaxial gas-pressure deformation of 20%Al_2O_3/YTZP and YTZP [62,63]. In the case of gas-pressure deformation, the key features of the equipment are shown in Fig. 7. High-purity argon gas was used to impose the deformation pressure during forming operations. In-situ deformation is measured when the diaphragm expands upwards to form a hemisphere; this displaces a silicon carbide sensor rod linked to an LVDT. The forming pressure was monitored both with a dial indicator as well as with an electronic DC strain gauge pressure transducer. The apparatus was inductively heated and fully enclosed within a vacuum chamber. The typical heat-up time was 30 minutes with a ten minute stabilization time prior to the application of forming pressure. The apparatus is capable of forming ceramic discs of 50 mm diameter discs. The discs were clamped at their periphery resulting in an unconstrained diaphragm with a diameter of 38 mm, producing a hemisphere of 19 mm in height.

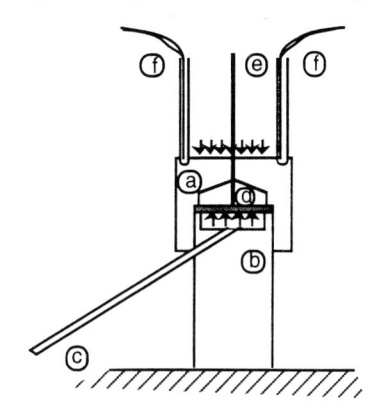

Fig. 7 Gas-pressure forming apparatus. Upper die (a) and lower die (b). Gas is admitted through integral pressure tube (c), ceramic diaphragm (d) deforms upwards causing movement of deflection sensor (e). Temperature is monitored by twin thermocouples (f).

Using the apparatus in Fig. 7, it is possible to make intricately shaped, net shaped parts from superplastic ceramic sheet. Examples of a cone-on-cylinder geometry, a hat section, and a hemisphere of YTZP are shown in Fig. 8. Various shape geometries are possible with this process – as determined by the shape of the die. Recently, Nieh and Wadsworth [64] also used the technology to successfully produce a sophisticated high temperature millimeter wave radome from Si wafers, as shown in Fig. 9. The radome was fabricated by co-deformation (superplastic forming/diffusion bonding) of a 10-cm-diameter silicon wafer with micromachined coolant slots and another silicon wafer. The structure in Fig. 9 would be extremely difficult, if not impossible, to make without using the gas-pressure forming process. In addition to Si, a hybrid YTZP/C103 (nominal composition: Nb-10Hf-1Ti.) ceramic-metal structure has been made using a superplastic forming/diffusion bonding technique [65]. Such a thin-walled engineered metal-ceramic structure could have great utility in high thermal flux applications. Other potential applications include manufacturing net shape turbine rotors directly from fine Si_3N_4 powders and the bulk forming of automotive components from superplastic Si_3N_4 [66].

Superplastic Forming of Intermetallics

There is only limited study on the superplastic forming of intermetallics. The aerospace application of titanium aluminides has not been widely revealed, although some demonstration parts has been made [67]. This is, in part, because of its proprietary or classified nature. Nonetheless, it is noted that super α_2 foils have been fabricated from sheet material by vacuum pack rolling in the temperature and strain rate ranges where the material is superplastic [68]. The foils can be used as facing sheet for further fabrication of intermetallic composites. Also noted is that, similar to that in the case of Si_3N_4, γ-TiAl are being actively evaluated to be used for turbine rotors and high-temperature gas valves [69]. As a result of the great difficulty in machining TiAl, net shape forming techniques would be attractive.

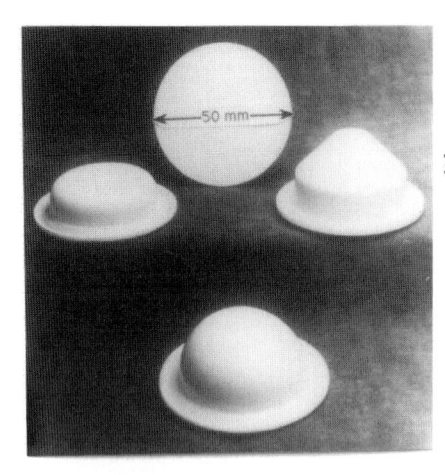

Fig. 8 Examples of a cone-on-cylinder geometry, a hat section, and a hemisphere superplastically-formed from YTZP and Al_2O_3/YTZP. (from Ref. [63])

Fig. 9 High temperature millimeter wave Si radome manufactured by a concurrent superplastic forming and diffusion bonding technique. Cooling channels are indicated. (from Ref. [64])

Conclusions

Superplasticity has been observed in many fine-grained (< 1 µm) ceramics, including YTZP, Al_2O_3, hydroxyapatite, β-spodumene glass ceramics, and Al_2O_3-zirconia, SiC-Si_3N_4, and Fe_3C-Fe composites) as well as intermetallics, including Ni_3Al, Ni_3Si, Ti_3Al, and TiAl. In the case of ceramics, the presence of glassy phases may be unnecessary but can strongly affect the optimum temperature for superplasticity. The tensile elongation of a superplastic ceramic is found to be inversely proportional to the flow stress. In the case of intermetallics, two-phase alloys, despite having a relatively large grain size (>10 µm), exhibit better superplastic properties than single-phase alloys. For some alloys such as Ni_3Si, superplasticity appears to be relatively rate insensitive. Superplastic forming techniques have been demonstrated in some ceramics and intermetallics, but applications are primarily performance-driven (e.g., aerospace). High cost is still the major obstacle for wide-spread, commercial applications of superplastic ceramics and intermetallics.

Acknowledgment

This work was performed, in part, under the auspices of the U.S. Department of Energy by Lawrence Livermore National Laboratory under contract No. W-7405-Eng-48.

References

1. R.M. Fulrath, *Ceram. Bull.*, **43**, 880 (1964).

2. R.W. Rice, *Ultrafine-Grain Ceramics* (edited by J.J. Burke, N.L. Reed, and V. Weiss), p. 203, Syracuse University Press, Syracuse, NY (1970).
3. F. Wakai, S. Sakaguchi, and Y. Matsuno, *Adv. Ceram. Mater.*, **1**, 259 (1986).
4. R.B. Day and R.J. Stokes, *J. Am. Ceram. Soc.*, **49**, 345 (1966).
5. P.E.D. Morgan, "Superplasticity in Ceramics," in *Ultrafine-Grain Ceramics*, edited by J.J. Burke, N.L. Reed, and V. Weiss, Syracuse University Press, New York (1970), p. 251.
6. K.R. Venkatachari and R. Raj, *J. Am. Ceram. Soc.*, **69**, 135 (1986).
7. C.K. Yoon and I.W. Chen, *J. Amer. Ceram. Soc*, **73**, 1555 (1990).
8. T.G. Nieh, C.M. McNally, and J. Wadsworth, *Scr. Metall.*, **22**, 1297 (1988).
9. P. Gruffel, P. Carry, and A. Mocellin, *Science of Ceramics, Volume 14* (edited by D. Taylor), p. 587, The Institute of Ceramics, Shelton, Stoke-on-Trent, UK (1987).
10. F. Wakai, Y. Kodama, S. Sakaguchi, and T. Nonami, *J. Am. Ceram. Soc*, **73**, 257 (1990).
11. J.-G. Wang and R. Raj, *J. Am. Ceram. Soc.*, **67**, 399 (1984).
12. F. Wakai and H. Kato, *Adv. Ceram. Mater.*, **3**, 71 (1988).
13. T.G. Nieh and J. Wadsworth, *Acta Metall. Mater.*, **39**, 3037 (1991).
14. F. Wakai, *Superplasticity in Metals, Ceramics, and Intermetallics, MRS Proceeding No. 196R* (edited by M.J. Mayo, J. Wadsworth, and M. Kobayashi), p. 349, Materials Research Society, Pittsburgh, PA (1990).
15. T. Rouxel, F. Wakai, K. Izaki, and K. Niihara, *Pro. 1st Inter. Symp. on Science of Engineering Ceramics* (edited by S. Kimura and K. Niihara), p. 437, Ceramic Society of Japan, Tokyo. Japan (1991).
16. W.J. Kim, G. Frommeyer, O.A. Ruano, J.B. Wolfenstine, and O.D. Sherby, *Scr. Metall.*, **23**, 1515 (1989).
17. I.-W. Chen and L.A. Xue, *J. Am. Ceram. Soc.*, **73**, 2585 (1990).
18. T.G. Nieh, J. Wadsworth, and F. Wakai, *Inter. Mater. Rev.*, **36**, 146 (1991).
19. T.G. Nieh and J. Wadsworth, *Acta Metall. Mater.*, **38**, 1121 (1990).
20. T. Hermanson, K.P.D. Lagerlof, and G.L. Dunlop, *Superplasticity and Superplastic Forming* (edited by C.H. Hamilton and N.E. Paton), p. 631, TMS (1988).
21. L.A. Xue, X. Wu, and I.W. Chen, *J. Am. Ceram. Soc.*, **74**, 842 (1991).
22. F. Wakai, Y. Kodama, S. Sakaguchi, N. Murayama, H. Kato, and T. Nagano, *MRS Intl. Meeting on Advanced Materials Vol 7 (IMAM-7, Superplasticity)* (edited by M. Doyama, S. Somiya, and R.P.H. Chang), p. 259, Materials Research Soc., Pittsburgh, Pennsylvania (1989).
23. T. Nagano, H. Kato, and F. Wakai, *J. Am. Ceram. Soc.*, **74**, 2258 (1991).
24. X. Wu and I.-W. Chen, *J. Am. Ceram. Soc.*, **75**, 2733 (1992).
25. F. Wakai, Y. Kodama, S. Sakaguchi, N. Murayama, K. Izaki, and K. Niihara, *Nature (London)*, **334**, 421 (1990).
26. T.G. Nieh and J. Wadsworth, *J. Am. Ceram. Soc.*, **72**, 1469 (1989).
27. F. Wakai, S. Sakaguchi, and H. Kato, *J. Ceram. Soc. Japan (In Japanese)*, **94**, 72 (1986).
28. D.J. Schissler, A.H. Chokshi, T.G. Nieh, and J. Wadsworth, *Acta Metall. Mater.*, **39**, 3227 (1991).
29. B.J. Kellett, C. Carry, and A. Mocellin, *J. Amer. Ceram. Soc*, **74**, 1922 (1990).
30. C. Carry, *MRS Intl. Meeting on Advanced Materials Vol.7 (IMAM-7, Superplasticity)* (edited by M. Doyama, S. Somiya, and R.P.H. Chang), p. 251, Materials Research Soc., Pittsburgh, PA (1989).
31. Y. Ma and T.G. Langdon, *Superplasticity in Metals, Ceramics, and Intermetallics, MRS Proceeding No. 196* (edited by M.J. Mayo, J. Wadsworth, and M. Kobayashi), p. 325, Materials Research Society, Pittsburgh, PA (1990).
32. R. Duclos, J. Crampon, and B. Amana, *Acta Metall.*, **70**, 877 (1987).
33. T.G. Nieh and J. Wadsworth, *Scr. Metall. Mater.*, **24**, 763 (1990).
34. W.J. Kim, J. Wolfenstine, and O.D. Sherby, *Acta Metall. Mater.*, **39**, 199 (1991).
35. S.E. Dougherty, T.G. Nieh, J. Wadsworth, and Y. Akimune, *J. Mater. Res.*, (1995). – in press
36. T.G. Nieh, D.L. Yaney, and J. Wadsworth, *Scripta Metall.*, **23**, 2007 (1989).
37. F. Wakai, H. Okamura, N. Kimura, and P.G.E. Descamps, *Proc. 1st Japan Int'l SAMPE Symp.* (edited by N. Igata, K. Kimpara, T. Kishi, E. Nakata, A. Okura, and T. Uryu), p. 267, Society for the Advancement of Materials and Process Engineering (1989).
38. L.A. Xue, *J. Mater. Sci. Lett.*, **10**, 1291 (1991).
39. S. Hanada, T. Sato, S. Watanabe, and O. Izumi, *J. Jpn Inst. Metals*, **45**, 1293 (1981).

40. V.K. Sikka, C.T. Liu, and E.A. Loria, *Processing of Structural Metals by Rapid Solidification* (edited by F.H. Froes and S.J. Savage), p. 417, Am. Soc. Metals, Metals Park, OH (1987).
41. M.S. Kim, S. Hanada, S. Wantanabe, and O. Izumi, *Mater. Trans. JIM*, **30**, 77 (1989).
42. A. Choudhury, A.K. Muhkerjee, and V.K. Sikka, *J. Mater. Sci.*, **25**, 3142 (1990).
43. T.G. Nieh and W.C. Oliver, *Scr. Metall.*, **23**, 851 (1989).
44. T. Takasugi, S. Rikukawa, and S. Hanada, *Acta Metall. Mater.*, **40**, 1895 (1992).
45. D. Lin, A. Shan, and D. Li, *Scr. Metall. Mater.*, **31**, 1455 (1994).
46. S.C. Cheng, J. Wolfenstine, and O.D. Sherby, *Metall. Trans.*, **23A**, 1509 (1992).
47. T. Tsujimoto, K. Hashimoto, and M. Nobuki, *Mater. Trans., JIM*, **33**, 989 (1992).
48. W.B. Lee, H.S. Yang, Y.-W. Kim, and A.K. Muhkerjee, *Scr. Metall. Mater.*, **29**, 1403 (1993).
49. A. Dutta and D. Banerjee, *Scr. Metall. Mater.*, **24**, 1319 (1990).
50. H.S. Yang, P. Jin, E. Dalder, and A.K. Muhkerjee, *Scr. Metall. Mater.*, **25**, 1223 (1991).
51. S.L. Stoner and A.K. Muhkerjee, *International Conference on Superplasticity in Advanced Materials (ICSAM-91)* (edited by S. Hori, M. Tokizane, and N. Furushiro), p. 323, The Japan Society for Research on Superplasticity (1991).
52. O.D. Sherby and J. Wadsworth, *Deformation, Processing and Structure* (edited by G. Krauss), p. 355, ASM, Metal Park, Ohio (1984).
53. J.K. Gregory, J.C. Gibeling, and W.D. Nix, *Metall. Trans.*, **16A**, 777 (1985).
54. T. Takasugi, S. Rikukawa, and S. Hanada, *Scr. Metall. Mater.*, **25**, 889 (1991).
55. N. Ridley, M.F. Islam, and J. Pilling, *Structural Intermetallics* (edited by R. Darolia, J.J. Lewandowski, C.T. Liu, P.L. Martin, D.B. Miracle, and M.V. Nathal), p. 63, TMS, Warrendale, PA (1993).
56. R.M. Imayev, O.A. Kaibyshev, and G.A. Salishchev, *Acta Metall. Mater.*, **40**, 581 (1992).
57. D. Vanderschueren, M. Nobuki, and M. Nakamura, *Scr. Metall. Mater.*, **28**, 605 (1993).
58. M. Tokizane, T. Fukami, and T. Inaba, *ISIJ Inter.*, **10**, (1991).
59. S.L. Kampe, J.D. Bryant, and L. Christodoulou, *Metall. Trans.*, **22A**, 447 (1991).
60. F. Wakai, *Brit. Ceram. Trans. J.*, **88**, 205 (1989).
61. X. Wu and I.W. Chen, *J. Am. Ceram. Soc.*, **73**, 746 (1990).
62. J.P. Wittenauer, T.G. Nieh, and J. Wadsworth, *J. Am. Ceram. Soc.*, **76**, 1665 (1993).
63. T.G. Nieh and J. Wadsworth, *J. Mater. Eng. Performance*, **3**, 496 (1994).
64. T.G. Nieh, LMSC-F070359, pp.7-301-7-321, Lockheed Missiles and Space Co. (1986).
65. T.G. Nieh and J. Wadsworth, *MRS Intl. Meeting on Advanced Materials Vol 7 (IUMRS-ICAM-93, Symp. E, Superplasticity)* (edited by M. Doyama, S. Somiya, and R.P.H. Chan), Pergamon Press, Netherland (1994). – in press
66. K. Watanabe, private communication, NGK Insulators, Ltd., Nangoya, Japan (1994).
67. D. Shih, private communication, McDonnell Douglas, St. Louis, MO (1994).
68. J.P. Wittenauer, C. Bassi, and B. Walser, *Scr. Metall.*, **23**, 1381 (1989).
69. Y. Nishiyama, T. Miyashita, S. Isobe, and T. Noda, *High Temperature Aluminides & Intermetallics* (edited by S.H. Whang, C.T. Liu, D.P. Pope, and J.O. Stiegler), p. 557, The Minerals, Metals, and Materials Society (1990).

Superplasticity in Al_2O_3-ZrO_2-Al_2TiO_5 Ceramics

John Pilling and James Payne
Metallurgical and Materials Engineering,
Michigan Technological University
Houghton MI 49931

Introduction

In order for ceramics to exhibit superplasticity, it has been shown[1] that a stable grain size less than 1µm must be present and that like metals, grain growth either prior to or during deformation can result in a loss of superplasticity, e.g. an increase in grain size from 0.4 to 3 µm resulted in a 5 fold increase in the flow stress of yttria stabilised zirconia during compression testing at 1450°C[2]. The primary use of superplastic ceramics is in the net shape forging of parts, where the closed die forging operation will result in a fully dense (and if ductile a crack free) ceramic part with little, if any, machining. Since up to 80% of the cost of ceramic parts is attributable to machining costs it is of benefit to develop ceramics that are superplastic and resistant to grain growth.

Grain growth in ceramics at high temperature occurs as a result of grain boundary migration where like grain boundaries are in contact, i.e. all grain boundaries in a single phase material such as zirconia. The addition of a second phase, e.g. alumina, has been shown to increase the extent of superplasticity and reduce the flow stress by reducing grain growth during deformation by the simple expedient of reducing the number of like grain boundaries in contact. It is our hypothesis that with the addition of a third, mutually insoluble phase, in this instance aluminium titanate, the number of like grain boundaries can be significantly reduced, Fig. 1. The material will be more resistant to grain growth and maintain its superplastic behaviour at higher strains than either the single or two phase ceramic alloys investigated to date[3].

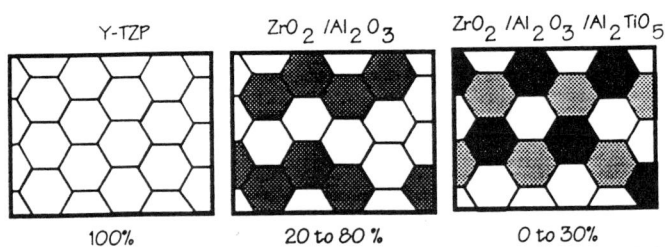

Y-TZP — 100%
ZrO_2 /Al_2O_3 — 20 to 80 %
ZrO_2 /Al_2O_3 /Al_2TiO_5 — 0 to 30%

Fig. 1. Reduction in phase contiguity associated with increasing the number of phases in a multiphase mixture.

Materials Preparation

Ceramic alloys were prepared from mixtures of the three component oxides, alumina[1], 3 mol% yttria stabilised, zirconia[2] and titania[3]. Each of the precursor powders contained crystals of 15 to 50nm in size but often in the form of coarse, >0.1mm, agglomerates. The powders were ball milled and electrostatically dispersed to form a slurry containing 3 vol% solids which was then ultrasonically

---

1. Linde B, Union Carbide, Indianapolis IN.
2. TZ-3Y, Tosho Soda/USA, Atlanta Ga.
3. UV-Titan, Kermira, Pori, Finland.

treated and allowed to settle. After returning any coarse agglomerates to the ball mill, the liquor was flocculated and the nanocrystalline ceramic particles collected. Each of the three slurries, containing approximately 25 vol% solids was then resuspended by adjusting the pH to 2 and mixed together using 3 mass flow controlled pumps via an ultrasonic mixing cell. After drying, the powders were die pressed at ~80 kPa into 25mm diameter cylinders and then cold isostatically pressed at 400MPa. The resulting blanks were flash sintered at 1500°C in air for 10 minutes. The compositions of the 4 alloys tested are summarised are shown in Fig. 1, and their properties listed in Table 1.

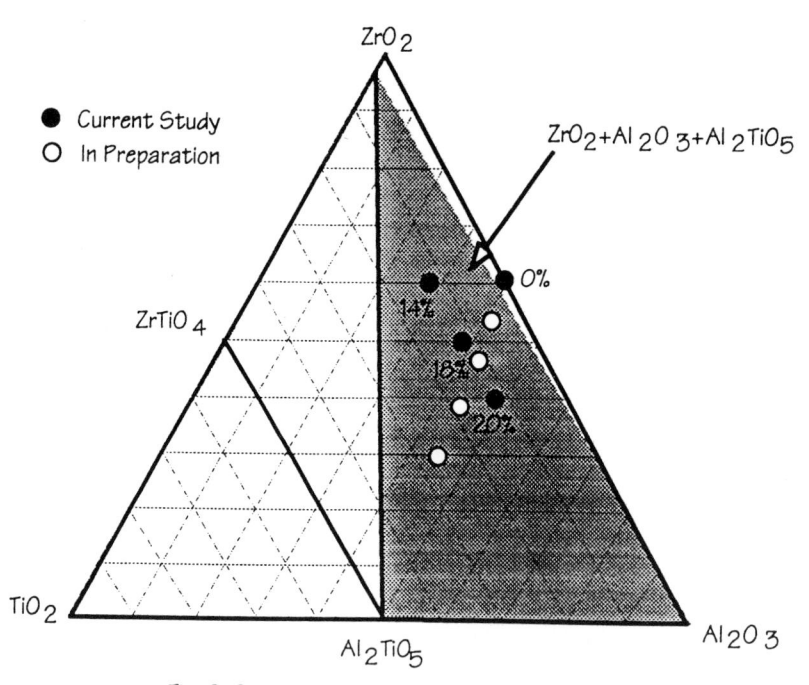

Fig. 2. Compositions of the alloys investigated.

Compression tests were carried out using an Instron model 4206 testing machine with a Centorr vacuum furnace equipped with molybdenum compression platens lubricated with BN[1]. Constant strain rate strain rate jump tests were carried out at 1420, 1470, 1500, 1520°C, the cross head speed being continuously adjusted by assuming a constant specimen volume after compensation for the (calibrated) elastic deflection of the loading system at temperature. Strain rate sensitivities were determined from the derivative of a polynomial regression of the $\log(\sigma)$- $\log(\dot{\varepsilon})$ data. Temperatures were controlled using an emissivity corrected accufiber optical pyrometer viewing the (stationary) molybdenum platen immediately below the test sample. Temperatures were measured and controlled to within +/- 3°C. Constant strain rate compression tests were carried out at 1500 and 1520°C on each of the 4 alloys at strain rates of 2×10^{-4}, 2×10^{-3} and 2×10^{-2} s^{-1}

The microstructures of each of the alloys were characterised in terms of grain boundary contiguity. In a given sample the total perimeter length of the grains of a given phase is measured together with the

---

1. ZYP Coatings Inc. Oak Ridge TN.

perimeter length in contact with grains of the same phase and the other phases. The ratio of the phase contact length to total length then expresses the phase contiguity. Thus in a single phase material the contiguity is 100% while in a perfectly mixed multiphase material with no like phases in contact the self-contiguity is zero.

Volume Fraction (%)[a]			Density (g/cm³)		%	Grain Size[b] (μm)	Phase Self-Contiguity (%)		
Al_2O_3	ZrO_2	Al_2TiO_5	Theoretical	Measured[c]			Al_2O_3	ZrO_2	Al_2TiO_5
24	62	14	5.10	4.98	98	0.73	11	57	12
36	46	18	4.78	4.74	99	0.65	38	11	18
42	38	20	4.64	4.57	98	0.63	17	24	22
50	50	-	4.86	4.71	97	1.41	23	52	-

Table 1: Physical Properties of AZT ceramic alloys

a. Ideally the ratio of alumina:zirconia should be 1:1, see Fig. 1
b. mean linear intercept of all phases.
c. Archimedes method

Results[1]

A microstructure typical of the AZT alloys is shown in Fig. 3 The self-contiguity of each phase in each of the alloys is summarised in Table 1, the lower the value for a given phase, the less that phase is likely to coarsen. Low values of self contiguity in multiphase materials can arise either from a low volume fraction or from homogeneous dispersion of that phase.

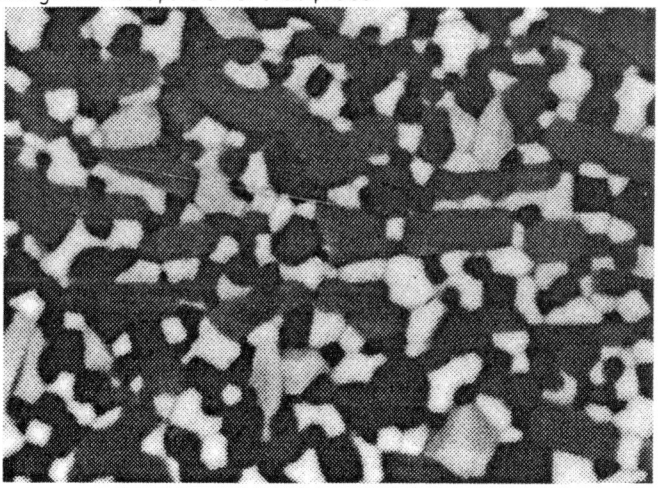

Fig. 3. Microstructure of 20% Al_2TiO_5 Alloy
(black=Al_2O_3; white=ZrO_2, gray=Al_2TiO_5

---

1. The Mechanical Properties Data presented in this paper is available via anonymous ftp through the Internet from austenite.my.mtu.edu. Filename is spAZT.data

The flow stress shows a strain rate dependence typical of superplastic materials, with the maximum strain rate sensitivities for each of the three ternary alloys peaking at m~0.70-0.74 at strain rates of $10^{-3} s^{-1}$. In general the flow stress decreases with increasing temperature. In the case of the 14% Al_2TiO_5 alloy the flow stresses at 1470 and 1500°C overlapped while in the 18% Al_2TiO_5 alloy the flow stresses at 1520°C were higher than the measured flow stresses at 1500°C, Figs. 4-6. The 20% Al_2TiO_5 alloy showed little grain growth at the higher test temperatures. The effect of the volume fraction of Al_2TiO_5 on the flow stress of the ceramic alloys is shown in Fig. 7 wherein the flow stress decreases with increasing amounts of the Al_2TiO_5 phase at a given strain rate and temperature.

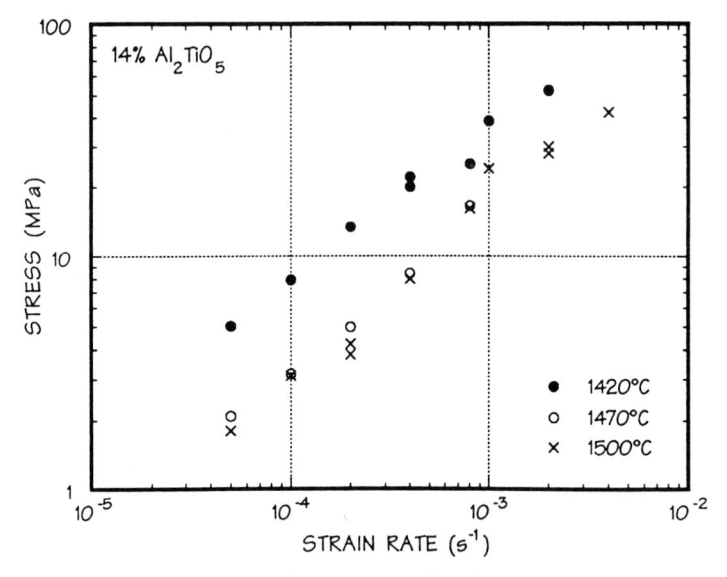

Fig. 4. Stress Strain Rate Values for 14% Al_2TiO_5.

The superplastic behaviour of the ternary ceramics is maintained to strain rates as high as $2 \times 10^{-2} s^{-1}$. The progressive change in section of the 18% Al_2TiO_5 alloy deformed at 1500°C and $2 \times 10^{-2} s^{-1}$ is shown in Fig. 8, where cracking due to circumferential tensile stresses is only evident after a strain of -1.1 (equivalent to 200% tensile strain). The flow stresses of the ternary alloys was significantly lower than those reported in the literature for binary Al_2O_3/ZrO_2 alloys in compression[4,5].

Conclusions

Ternary Al_2O_3-ZrO_2-Al_2TiO_5 ceramics have been shown to exhibit superplastic behaviour and to have much lower equivalent flow stresses than the single phase ZrO_2 and microduplex Al_2O_3-ZrO_2 ceramics investigated previously. High ductility is maintained to strain rates as high as $2 \times 10^{-2} s^{-1}$ and the alloys would appear suitable for a slow forging operation.

Acknowledgments

The Authors are grateful to Kermira for kindly donating the nanocrystalline TiO_2 used in this study. This work was supported by Caterpillar Inc.

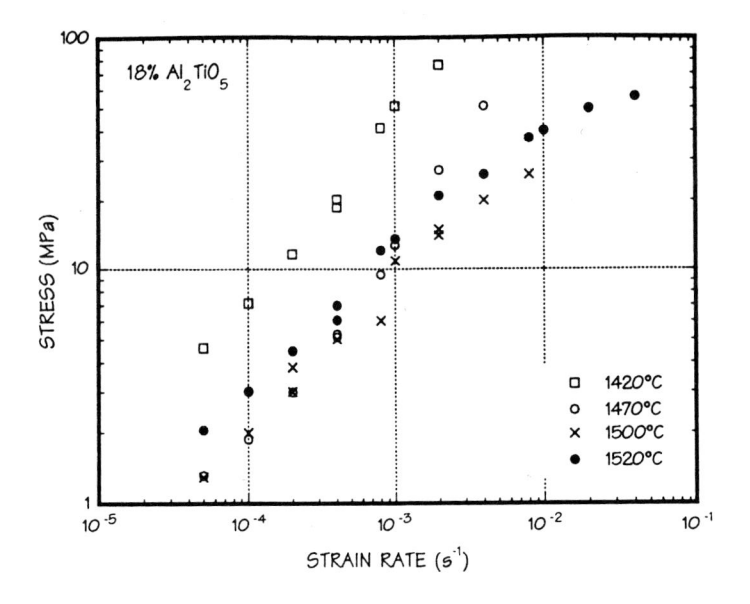

Fig. 5. Stress Strain Rate Values for 18% Al_2TiO_5.

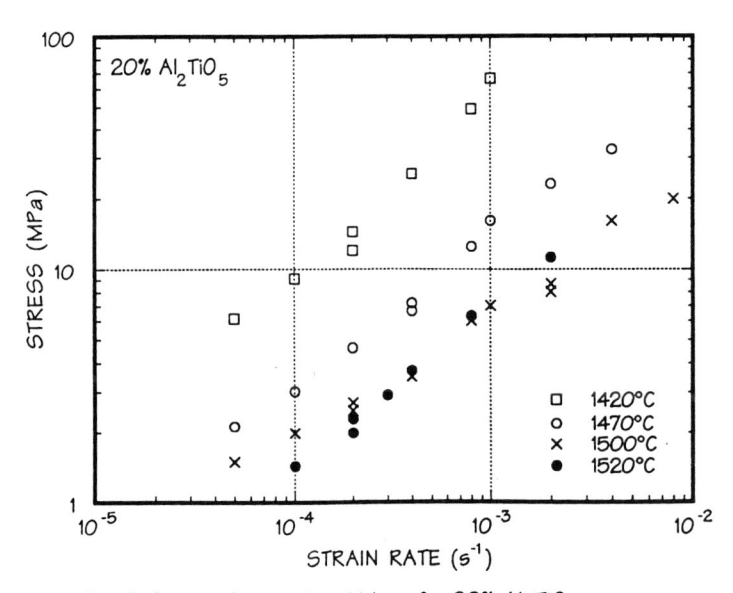

Fig. 6. Stress Strain Rate Values for 20% Al_2TiO_5.

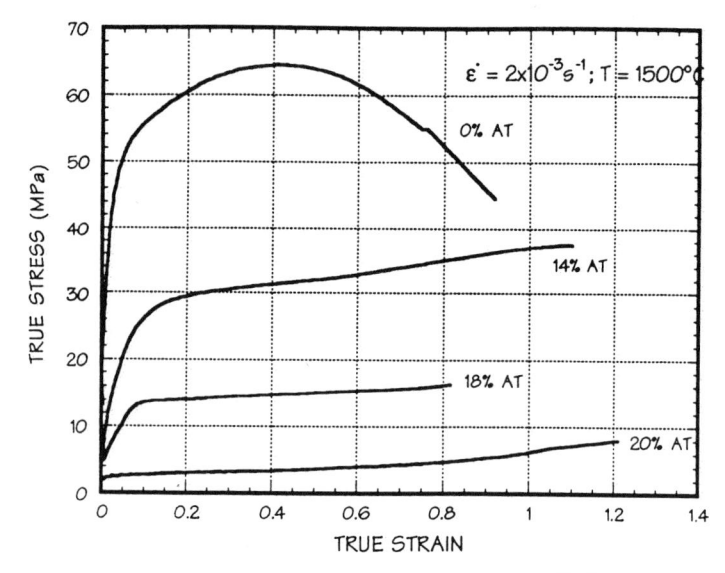

Fig. 7. Stress Strain Values for all alloys at $2 \times 10^{-3} s^{-1}$, 1500°C.

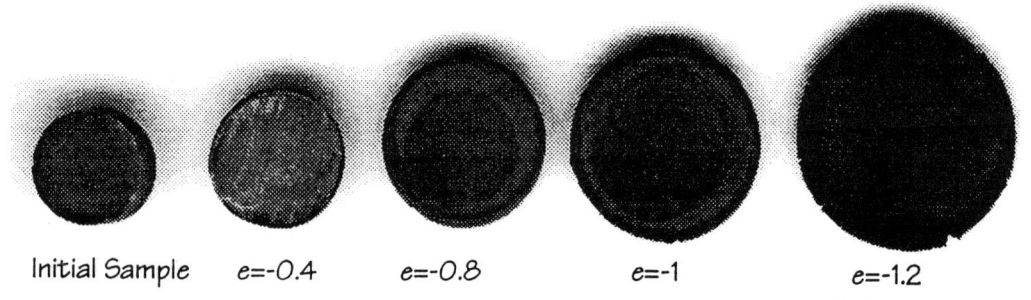

Fig. 8. Progressive deformation of 18%Al_2TiO_5 at 1500°C and $2 \times 10^{-2} s^{-1}$

References
(1) Nieh, T. G.; McNally, C. M.; Wadsworth, J. Scripta Metallurgica **1988**, 22, 1297-1300.
(2) Wakai, F. British Ceramic Society Transactions and Journal **1989**, 88, 205-208.
(3) Chen, I.-W.; Xue, L. A. Journal of the American Ceramics Society **1990**, 73, 2585-2609.
(4) Wakai, F.; Kodama, Y.; Sakaguchi, S.; Murayama, N.; Kato, H.; Nagano, T. In MRS International Meeting on Advanced Materials; Materials Research Society, Pittsburg PA: 1989; pp 259 - 266.
(5) Kellet, B. J.; Lange, F. F. Journal of Materials Research **1988**, 3, 545-551.

Electronic Contribution to the 'Superplastic Partition' in Ceramics

T.J. Davies, A.A. Ogwu, N. Ridley and Z.C. Wang
Manchester Materials Science Centre
Manchester University/UMIST
Grosvenor Street
Manchester M1 7HS, UK.

ABSTRACT

Recent developments in the understanding of the mechanical behaviour of ceramics assessed for superplastic deformation indicates that neither the grain boundary viscous flow mechanism nor the grain boundary space charge concept (due to elemental segregation) can satisfactorily account for the presence or absence of superplasticity in ceramics. It is our observation, based on the experimental work carried out in Manchester and elsewhere, that a 'superplastic partition' exists in ceramics that corresponds to a direct correlation with observed 'metallic behaviour' in certain ceramics. When this observation is linked with the Gifkins core-mantle concept and the findings of Mott, Cottrell and Gilman on insulator-metal transitions, this yields an improved appreciation of the superplastic partition in ceramics.

INTRODUCTION

The relationship expressed as

$$\sigma = K[\dot{\varepsilon} \exp(Q_c/RT)]^m \qquad (1)$$

is found to describe flow stress and strain rate relationships during superplastic deformation of metals and ceramics. In this relationship, σ is the flow stress, R is the gas constant, T is the absolute temperature, K is a material constant and m is the strain rate sensitivity index. The term $[\dot{\varepsilon} \exp(Q_c/RT)]$ is referred to as the Zener-Hollomon parameter, Q_c is the activation energy for superplastic flow and $\dot{\varepsilon}$ is the strain rate. One of the major differences between superplastic deformation in ceramics and metals is that superplasticity in ceramics is very sensitive to changes in the Zener-Hollomon parameter while superplastic deformation in metals is relatively insensitive to this parameter [1], provided that other requirements like ultra-fine grain sizes and a high strain rate sensitivity index is obtained. Fine grained ceramics are known to exhibit a drop in tensile ductility with an increase in $\dot{\varepsilon} \exp(Q_c/RT)$ even when the strain-rate sensitivity exponent remains high [1]. These observations may be explained by superplastic flow occuring at the tip of a crack, and inhibiting its propagation and fracture [1], consistent with the Gifkins [2] core-mantle concept. If the strain rate is increased or temperature is decreased, i.e. as $\dot{\varepsilon} \exp(Q_c/RT)$ is increased, the amount of recovery at the crack tip through superplastic flow is diminished and this results in a rapid propagation of cracks leading to early (tensile) failure. Ceramic materials assessed for superplastic deformation can be partitioned into two categories, namely those that undergo extensive superplasticity and those that are relatively brittle. The argument in this paper is that this 'superplastic partitioning' in ceramics could be associated with the fact the ceramic materials observed (to date) to exhibit extensive superplastic deformation have an intrinsic 'metallic' character in their bonds. This could lead to less sensitivity to changes in the Zener-Hollomon parameter, which would produce behaviour similar to that found in pure metals, showing extensive deformation, whilst the brittle ceramics that lack this 'metallic' character, and are generally insulator type materials, would yield a high sensitivity to changes in the Zener-Hollomon parameter and promote brittle behaviour. A comparison of the tensile ductility behaviour of a metallic alloy with a ceramic alloy as a function of the Zener-Hollomon parameter is shown in Fig. 1.

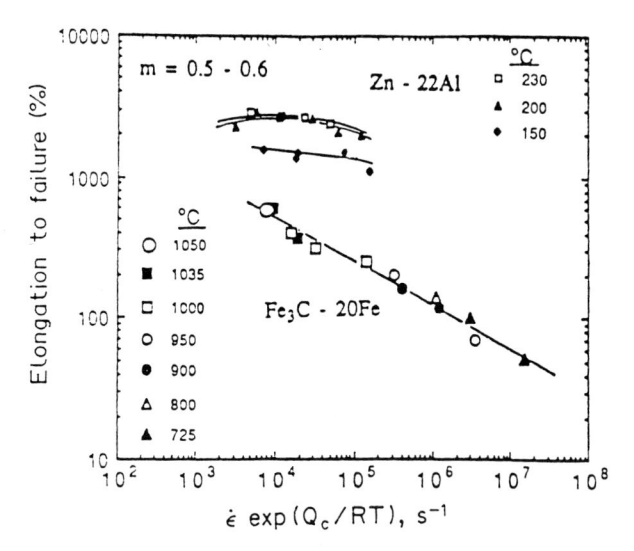

Fig. 1. Comparison of the tensile ductility behaviour of a metallic alloy (Zn-20Al) with a ceramic alloy (Fe$_3$C-20Fe) as a function of $\dot{\varepsilon}$ exp (Q_c/RT) in the range where m = 0.5 - 0.6 (After Kim et al [1]).

THEORETICAL CONSIDERATIONS

A. 'Metallic' Behaviour

The mechanical and electrical properties of a solid are both related to the nature of the bonding in the solid. The bonding in a solid determines the type of cohesion that would exist in the solid, as well as whether the solid would be an electrical insulator, semi-conductor or metallic conductor. A good example of the inter-relationship between the mechanical state and electrical properties of a material can be found in the report of shear induced metallization of silicon, observed experimentally by Pharr et al [3], and discussed by Gilman [4] and Cahn [5]. Gilman's proposal is that the transformation of silicon from the insulating to the metallic state under an indenter load as observed by Pharr et al [3], occurred because compression causes narrowing of the band gap until electrons can tunnel from the non-conducting valence band into the conduction band, leading to metallization. Another way of expressing the above description is that changes occur in the overlap of atomic wave functions as compression takes place leading to electron delocalization. This is consistent with an earlier finding by Gilman [6] that a linear correlation exists between glide activation energies (i.e. dislocation motion) and the average band gap energies in solids.

In summary, Gilman [4] suggests that since the metallisation process has an electronic origin (i.e. tunnelling of electrons from the valence band to the conduction band), it can be influenced by photons, doping (donors enhance whereas acceptors inhibit the transfer), currents, surface states, etc. The underlying bonding principles behind insulator - metal transitions have also been a subject of study by Edwards and Sienko [7], Mott [8] and Herzfeld [9].

Based on the findings of Gilman and others, and our previous work [10,11,12,13], the present authors suggest that the extensive superplastic deformation (up to 800% tensile elongation) in yttria stabilized tetragonal zirconia polycrystals reported by Nieh et al [14] and Wakai et al [15], could be due to the fact that yttria stabilized tetragonal zirconia has an intrinsic metallic bonding, as described below, which imparts the ability to resist cavitation, i.e. an accommodation process would be present at grain junctions by dislocation motion [16] due to the metallic bonding, leading to continuous elongation without fracture. The difficulty of achieving such extensive deformation in Si_3N_4/SiC composites [16] and alumina [17] could be due to the absence of this metallic character in their bonds.

B. Bonding Mechanism

It is generally agreed that the band theory of solids [18,19] cannot be applied to certain compounds of transition/rare earth metals with partly filled d and f shells, including oxides such as yttria stabilized tetragonal zirconia; such compounds have properties that range from metallic to semi-conducting and insulating. The bonding behaviour of transition/rare earth metal compounds has been accounted for in a model developed by Hubbard [20,21,22] which is based on a concept that the partial filling of the d and f orbitals in transition/rare earth metal compounds (sufficient for metallic behaviour according to the band theory) must be accompanied by a considerable overlap of the orbitals of the metal ions to give the metallic character; when the overlap is absent or severely limited, electrons are localized on individual metal ions and the materials are insulators. Oxides of transition metals often adopt a NaCl structure; the 3d band in transition metal ions split into two states, the lower designated t_{2g} (i.e. d_{xy}, d_{xz}, d_{yz}) can take 6N electrons and the upper one designated e_g (i.e. d_{z^2}, $d_{x^2-y^2}$) can take 4N electrons. For example, early transition metal oxides, such as TiO and VO, show a metallic conductivity, because the symmetry of the NaCl structure allows three of the five d orbitals on different metal atoms to overlap, i.e. d_{xy}, d_{xz} and d_{yz}, overlap as shown in Fig. 2a, to form a broad t_{2g} band. On moving towards the right-hand end of the transition metal series in the periodic table of elements, a contraction of the d orbitals occurs as the nuclear charge increases, leading to a localization of the d-orbital electrons on the metal ions, i.e. the orbitals $d_{x^2-y^2}$ would point directly at the oxygen ions, making orbital overlap difficult, with a resultant low electrical conductivity, as shown for NiO in Fig. 2b; oxides such as MnO, CoO and NiO do not show a metallic character.

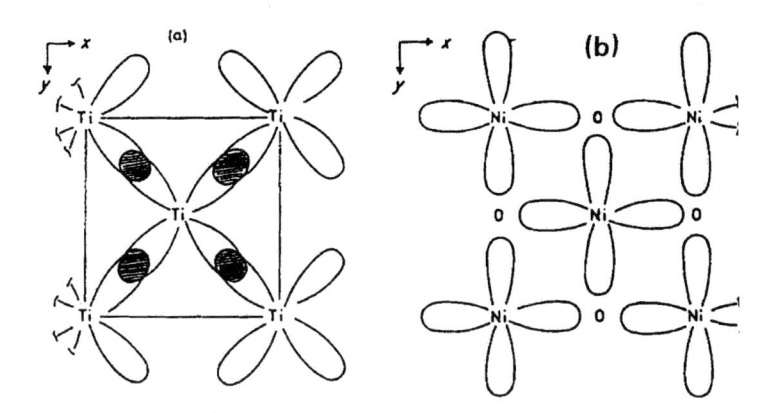

Fig. 2. (a) Section through TiO structure, parallel to unit cell face showing Ti^{2+} positions only - overlap of d_{xy} orbitals on adjacent Ti^{2+} ions, with similar overlap of d_{xz} and d_{yz} orbitals leads to t_{2g} band. (b) Structure of NiO, showing $d_{x^2-y^2}$ orbitals pointing directly at oxide ions and therefore unable to overlap and form e_g band (after Ref. [23]).

The Hubbard model can also be applied to rare earth metals and their compounds. The 4f orbitals are highly contracted in rare earth compounds (such as their oxides) with little or no overlap between them. However, for the early rare earths, e.g. cerium and it's compounds, the 4f orbitals are more likely to overlap. Empirical rules for d orbital overlap in transition metal compounds have been proposed by Phillips and Williams in [23] as follows:

(i)   The formal charge on the cations is small,
(ii) The cation occurs early in the transition metal series,
(iii) The cation is in the second or third transition metal series,
(iv) The anion is reasonably electropositive.

It is interesting that yttria and zirconia, the two compounds associated with extensive superplastic deformation in structural ceramics, fall into the group of compounds covered by these empirical rules; the temperature and pressure dependencies of insulator-metal transitions in early transition/rare earth metal oxides has been discussed by Mott [8]. The metallic behaviour of such materials has been shown by the evident plasticity revealed in recent transmission electron microscopy observations of dislocation structures around Vickers indentations in a 9.4 mol-% Y_2O_3 stabilised cubic ZrO_2 single crystal [24] at elevated temperatures, as shown in Fig. 3.

It is known that dislocation loops can form in ceramics and other materials by processes such as the condensation of a super-saturated concentration of vacancies [25]. However, dislocations observed in yttria stabilized tetragonal zirconia (quoted above) occurred under an indenter, i.e. under high local pressure. Materials frequently crack when an indenter is forced into them; the case of whether elastic strains generated in the vicinity

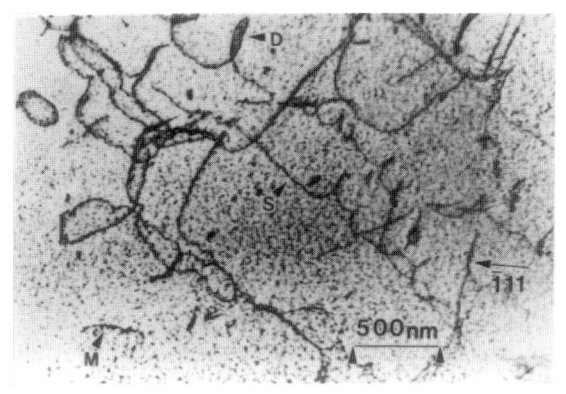

Fig. 3. Bright field transmission electron micrograph of dislocation structure 50 μm below surface of {111} surface indentation in cubic zirconia single crystal: Vickers hardness test carried out at 800°C using 4.9N load (after Ref. [24]).

of such cracks are relieved by the movement of dislocations or by crack extension relates to a classification of materials into ductile and brittle categories. The peierls stress for dislocation motion in solids is relatively high for bound electron systems with directional bonding (e.g. insulators) and low for delocalized electron systems (e.g. metals) i.e. at the insulator-metal transition point a mixed mode or transitional behaviour occurs. Insulator-metal transitions are well documented in ceramics [8] and as stated by Cottrell [26], all elements in the periodic table are expected to be metallic when the right conditions exist, such as sufficiently high pressures; Cottrell [26] cites the example of hydrogen and iodine becoming metals at appropriate pressures. Some other conditions that could lead to insulator-metal transitions have been identified by Mott [8] and Gilman [4].

The good electrical conductivity found in the early rare earth and transition metal oxides is further evidence for the existence of metallic type bonding in these compounds. The compounds are referred to as mixed ionic and electronic conductors as discussed by Weber et al [27] and Kaneko et al [28]. Recently, Reidy and Simkovich [29] based on electrical conductivity measurements on ceria stabilized zirconia, found relatively large electronic conduction compared with the ionic conduction component; an electron hopping mechanism is thought to dominate the electrical conduction in the ceria doped zirconia at low oxygen partial pressures. The proposition regarding the electronic contribution to conduction is consistent with the arguments on orbital overlap and electron delocalization in the early rare earth and transition metal oxides.

Contrary to a fairly commonly held opinion, the presence of a glassy grain boundary phase is not an absolute requirement for superplastic deformation in ceramics, although in certain conditions it may contribute to it. Sherby and Wadsworth [30] reporting on high resolution analytical transmission electron microscopy studies on yttria stabilized tetragonal zirconia ceramics, stated that superplastic deformation occurred when they did not contain a glassy grain boundary phase. Similarly, Lakki et al [31] based on creep tests in compression and internal friction measurements in yttria stabilized tetragonal zirconia ceramics, have stressed that in pure compositions that do not contain an amorphous glassy grain boundary layer, the accommodation of grain boundary sliding (which is the main mechanism involved in superplastic deformation) could be explained in terms of a substantial increase in the grain boundary dislocation mobility; this is in agreement with transmission electron microscopy studies which revealed the presence of dislocations under an indenter (see Fig. 3).

Further evidence for metallic deformation behaviour was found in the work of Davies et al [32,33,34] on superplastic deformation in ceramics, in which yttria stabilized tetragonal zirconia polycrystals were found to deform with cavities aligned parallel to the tensile axis (see Fig. 4), a behaviour typical of superplastic deformation in metals. Ceramics usually have their cavities aligned perpendicular to the tensile axis as shown in Fig. 4 for Al_2O_3 doped with CuO and MgO.

It is evident that the materials that acquire a metallic behaviour during superplastic deformation do so owing to an intrinsic property of their bonding. An example of an 'externally' induced metallic character in superplastic ceramics can be found in a magnesium aluminate spinel ($MgO.nAl_2O_3$). Lappalainen et al [35] recently observed that thin film tensile specimens of this spinel failed at 3% tensile strain in a superplastic deformation test conducted at 1200°C and at a strain rate of 1×10^{-5} s^{-1}. However, when the magnesium aluminate spinel was doped with platinum (which has a relatively high electron density at the boundaries of its Wigner-Seitz cell [36]), superplastic flow (of a type previously associated with yttria stabilized tetragonal zirconia) was observed, including the presence of a serrated stress-strain curve, as shown in Fig. 5.

Based on an earlier observation by Chiang and Kingery [37] that a space-charge segregation of intrinsic lattice defects occurs in the grain boundaries in the magnesium aluminate spinel resulting in a net grain boundary charge, Lappalainen et al [35] proposed that the possible existence of an attractive stress at the grain boundaries in $MgO.nAl_2O_3$ spinel (due to the electrical potential difference at the grain boundaries, relative to the interior of the grains) could be responsible for an increased grain boundary resistance to cavitation which would allow extensive superplastic flow. However, Lappalainen et al stated that although they thought platinum could influence the development of a cavitation resistant stress at the grain boundaries, they could not identify the mechanism, since the results of Chiang and Kingery had previously indicated that the ratio of Al/Mg in the grain boundary regions of $MgO.nAl_2O_3$ spinel is relatively insensitive to variations in stoichiometry and that pure $MgO.nAl_2O_3$ is brittle.

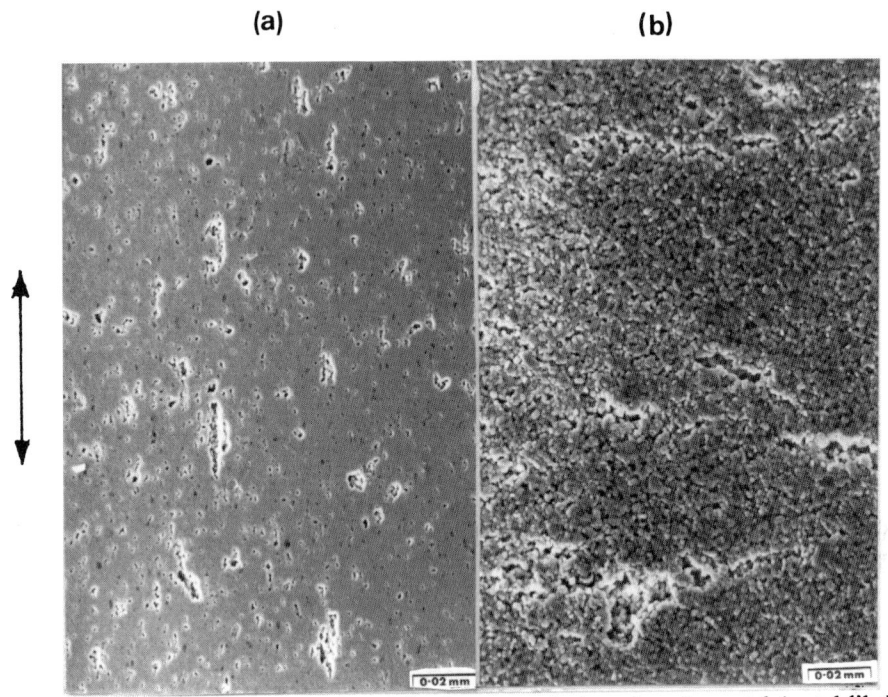

Fig. 4. (a) Cavity alignment in a yttria stablized tetragonal zirconia polycrystal (metal-like) and (b) Al_2O_3 doped with CuO and MgO (ceramic-like) : double arrow indicates tensile axis.

The present authors suspect that a metallisation of the $MgO.nAl_2O_3$ spinel could occur in the presence of platinum (with a relatively high electron density), as described above. The possibility of electron transfer into antibonding levels in the spinel are feasible according to recent proposals by Li [38] based on his studies of the wettability of ceramics by metals. Gilman's recent association of the activation of dislocation glide in silicon with the tunnelling of electrons into antibonding levels by processes that included alloying would also seem to be consistent with metallisation of the $MgO.nAl_2O_3$ spinel by the addition of platinum. It is suggested that the processes responsible for resistance to grain boundary crack growth in metals, such as dislocation glide activated processes which are known to contribute to superplastic flow in metals, become operative in platinum doped $MgO.nAl_2O_3$ spinel.
On the other hand, insulator type ceramics, like MgO with a band gap energy of 7.3 eV, and Al_2O_3 with 8.3 eV [39], show limited superplastic deformation even when they meet other requirements for superplasticity, e.g. ultrafine grain size. A similar reason will account for the limited superplasticity obtained in fine grained Si_3N_4 and SiC. It should be mentioned

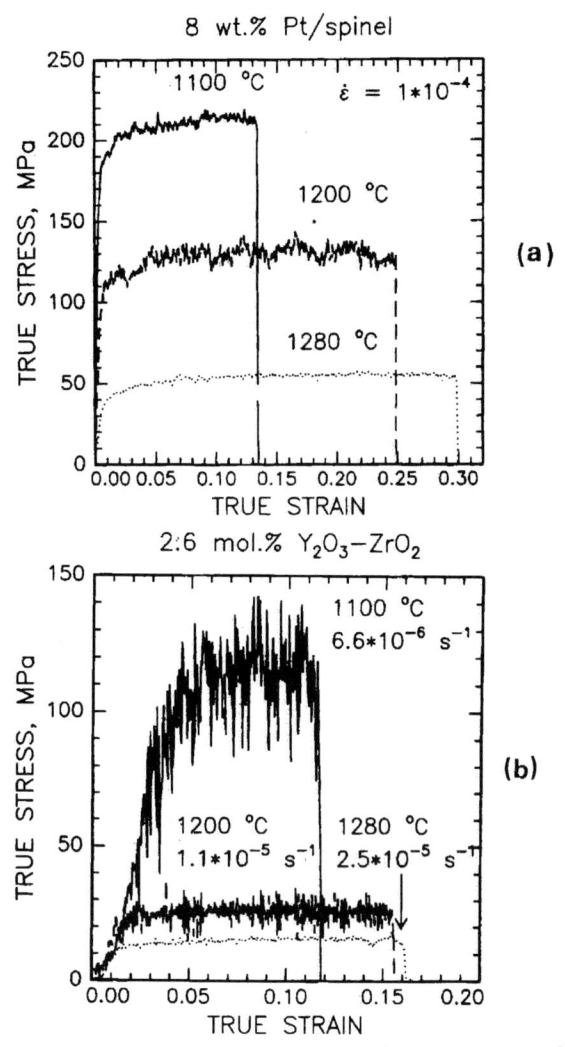

Fig. 5. Stress-strain curves for (a) a platinum doped magnesium aluminate spinel, (b) yttria stabilized zirconia (after Ref. [35]).
(a) grain size of specimens in range 35-90 nm - curves are serrated and exhibit no strain hardening; (b) grain size of specimens in range 70-290 nm - curves are very similar to those in (a).

that SiC has a structure consisting of a layered tetrahedral stacking sequence, with strong covalently bonded tetrahedral units with minimal possibility for plasticity. Typical examples of ceramic materials that exhibit a metallic character in their superplastic behaviour compared with insulating ceramics are presented in Table 1. A simple comparison will illustrate the presence of the metallic character: whereas Nieh et al obtained tensile elongations of up to 800% (Table 1) for yttria stablized zirconia, confirming its metallic character, Al_2O_3 + MgO (both with large band gap energies) would only give 38-54% tensile ductility at 1450°C. It is noticeable (Table 1) that when yttria stablized zirconia is alloyed with 20, 40, 60 and 80 wt% - Al_2O_3 and tested in a narrow temperature range, there is a corresponding decrease in the total tensile elongation obtainable with increasing alumina content. Alternatively, the addition of Cr_2O_3 and Y_2O_3 to Al_2O_3 + MgO leads to slight increments in the tensile elongations obtained at 1450°C (Table 1), indicating that the selection of alloying compounds from the left-hand end of the transition metal periods in the periodic table of elements contributes to an increase in the metallic character of the compounds.

Experimental verification of the data in Table 1 is shown in Fig. 6 below based on recent work in Manchester. Interestingly, improvements in the ductility and resistance to cavitation have been reported for Al_2O_3 doped with the early rare earth and transition metal oxides such as yttria, zirconia or hafnia [40,41,42,43,44] which satisfy the rules proposed by Phillips and Williams for orbital overlap, leading to metallic type bonding [23].

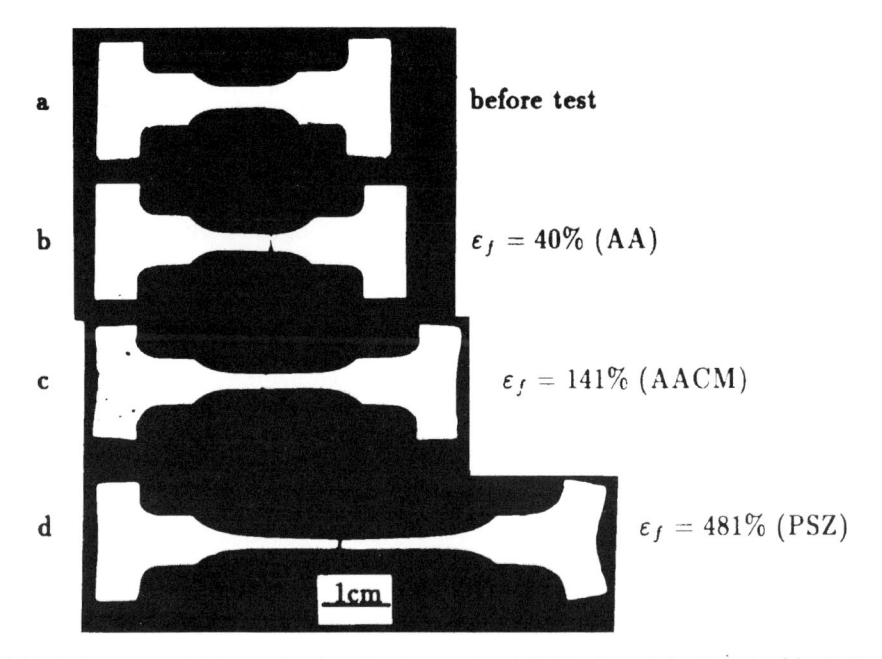

Fig. 6: (a) Before test; (b) Pure alumina (AA) tested at 1573K; (c) Al_2O_3 doped with CuO (AACM) tested at 1673K; (d) PSZ specimen tested at 1823K. Strain rate $1 \times 10^{-4} s^{-1}$

CONCLUSION

The superplastic flow observed in ceramics seems to be directly related to their electronic state. Extensive superplastic flow can be associated with the fact that some ceramics may be classified as having an intrinsic 'metallic' character, e.g. yttria stablized tetragonal zirconia polycrystal, either based on the nature of the orbitals responsible for their bonding, or due to external influences (extrinsic) such as alloying, e.g. platinum doped $MgO.nAl_2O_3$ spinel. The presence of the metallic character is likely to affect the sensitivity of the ceramics to changes in the Zener-Hollomon parameter, thereby assisting dislocation motion, which is active in the cavity accommodation process, and promoting extensive deformation.

REFERENCES

1. W.J. Kim, J. Wolfenstine and O.D. Sherby: Acta Metall. Mater. 1991, 39 (2), 199-208.
2. R.C. Gifkins: Metall. Trans. 7A (1976) 1225-1232.
3. G.M. Pharr, W.C. Oliver and D.S. Harding, J. Mater. Res., 1991, 6 (6), 1129-1130.
4. J.J. Gilman: J. Mater. Res. 1992, 7 (3) 535-538.
5. R.W. Cahn: Nature, 1992, 537, 535.
6. J.J. Gilman: J. Appl. Phys. 1975, 46, 5110.
7. P.P. Edwards and M.J. Sienko: Int. Rev. Phys. Chem. 1982, 3, 83.
8. N.F. Mott: metal-insulator transitions; 1974, London, Taylor & Francis.
9. K.F. Herzfeld: Phys. Rev., 1927, 29, 701.
10. T.J. Davies and A.A. Ogwu, Mat. Sci. and Tech., Vol. 10 (1994) 669-673.
11. A.A. Ogwu and T.J. Davies, Mat. Sci. and tech., 9 (3) (1993) 213-217.
12. A.A. Ogwu and T.J. Davies, J. Mat. Sci. (1993) 28, 847-852.
13. A.A. Ogwu and T.J. Davies, J. Mat. Sci. (1992) 27, 5382-5388.
14. T.G. Nieh, C.M. McNally and J. Wadsworth: Scr. Metall., 1988, 22, 1297.
15. F. Wakai, S. Sakaguchi and Y. Matsuno: Adv. Ceram. Mater., 1986, 1, 259.
16. F. Wakai, Y. Kodama, S. Sakaguchi, N. Murayama, K. Izaki and K. Niihara: Nature, 1990, 344, 431.
17. P. Gruffel, P. Carry and A. Mocellin: In 'Science of Ceramics' (ed. D.Taylor), Vol. 14, 587, 1988, Stoke-on-Trent, The Institute of Ceramics.
18. J.B. Goodenough: Prog. Solid State Chem., 1971, 5, 143.
19. P.A. Cox: 'The electronic structure and chemistry of solids", 77, 1987, Oxford, Oxford University Press.
20. J. Hubbard: Proc. R. Soc., 1963, A276, 238.
21. J. Hubbard: Proc. R. Soc., 1964, A277, 237.
22. J. Hubbard: Proc. R.Soc., 1964, B281, 401.
23. A.R. West: in 'Basic solid state chemistry' 1988, Chichester, Wiley, 116-118.
24. D. Holmes, A.H. Heuer and P. Pirouz: Philos Mag.A, 1993, 67 (2) 325-342.
25. Chong-Min Wang and F.L. Riley: J. Am. Ceram. Soc. 1993, 76 (8), 2136-2138.
26. A.H. Cottrell: 'Introduction to the modern theory of metals' Chapter 1; 1988, London, The Institute of Metals.
27. W.J. Weber, H.L. Tuller, T.O. Mason and A.N. Cormack: Mater. Sci. Eng. 1993, B18, 62.
28. H. Kaneko, F. Jin and H. Taimatsu: J. Am. Ceram. Soc. 1993, 76 (3), 793-795.
29. R.F. Reidy and G. Simkovich: Solid State Ionics, 1993, 62, 85-97.

30. O. Sherby and J. Wadsworth: Prog. Mater. Sci., 1989, 33, 181.
31. A. Lakki, R. Schaller, M. Nauer and C. Carry: Acta Metall. Mater., 1993, 41 (10), 2852.
32. T.J. Davies, Z.C. Wang and N. Ridley: in Proc. 5th Int. Conf. on 'Creep and fracture of engineering materials and structures', University College, Swansea, March-April 1993, The Institute of Materials, 325-334.
33. Z.C. Wang, T.J. Davies and N. Ridley: Scr. Metall. Mater., 1993, 28, 301-306.
34. Z.C. Wang, T.J. Davies and N. Ridley: in Proc. 3rd European Ceramic Society Conference (ECRS), (ed. P. Duran and J.F.Fernandez) Vol. 1, 681-688, 1993, Madrid, Spain, European Ceramic Society.
35. R. Lappalainen, A. Pannikat and R. Raj: Acta Metall. Mater., 1993, 41 (4) 1229-1235.
36. J.A. Alonso and N.H. March: Electrons in metals and alloys' 35: 1989, London, Academic Press.
37. Y.M. Chiang and W.D. Kingery: J. Am. Ceram. Soc. 1990, 73, (5) 1153-1158.
38. J.G. Li: J. Am. Ceram. Soc. 1992, 75 (11) 3118-3126.
39. J.A. Duffy: 'Bonding, energy levels and bands in inorganic solids', 169-170, 1990, London, Longman.
40. T. Hermansson, K.P.D., Lagerlof and G.L. Dunlop: in 'Superplasticity and Superplastic Forming', (ed. C.H. Hamilton and N.E. Paton), 1988, Warrendale, Pa.,TMS.
41. F. Wakai and H. Kato: Adv. Ceram. Mater., 1988, 3, 71.
42. T.G. Nieh, C.M. McNally and J. Wadsworth: Scr. Metall., 1989, 23, 457.
43. F. Wakai: PhD dissertation, Kyoto University, Japan, 1988.
44. J. Wang and R. Raj: Acta Metall. Mater., 1991, 39, 2909.

Table 1. Experimental data for fine grained ceramics [1] (strain rate hardening exponent m ≈ 0.5 for all materials)

Material	Investigators	Grain Size (μm)	m	Strain Rate range (s^{-1})	Q_c (kJ·mol^{-1})	Temperature range (°C)	Tensile Elongation range (%)
Y.TZP	Wakai et al. [5]	0.3-0.4	~ 0.5	1.11×10^{-4}-5.56×10^{-4}	580	1400-1500	60-180
Y.TZP	Nieh et al. [3]	~ 0.3	~ 0.5	8.3×10^{-3}-2.7×10^{-4}	580	1500-1530	400-800
Y.TZP	Hermansson [9]	~ 0.3	~ 0.5	4.8×10^{-3}	580	1450	250
Y.TZP + 20 wt% Al_2O_3	Wakai et al. [6]	ZrO_2: 0.5 Al_2O_3: 0.5	~ 0.5	1.11×10^{-4}-1.11×10^{-3}	620	1350-1500	30-220
Y.TZP + 20 wt% Al_2O_3	Nieh et al. [4]	ZrO_2: 0.5 Al_2O_3: 0.5	~ 0.5	8.33×10^{-4}	620	1650	500
Y.TZP + 40 wt% Al_2O_3	Wakai et al. [5]	ZrO_2: 0.51 Al_2O_3: 0.61	~ 0.5	1.11×10^{-4}-2.78×10^{-4}	720	1450-1550	145-250
Y.TZP + 60 wt% Al_2O_3	Wakai et al.	ZrO_2: 0.59 Al_2O_3: 0.99	~ 0.5	1.11×10^{-4}-2.78×10^{-4}	700	1450-1530	70-140
Y.TZP + 80 wt% Al_2O_3	Wakai et al.	ZrO_2: 0.47 Al_2O_3: 1.0	~ 0.5	1.11×10^{-4}-2.78×10^{-4}	753	1450-1550	60-110
Al_2O_3 + MgO	Gruffel et al. [7]	0.77-1.51	~ 0.5	1.2×10^{-4}	400	1450	38-54
Al_2O_3 + MgO + Cr_2O_3	Gruffel et al. [7]	0.83	~ 0.5	8.8×10^{-3}	400	1450	55
Al_2O_3 + MgO + Y_2O_3	Gruffel et al. [7]	0.66	~ 0.5	3.4×10^{-3}	400	1450	65
Al_2O_3 + MgO + Ti_2O_3	Gruffel et al. [7]	0.72	~ 0.5	5.5×10^{-4}-1.2×10^{-4}	400	1250	25-30

EFFECT OF Fe AND Co ADDITIONS ON THE SUPERPLASTIC BEHAVIOUR IN THE NiAl-Ni$_3$Al TWO PHASE ALLOY

S.OCHIAI[a], M.KOBAYASHI[b]

[a] The Nishi Tokyo University, Uenohara-cho, Yamanashi, 409-01, JAPAN
[b] Chiba Inst. of Technology, Tsudanuma, Narashino, 275, JAPAN

Abstract

Fe and Co, representative substitutional elements for Ni sites in NiAl and Ni$_3$Al, were added to NiAl-Ni$_3$Al two phase alloy in order to investigate microstructures and mechanical properties at high temperatures. Alloys were arcmelted and subjected to heat refining treatment which was composed of oil quenching from 1573K and tempering at 1073K. Very fine lamellar structure was observed for the base alloy. Iron added alloys with 2 and 4 mol%Fe showed coarser two phase structures with increasing iron content. On the contrary, Co added alloys with 2 and 4 mol%Co exhibited fine lamellar structures irrespective of Co content. The result of compression tests demonstrated that while the strain rate sensitivity exponent of flow stress, m value, of the base alloy was about 0.4 at 1073K, iron added alloys exhibited smaller m value. On the other hand, the m value of about 0.4 was maintained for cobalt added alloys. it was found that although the elongation-to-failure value of more than 150% was confirmed for the base alloy, it was decreased with increasing iron content. But, the cobalt addition showed larger elongation- to failure than the base alloy. These facts suggest that the iron addition suppressed the superplastic deformation, whereas cobalt addition can enhance it.

1. Introduction

It has been known that the Ni-rich NiAl(β) intermetallic phase transforms into martensite, when quenched from high temperature in the single β-phase region[1]. We have reported that the very fine lamella structure consisting of β and γ'(Ni$_3$Al) phases can be obtained, particularly in the Ni-34mol%Al alloy, by tempering the martensite at the intermediate temperature of 1073K in the (β+γ') two phase field[2]. According to the compression test result at high temperature, such heat refined alloy with 34mol%Al deformed without fracture and showed high strain-rate sensitivity exponent (namely m value) over 0.4. Also, it has been revealed that the boron micro-alloyed alloy with lamellar structure exhibited high fracture strain over 200% at 1073K by tension test[3]. These results are believed to be indicating the development of superplastic behavior in the alloy.

Applying the superplastic behavior, the significant improvement of deformability for this (β+γ') two phase alloy can be expected. On the other hand, the further strengthening, such as solid solution hardening, will be necessitated in this alloy for structural applications. From this point of view, iron and cobalt can be selected as additional elements, because their solid solubility for β or γ' phase is large and they are expected not to change the transformation temperature so much compared with the base alloy [4,5]. It is the purpose of this study to clarify the effects of iron and cobalt addition on the microstructures and mechanical properties at high temperature in (β+γ') two phase alloys.

2. Experimental

The compositions of specimens investigated are Ni-33.9mol%Al-0.1mol%B (designated the base alloy) and Ni-33.9mol%Al-0.1mol%B-Xmol%Fe (or Co) (X=2,4, designated 2Fe etc.). Using pure Ni(99.9%), Al(99.99%), Fe(99.9%) and Co(99.9%) as raw materials, these five alloys were produced by arc-melting method in an argon atmosphere. Button ingots were remelted in an induction furnace and centrifugally cast into the molds. Obtained ingots were machined into compression test specimens with length of 12mm and diam. of 6mm and into tensile specimens with the gage length of 10mm and the diam. of 3mm.

Specimens were subjected to the heat refining treatment. Namely, they were solutionized at 1573K for 7.2ks and quenched from this temperature into oil bath at 373K. Oil quenching was employed to avoid the formation of microcracks during quenching. Then, they were tempered at 1173K for 36ks. Following heat treatment to obtain lamella structure consisting of β and γ' phases, mechanical properties at high temperatures were studied. Compression tests were conducted using the universal testing machine named "Thermecmaster" at 1023K ,1073K and 1123K and the strain rate change test was adopted in order to investigate the strain rate sensitivity of the stress. On the other hand, tensile tests were carried out using an Instron type testing machine at 1073K and the initial strain rate of 8.3×10^{-5} s^{-1} in an argon atmosphere in order to avoid oxidation.

Optical microscopy (OM) was employed to examine microstructures and X-ray diffraction was utilized to identify the phases present.

3. Results and Discussion

3.1. Microstructures

Fig.1 shows the microstructures of quenched specimens from 1573K. The twin structures inherent in martensite formation can be observed definitely in 2Fe, 2Co and 4Co alloys as well as base alloy. Since the martensite in NiAl has 3R structure and shows the shape memory effect, the transformation is known as the thermoelastic type one. On the other hand, β phase containing fine γ' was observed instead of martensite structure for 4Fe, indicating that iron addition has the tendency to suppress the martensite transformation.

By tempering the quenched specimens, very fine lamellar

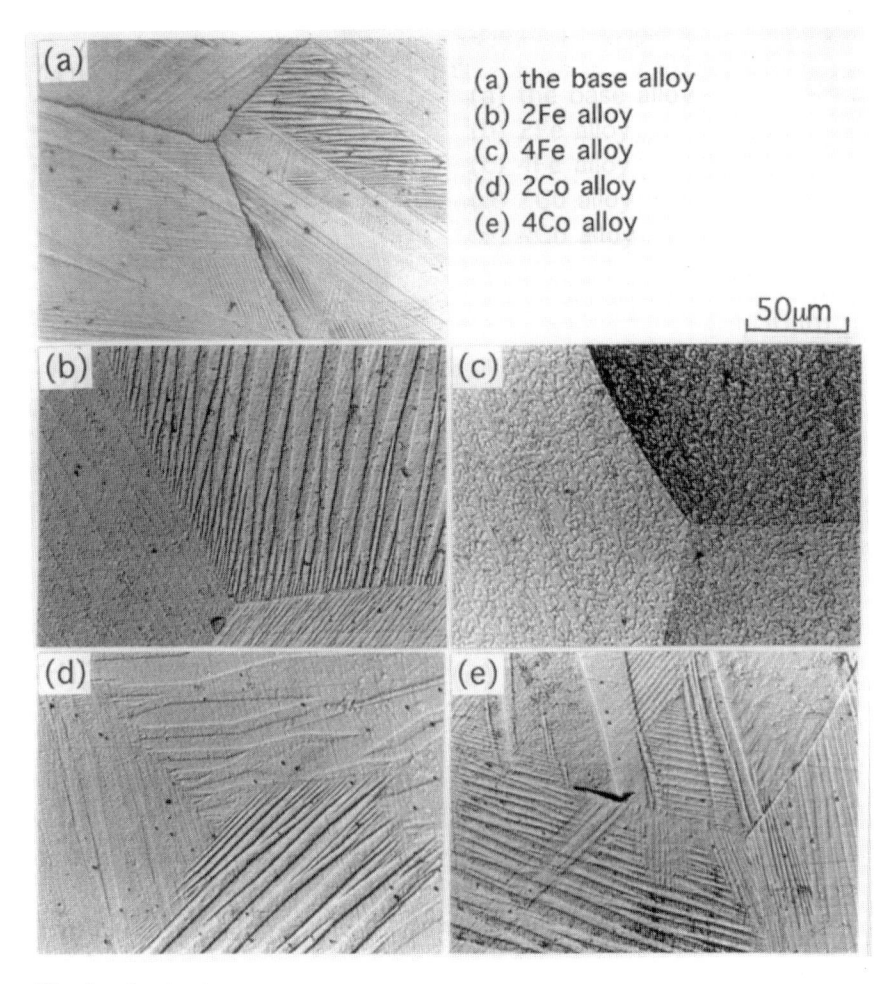

Fig.1 Optical microstructures of alloys quenched from 1573K.

(a) the base alloy
(b) 2Fe alloy
(c) 4Fe alloy
(d) 2Co alloy
(e) 4Co alloy

structures are clearly seen in 2Co and 4Co alloys as presented in Fig.2. This lamellar structure is consisting of β and γ' phases which are layered alternately having the intervals of less than 0.4 μm and it has been acknowledged that such structure is able to be expected to form when the volume of the phase is equal to each other. On the contrary, in iron containing alloys including 2Fe alloy, precipitation of massive γ' phase with acicular shape was observed. In all alloys, continuous precipitates of γ' phase were visible on the prior β grains, though the width of precipitate is very small like film.

3.2. Compressive test

In Fig.3, the representative true stress vs. true strain curves for the base alloy obtained at 1023K is illustrated. During testing, strain rate was changed in a variety of 5 steps. Any alloy showed the maximum stress at the initial stage of deformation and then the

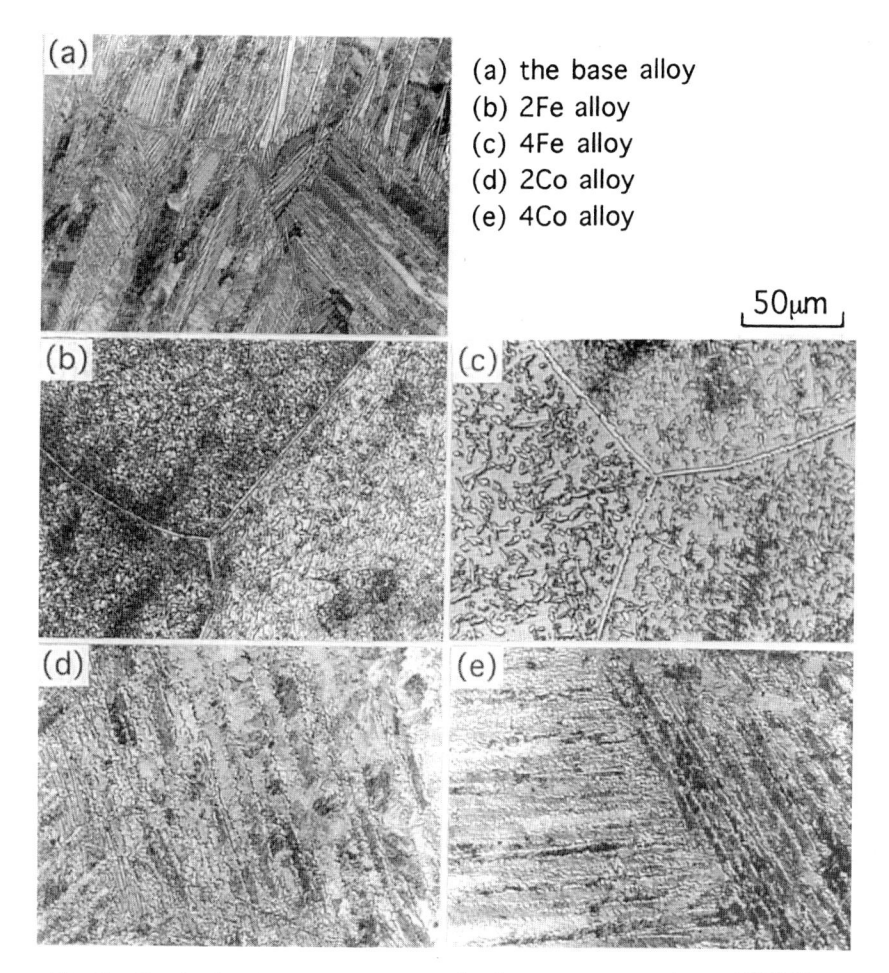

Fig.2 Optical microstructures of alloys tempered at 1073K.

(a) the base alloy
(b) 2Fe alloy
(c) 4Fe alloy
(d) 2Co alloy
(e) 4Co alloy

stress decreased gradually with the increase of strain. Hereafter, the stress became constant, suggesting the steady state deformation.

The relations between the 0.2%flow stress, the maximum stress and the stresses at the strain of both 0.2 and 0.45 vs. the concentration of Fe and Co additives are depicted in Fig.4 and Fig.5, respectively. The 0.2%flow stress and the maximum stress which are in the initial stage of deformation and at the strain rate of 7.4×10^{-4} s^{-1} tend to decrease with increasing additive concentration regardless of the kind of addition. But, stresses at the strain of both 0.2 and 0.45 which are in the steady state stage of deformation and at the strain rate of 1.0×10^{-4} s^{-1} have the tendency to increase with the incerase of Fe addition and to be almost constant for Co addition, respectively.

Therefore, in terms of the steady state deformation, iron addition is considered to be more effective on the strengthening of ($\beta + \gamma'$) two phase alloy than cobalt addition. These trends were also recognized at 1023K and 1123K. Furthermore, It is worth to note that there is a large difference in the stress level between the 0.2%flow stress, the maximum stress and the stresses at the strain

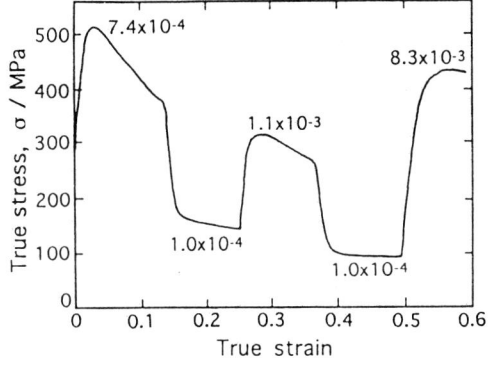

Fig.3 True stress - true strain curve for the base alloy at 1023K.

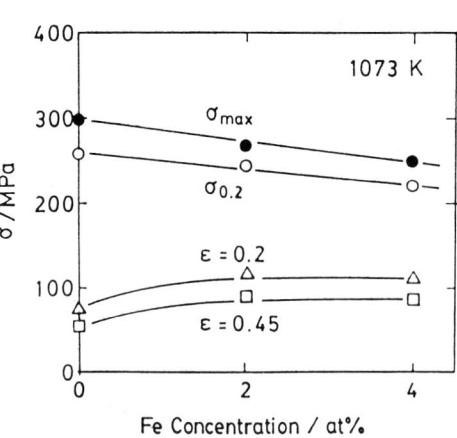

Fig.4 Effect of iron addition on the deformation stresses at 1073K.

Fig.5 Effect of cobalt addition on the deformation stresses at 1073K.

of both 0.2 and 0.45. This is due to the difference of strain rate, indicating the strong sensitivity of the stress for strain rate.

Strain rate sensitivity exponent, m value, can be estimated using following equation.

$$\dot{\varepsilon} = C \sigma^{1/m} \quad (1)$$

where σ is the flow stress, ε is the strain rate and C is a constant. Backofen et al. regarded m value of above 0.3 as one of the measures for the superplasticity[6]. In this case, the necking is restrained so that the large and homogeneous elongation comes to occur.

Fig.6 and Fig.7 show the mean m value as a function of test temperature for Fe added and Co added alloys, respectively. It has been revealed that the m value inclines to increase with the increase of temperature. While the base alloy without Fe or Co addition exhibits high m value of about 0.4, iron added alloys show sufficiently lower m value of about 0.27. But, the alloys containing cobalt exhibit almost the same value as the base alloy. This seems to suggest that iron addition is detrimental to the superplastic phenomenon. This alloying effect on the mechanical properties seems

Fig.6 Temperature dependence of the average m value for alloys containing various iron additions.

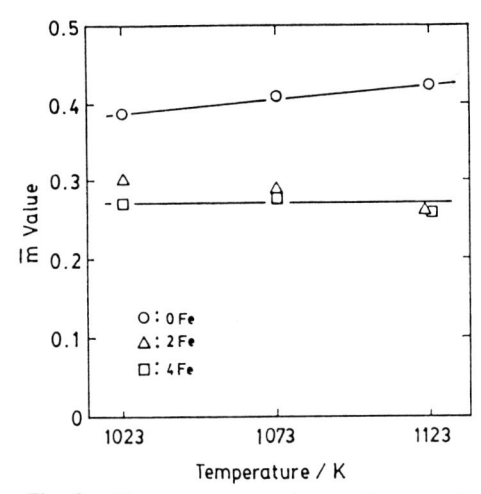

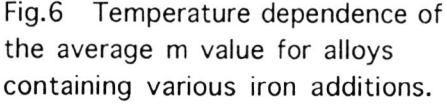

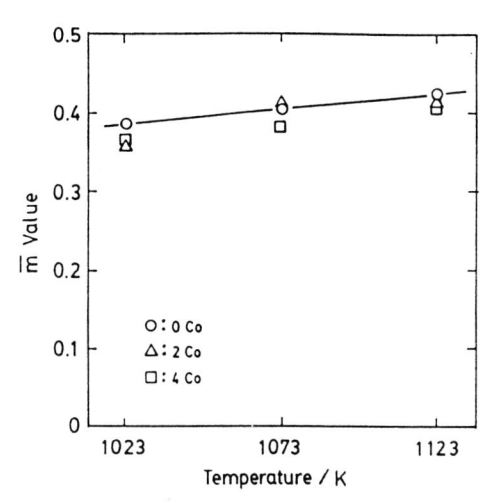

Fig.7 Temperature dependence of the average m value for alloys containing various cobalt additions.

to be tightly connected with the microstructural change; iron addition changed microstructure from fine to coarse, whereas cobalt did not markedly.

Based on the following equation, we can estimate the apparent activation energy, Q, for the high temperature deformation:

$$\dot{\varepsilon} = C \sigma^{1/m} \exp(-Q/RT) \quad (2)$$

where R is the gas constant and T is the absolute temperature. Fig.8 and Fig.9 show the logarithmic plots of the flow stress vs reciprocal temperature for Fe added and Co added alloys, respectively. In these figures, it is confirmed that there exists linear relation for each alloy. The apparent activation energies resulted from the gradients are 224kJ/mol for the base alloy, 251kJ/mol for 2Fe, 262kJ/mol for 4Fe, 230kJ/mol for 2Co and 230kJ/mol for 4Co. Therefore, it is clear that while the iron addition increases the activation energy, the value is held almost constant despite of the cobalt addition.

The activation energy of the deformation at high temperatures for γ' phase has been reported to be 315 ± 30 kJ/mol[7] and 326 ± 9 kJ/mol[8]. Also, 303 kJ/mol is obtained for self diffusion in γ'[9]. On the other hand, as the activation energy of the deformation for β phase, the values of 310 kJ/mol[10] and 326 kJ/mol[11] were presented along with the value of 310 kJ/mol[12] for the self diffusion. Since the values of activation energy measured in this work are substantially less than those for reported values, it is difficult to think that the lattice diffusion controls the high temperature deformation of (β+γ') two phase alloys tested. Instead, there is a possibility that the grain boundary sliding mechanism operated.

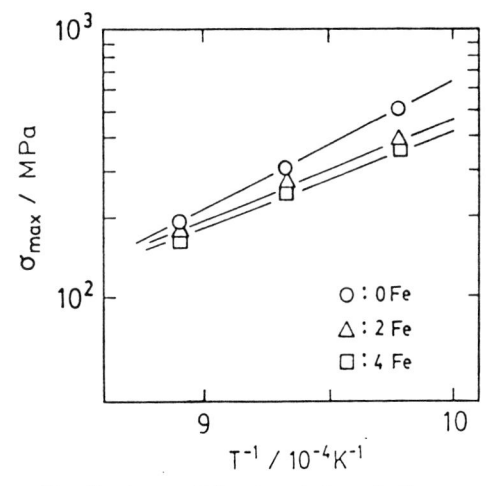

Fig.8 Logarithmic plots of the maximum stress vs. reciprocal temperature for alloys with and without iron addition.

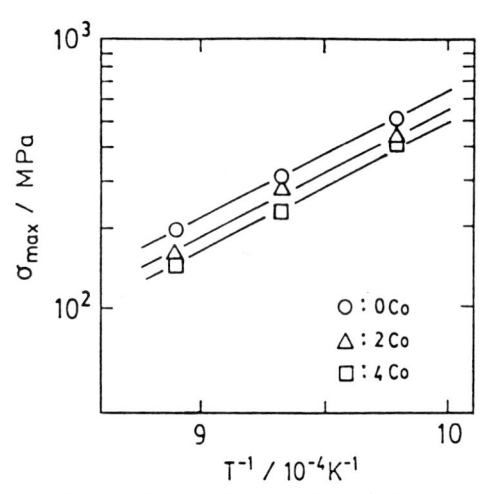

Fig.9 Logarithmic plots of the maximum stress vs. reciprocal temperature for alloys with and without cobalt addition.

3.2 Tensile Test

High temperature tensile test was conducted at 1073K and at the initial strain rate of 8.3 x 10^{-5} s^{-1}. Resultant engi. stress vs. engi. strain curves are shown in Fig.10 and Fig.11. It is clear that Fe addition increases the maximum stress from 75MPa of the base alloy to 120MPa of 4Fe alloy. However, iron decreases the elongation-to-failure from 180 % of the base alloy to 45% of 4Fe alloy. On the contrary, Co addition tends to decrease the maximum stress; 60MPa for 4Co alloy. But, the elongation-to-failure was increased by the addition of cobalt and no less than the value of 250% was obtained for 4Co alloy.

Above mention clearly indicates that cobalt addition for ($\beta+\gamma'$) two phase alloy is effective for the development of the superplasticity, whereas the iron deteriorate the superplastic phenomenon. These tendency is consistent with the observation of microstructure and the result of compression test.

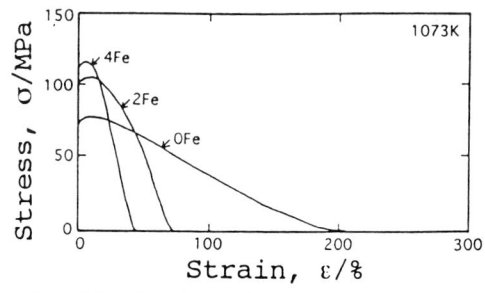

Fig.10 Engineering stress - strain curves for alloys with various iron content at 1073K.

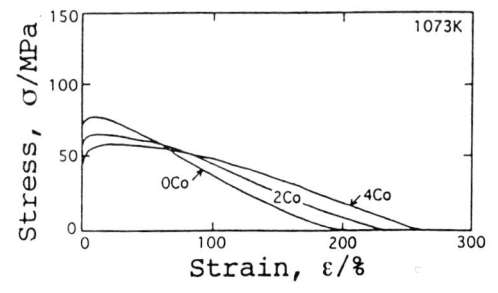

Fig.11 Engineering stress - strain curves for alloys with various cobalt content at 1073K.

4. Summary

Transformation of NiAl(β) phase into martensite on quenching was confirmed for the base alloy, 2Fe, 2Co and 4Co alloys. After tempering at 1073K, very fine lamellar structure consisting of NiAl and Ni$_3$Al(γ') two phases was observed in the base, 2Co and 4Co alloys, whereas coarse structure of β phase containing γ' in it was obtained for iron added alloys. Results of compression test showed that the base alloy and cobalt added alloys exhibited high strain rate sensitivity exponent, m value, as 0.4, while iron added alloys exhibited lower m value as 0.27. The apparent activation energies for high temperature steady state deformation were estimated to be 224kJ/mol for the base alloy, 230kJ/mol for cobalt added alloys and about 255kJ/mol for iron added alloys. Those values seem to suggest the operation of grain boundary sliding mechanism. Cobalt addition enhanced the elongation-to-failure and no less than 250% was exhibited for 4Co alloy. Iron addition, however, had the tendency to reduce the elongation. This large elongation value and high m value shown in the base and cobalt added alloys are believed to be the proofs for the development of the superplasticity.

References

1. S.Chakravorty and C.M.Wayman, Metall.Trans., 7A, 1976, 555.
2. S.Ochiai, I.Yamada and Y.Kojima, J.Japan Inst.Metals, 54, 1990, 301.
3. S.Ochiai, Y.Doi, I.Yamada and Y.Kojima, J.Japan Inst.Metals, 57, 1993, 214.
4. S.M.Russell, C.C.Law and M.J.Blackburn, Mat.Res.Soc.Symp.Proc., 133, 1989, 627.
5. M.Rudy and G.Sauthoff, Mat.Res.Soc.Symp.Proc., 39, 1985, 327.
6. W.A.Backofen, I.R.Turner and D.H.Avery, Trans.ASM, 57, 1964, 980.
7. J.R.Nicholls and R.D.Rawlings, J.Mater.Sci., 12, 1977, 2456.
8. P.A.Flinn, Trans.Met.Soc.AIME, 218, 1960, 145.
9. G.F.Hancock, Phys.Status.Solidi., A7, 1971, 535.
10. J.D.Whittenberger, J.Mater.Sci., 22, 1987, 394.
11. R.R.Vandervoort, A.K.Mukherjee and J.E.Dorn, Trans.ASM, 59, 1966, 930.
12. G.F.Hancock and B.R.McDonnell, Phys.Status Solidi (a), 4, 1971, 143.

Microstructural Evolution during Superplastic Flow

Microstructural and Textural Changes During the Superplastic Deformation of a Modified IMI550 Titanium Alloy (Ti-4Al-3Mo-2Sn-1Fe-0.5Si)

M. Tuffs and C. Hammond
School of Materials, The University of Leeds.

Abstract

The microstructural and textural changes that occur during the superplastic deformation of a two phase titanium sheet alloy IMI550 that has been modified by the replacement of 1%Mo with 1%Fe (Ti-4Al-3Mo-2Sn-1Fe-0.5Si) have been studied. The microstructure of the alloy consists of fine, relatively equiaxed primary grains of α surrounded by β in which a secondary α phase is distributed. The microstructure also contains a significant number of coarser features such as elongated and blocky α. The mean linear intercept value of the primary α grains was approximately 3.0 µm and the total $V_\alpha \sim 0.60$. The alloy possesses weak textures of the $\{11\bar{2}0\}<0001>$ and $\{100\}<110>$ types in the α and β phases, respectively. The addition of iron has reduced the flow stresses at all strain rates within the superplastic range at 800 and 850°C by more than 20%. The optimum temperature for superplastic deformation, in terms of maximum elongation, was 850°C and hence this temperature was chosen to investigate in detail the effect of superplastic deformation on the modified alloy. At 850°C the alloy showed superplastic properties at all the imposed strain rates between approximately 2×10^{-5} and 5×10^{-3} s^{-1}. The microstructures after deformation consisted of relatively equiaxed grains of α and β, the coarser α features having been eliminated, with $V_\alpha \sim 0.45$. The softer β phase also existed as a thin layer separating some α grains. Strain induced grain growth was observed after 200% strain at all but the highest strain rate of approximately 4×10^{-3} s^{-1}. The specimens deformed at the lower strain rates exhibited the greatest amount of growth, which lead to significant strain hardening. Superplastic deformation resulted in the randomisation of well developed annealing textures of the $\{11\bar{2}0\}<0001>$ and $\{100\}<110>$ types in the α and β phases, respectively. The observations are consistent with the superplastic flow being attributed to grain boundary sliding with grain rotation.

1. Introduction

IMI550 is a high strength alpha+beta titanium alloy with a nominal composition of (wt %) Ti-4Al-4Mo-2Sn-0.5Si. The tin and aluminium additions stabilise and solid solution strengthen the alpha phase. The molybdenum stabilises the beta phase and widens the alpha+beta phase field to improve forgeability and heat treatment response. The silicon gives an increase in tensile and creep strength and helps to refine grain size [1]. Previous studies have shown that IMI550 can be superplastically deformed at temperatures above 850°C providing that the microstructure of the alloy consists of fine grained equiaxed α and β [2,3]. The reduction of superplastic forming temperatures for this alloy would be beneficial for a number of reasons such as reduced oxidation and microstructural variations.

It has been well documented that the superplastic response of another $\alpha+\beta$ alloy, Ti-6Al-4V, can be improved by small additions of alloying elements such as cobalt, iron and nickel [4-7]. The alloy additions (Fe, Ni, Co) were chosen for two reasons:
 (i) they have high diffusivities in the β phase which is expected to increase the value of the appropriate diffusion coefficient for creep in the β phase.
 (ii) they partition to, and stabilise the β phase leading to an increased proportion of the softer β phase at lower temperatures.

The same philosophy has been applied to IMI550, and as a result, trial quantity of a modified alloy based on IMI550 but with 1% Mo replaced with 1% Fe has been produced in the form of 2 mm thick sheet [8], the composition of which is given in table 1 below.

Table 1 Composition of IMI550-1Mo+1Fe alloy (wt %, Balance Titainium)

Al	Mo	Sn	Fe	Si	O
4.17	2.99	1.91	98	0.50	0.14

The microstructure of the as received IMI550-1Mo+1Fe sheet, as shown in fig. 1, consists of fine and relatively equiaxed primary α grains surrounded by β (the lighter phase on the SEM photographs) in which secondary α is distributed. The microstructure also contains a significant amount of coarser features such as elongated and blocky α, examples of which are given in fig. 2. The mean linear intercept and volume fraction of the primary α grains were determined as 3.0 μm and 0.60, respectively.

A comparison between stress-strain rate and strain rate sensitivity behaviour of IMI550 and IMI550-1Mo+1Fe sheet alloys tested at 800 and 850°C is shown in fig. 3. The testing was carried out using a cross head speed cycling procedure on 2 mm thick sheet specimens with a 25 mm gauge length and a 5 mm gauge width. The data was obtained after each of the specimens had received sufficient deformation to achieve a refined microstructure and overcome any transient stress-strain rate behaviour. It can be seen that the alloy modification has led to the reduction in flow stresses, by more than 20%, at all the imposed strain rates at both temperatures. The strain rate sensitivity data illustrates that the maximum strain rate sensitivities are higher for the modified alloys.

Superplastic elongation to failure and β volume fraction data at various temperatures for both alloys are shown in fig. 4. The elongation to failure tests were carried out on specimens of IMI550 and IMI550-1Mo+1Fe with 25 and 20 mm gauge lengths, respectively. The specimens were deformed at a cross head speed of 1.0 mm min^{-1}, which translates into initial strain rates of 6.7x10^{-4} and 8.3x10^{-4} s^{-1} for the base and modified alloy, respectively. During the course of the test the cross head speed was periodically increased, when necessary, to return the strain rate to its initial value. The maximum elongation for the modified alloy is achieved at 850°C with V_β= 0.55. For the base IMI550 alloy V_β= 0.55 at approximately 880°C, the temperature at which IMI550 is superplastically formed commercially. At 800°C the modified alloy achieves a greater elongation to failure than the base alloy whilst at 900°C this situation is reversed.

The aim of the present work is to study the stress-strain rate behaviour and the microstructural and textural changes that occur during the superplastic deformation of IMI550-1Mo+1Fe sheet alloy at 850°C, the optimum temperature for superplastic deformation in terms maximum elongation to failure.

2. Experimental Procedures

2.1 Mechanical Testing

The high temperature testing was carried out using an Instron tensile testing machine on which was mounted a three zone tube furnace accurate to ± 2°C over a 300 mm zone. The ends of the furnace were sealed with insulating mineral wool. High purity argon was passed through the furnace during testing to reduce oxidation of the specimen. Sheet specimens with a thickness of

2 mm, a gauge length of 20 mm and a gauge width of 5 mm were used for the uniaxial deformation, with the gauge lengths parallel to the longitudinal direction of the sheet. Each specimen was held at the test temperature for 30 minutes prior to deformation to allow temperature equilibration. During this period the as received microstructure changed to generally fine and equiaxed grains of α and β, i.e. homogenisation, and the volume fraction of the β increased to approximately 0.55, as shown in fig. 5. However, the presence of coarser features such as elongated and blocky α were not eliminated.

Two types of testing procedures were employed. The first was a strain rate cycling procedure carried out on a single tensile specimen. This involved initially straining the specimen at a cross head speed that corresponded to a relatively high strain rate of approximately 2×10^{-3} s^{-1} until the work hardening plateau had been overcome and the load stabilised. The cross head speed was then reduced in stages to give the lowest strain rate of approximately 2×10^{-5} s^{-1}; the load being allowed to stabilise at each stage. After the load had stabilised at the lowest strain rate the cross head speed was then increased in stages to give the highest strain rate of approximately 4×10^{-3} s^{-1}; the load being allowed to stabilise at each stage. This sequence of consecutively cycling between high and low strain rates was repeated several times.

The second test procedure involved deforming five separate specimens to 200% strain at particular cross head speeds that corresponded to strain rates throughout the superplastic range. Details of the strain rate schedules for each specimen are given in table 2. For the tests that involved initial cross head speeds of below 2.0 mm min^{-1} an initial period of deformation, amounting to less than 10% strain, was allowed at a cross head speed of 2.0 mm min^{-1} to remove slack from the grip and specimen arrangement and to reach the plateau of the load versus extension curve with minimum delay. During deformation the true strain rate of each specimen will decrease slightly. Therefore, to ensure the strain rate range for each specimen remained as narrow as possible then the cross head speed was increased after a specific strain to return the strain rate to its initial value. By using this second procedure the microstructural development and the variation in flow stress with strain during superplastic deformation at high, intermediate and low strain rates could be assessed.

Table 2. Strain rate schedules for specimens with a gauge length of 20 mm deformed at 850°C.

Initial strain rate	Initial cross head speed	Duration	Elongation at change in CHS	Second cross head speed	Duration	Total elongation
(s^{-1})	(mm min^{-1})	(mins)	(%)	(mm min^{-1})	(mins)	(%)
8.3×10^{-5}	0.1	200	100	0.2	100	200
4.2×10^{-4}	0.5	40	100	1.0	20	200
8.3×10^{-4}	1.0	20	100	2.0	10	200
1.7×10^{-3}	2.0	15	150	5.0	2	200
4.2×10^{-3}	5.0	4	100	10.0	2	200

After deformation had ceased, using both procedures, the specimens were removed from the furnace rapidly to minimise α grain growth that would occur excessively, and increase the α volume fraction, if slow or furnace cooling was permitted. Dimensional measurements were taken along the gauge length of the cooled specimens using a pointed jaw micrometer to determine;
  a) the level of strain anisotropy between the transverse and short transverse directions, and,
  b) the amount of necking i.e. the ratio of maximum to minimum gauge area.

2.2 Metallography

Microstructural changes were evaluated using a Camscan series 4 scanning electron microscope. The microstructure of the as received alloy was studied together with material taken from the gauge length and grip section of the deformed specimens. Microspecimens were dip etched in a solution of 10% HNO_3 and 2% HF for an appropriate time to reveal the microstructure. The mean linear intercept values for the α phase were determined in the longitudinal, transverse and short transverse directions at a magnification of 1400 times with measurements taken over at least 300 grains. The proportions of α and β phases were determined using a 25 point grid over at least 20 fields, i.e. at least 600 points.

2.3 Textural Analyses

Textural analyses were carried out using a Philips diffractometer with filtered Mo K_α radiation, which was chosen because it has a low mass absorption coefficient with titanium and would generate X-ray data with correspondingly low levels of noise. However, Mo K_α has a relatively low wavelength of 0.7093 Å that results in a low resolution. Therefore, the α and β textures had to be determined from planes that gave reasonably strong reflections well separated from others. The reflections from the $\{11\bar{2}0\}$ planes for the α phase and the $\{200\}$ planes for the and β phase fulfilled these requirements and were used to generate pole figures from superplastically deformed, as received and heat treated material.

3. Results

The stress-strain rate and corresponding derived strain rate sensitivity data obtained by using the strain rate cycling procedure on a tensile specimen, which is shown in fig. 6, of the as received modified sheet at 850°C are shown in figs 7 and 8 for the increasing and decreasing strain rate cycles, respectively. The specimen, had been deformed to a total strain of 276% (approximately 30% strain per cycle) with a good resistance to neck formation, the ratio of maximum to minimum area along the gauge length being only 1.17. The first decreasing cycle was characterised by the lowest strain rate sensitivities with a maximum value of only 0.45. As the test progressed into the second and subsequent cycles then higher strain rate sensitivities were recorded which stabilised and reached a maximum of 0.55 and 0.61 for the decreasing and increasing cycles, respectively, at a strain rate of approximately 5×10^{-4} s^{-1}. The stress-strain rate data also shows that the flow stress gradually increased with strain suggesting the occurrence of strain enhanced grain growth, as is shown to be the case by a comparison between the microstructures in the specimen grip (annealed) and gauge length (deformed), shown in figs 9 and 10, respectively. It is also clear that the coarser α features within the gauge length have been broken down by the deformation and a relatively equiaxed $\alpha+\beta$ structure has been achieved with the β phase separating some α grains.

The mechanical test results obtained by using the constant cross head speed procedure on five separate tensile specimens are shown in fig. 11, the stress and strain rate values obtained from each tensile specimen after 20, 40, 100 and 200% strain are plotted. The specimen deformed at an initial strain rate of 1.6×10^{-3} s^{-1} is shown in fig. 6. It can be seen that each specimen exhibits some degree of strain hardening; the greatest amount of hardening being associated with the specimens deformed at th. lowest strain rates. The tensile tests involving the three highest initial strain rates were repeated with almost identical results being obtained thus confirming the stress-strain rate behaviour characterised in fig. 11.

A detailed microstructural analysis was carried out on material taken from the grip sections of the specimens deformed at the lowest and highest initial strain rates. The results, shown in table 3, illustrate that the amount of grain growth induced by static annealing at 850°C is small and that before deformation begins the grains in the longitudinal-transverse plane are equiaxed.

The results of the microstructural analyses that were carried out on material taken from the gauge lengths of the deformed specimens are given in table 4. V_α for all of the specimens is close to 0.45 indicating that the volume fraction of the α and β phases is independent of strain rate. The consistency of the V_α values also suggests that the any α grain growth that occurred during the cooling of each specimen was insignificant. It is evident that grain growth and/or changes in grain shape have occurred in all specimens as a result of the deformation. In all cases the ratio of grain size in the longitudinal direction to grain size in the transverse direction (L/w) has increased, i.e. the grains have elongated along the tensile axis, the greatest elongations being associated with the highest strain rates.

Table 3 Microstructural analyses from specimens of IMI550-1Mo+1Fe annealed at 850°C.

Duration of test (mins)	V_α	α grain size (μm)[*]				L/w
		L	w	t	average	
6	0.47	3.5	3.5	2.4	3.1	1.0
300	0.46	3.7	3.7	2.5	3.3	1.0

Table 4 Microstructural and dimensional measurements from specimens of IMI550-1Mo+1Fe deformed to 200% strain at 850°C.

Initial strain rate (s^{-1})	V_α	α grain size (μm)[*]				L/w	$\varepsilon_t/\varepsilon_w$ [**]	A_{max}/A_{min} [***]
		L	w	t	average			
8.3x10^{-5}	0.46	5.3	4.7	4.1	4.7	1.1	2.2	1.25
4.2x10^{-4}	0.48	5.2	4.7	3.4	4.4	1.1	1.8	1.12
8.3x10^{-4}	0.47	4.6	3.9	3.0	3.8	1.2	1.7	1.05
1.7x10^{-3}	0.45	4.7	4.0	2.9	3.9	1.2	1.6	1.08
4.2x10^{-3}	0.46	4.0	3.1	2.3	3.1	1.3	1.6	1.17

* Grain sizes were determined using the mean linear intercept method with $V_\alpha = 0.45$. The values for the three principal directions are given, namely;
    L - Longitudinal to the stress.
    w - Transverse to the stress across the width of the specimen.
    t - Transverse to the stress through the thickness of the specimen (short transverse).
** Ratio of true strain across thickness to true strain across width of gauge length.
*** Ratio of maximum to minimum area along gauge length.

Examples of the microstructures after 200% strain at high, intermediate and low strain rates are given in figs 12, 13 and 14, respectively. At all strain rates a microstructure consisting of equiaxed grains of α and β has been developed and the coarser α features have been eliminated. It can be seen that the β phase exists as generally equiaxed grains as well as forming a thin layer separating some α grains. It is evident that the greatest amount of strain enhanced grain growth is associated with the specimens deformed at the lower strain rates. The amount of strain enhanced grain growth experienced by each specimen is quantified in table 5 which also gives the average grain growth rate during deformation. The growth rate experienced at the lowest strain rate was substantially lower than those experienced at intermediate strain rates and at the highest strain rate no grain growth has occurred. A further hardening mechanism, in excess of that due to concurrent grain growth, that occurs during the course of a test is the build up of α case on the specimen surface due to oxidation that would affect the specimens deformed at the lowest strain rates in particular.

The results of the dimensional measurements carried out on the deformed specimens, also given in table 4, indicate that the strain across the thickness was greater than that across the width of the gauge length, i.e. the deformation was approaching plane strain. It is noticeable that the lower strain rates produce the most anisotropy, which may be associated with the build up of α case. The dimensional measurements also reveal that the least amount of necking has occurred in the specimens deformed at the intermediate strain rates.

Table 5 Grain growth characteristics of IMI550-1Mo+1Fe deformed to 200% strain at 850°C.

Initial Strain rate (s^{-1})	Duration of test (mins)	Average α grain size after test (μm)		Grain growth (%)	Average growth rate (μm min^{-1})
		Grip	Gauge		
8.3×10^{-5}	300	3.3	4.7	42	4.7×10^{-3}
4.2×10^{-4}	60	3.2(*)	4.4	38	2.0×10^{-2}
8.3×10^{-4}	30	3.2(*)	3.8	19	2.0×10^{-2}
1.7×10^{-3}	17	3.2(*)	3.9	22	4.1×10^{-2}
4.2×10^{-3}	6	3.1	3.1	0	0

* These values were calculated by taking the mean of the grain sizes in the grips of the specimens deformed at the lowest and highest initial strain rates.

$\{11\bar{2}0\}$ and $\{200\}$ pole figures for the α and β phases, respectively, for as received, statically annealed and superplastically deformed material are given in figs 15 and 16. The as received material possessed a weak texture of the type $\{11\bar{2}0\}<0001>$ along the longitudinal and transverse directions in the α phase; the dual directionality of this texture occurred as a likely consequence of the cross rolling employed during the sheet manufacture. A similarly weak $\{200\}<100>$ texture was observed in the β phase. Static annealing at 850°C for 47 minutes sharpened these textures considerably. This heat treatment was chosen because it represents period of 30 minutes allowed for temperature equilibration and the duration of the test carried out at an initial strain rate of 1.7×10^{-3} s^{-1} and , therefore, the textures obtained reflect the annealing textures that would be present in the absence of deformation. After superplastic deformation to 200% strain at 1.7×10^{-3} s^{-1} and 850°C the strong texture developed during annealing was randomised with only a small amount of the annealed texture still apparent. After superplastic deformation at 8.3×10^{-5} s^{-1} the annealed texture had been similarly randomised.

4. Discussion

The strain rate cycling tests carried out at 850°C revealed that the initial microstructure of IMI550-1Mo+1Fe sheet alloy, which contained significant amounts of elongated and blocky α, was not ideal for superplastic deformation. During the first strain rate cycles, which amounted to approximately 60% strain, the coarser α features were broken down and a equiaxed $\alpha+\beta$ microstructure, more conducive to superplastic deformation, was evolved. The strain rate sensitivities for the subsequent cycles increased and stabilised with maximum values of 0.55 and 0.61 at a strain rate of approximately 5×10^{-4} s^{-1} for the decreasing and increasing cycles, respectively. The flow stresses increased with strain due to strain enhanced grain growth. However, this did not lead to a reduction in the strain rate sensitivities measured in the latter strain rate cycles. The slight difference between the maximum strain rate sensitivities for the increasing and decreasing cycles can be explained by the occurrence of strain enhanced grain growth since as the test proceeds a finite amount of grain growth will occur at each stage, particularly at low and intermediate strain rates, resulting in a small additional stress between strain rate increments. For a decreasing cycle this will lead to slightly lower strain rate sensitivities and for an increasing cycle to slightly higher strain rate sensitivities, as is the case. The true strain rate sensitivity will be close to the average of the values obtained during increasing and decreasing cycles, giving a maximum of ~0.58 in this case. The log σ-log ε plots, for both increasing an decreasing strain rate cycles, have a sigmoidal form with strain rate sensitivities dropping off at low (region I) and high (region III) strain rates. The occurrence of low strain rate sensitivities in region I have been explained by either the presence of a threshold stress [2] or by concurrent grain growth [9]. It is difficult to see how concurrent grain growth could account for the low strain rate sensitivities encountered in this investigation since grain growth could only account for low strain rate sensitivities on a decreasing cycle. Reductions in strain rate sensitivities in region III are attributed to the onset of dislocation climb controlled creep [10] where m~0.3. Overall the results of the strain rate cycling test have revealed that m ≥ 0.4 across the range of employed strain rates, from ~3×10^{-5} to 3×10^{-3} s^{-1} and, therefore, the IMI550-1Mo+1Fe sheet alloy is superplastic within this range.

The results obtained from the five specimens deformed using the constant cross head procedure across the range of initial strain rates from 8.3×10^{-5} s^{-1} to 4.2×10^{-3} s^{-1} revealed that 200% elongation was achieved in each case with a good resistance to necking. The least amount of necking was associated with the specimens deformed at intermediate strain rates (Region II) where the maximum strain rate sensitivities were measured. The microstructural investigations revealed that the greatest amount of grain growth was associated with the specimens deformed at the lowest strain rates, which was not unexpected since these tests took the longest time to complete and exhibited the greatest amounts of strain hardening. However, it was found that the specimens deformed at intermediate strain rates (region II) exhibited the greatest growth rates. The enhanced grain growth rates in this region could be associated with enhanced diffusion due to dislocation activity, i.e. pipe diffusion, which would, presumably, also benefit the appropriate diffusion coefficient for creep in both the α and β phases.

Deformation at the highest initial strain rate of 4.2×10^{-3} s^{-1} resulted in structural refinement with no increase in the average grain size. This was not entirely unexpected since the occurrence of grain refinement was observed by Lee and Backofen during superplastic deformation at comparably high strain rates and higher temperatures in their pioneering work with Ti-6Al-4V [11]. Further scrutiny of the grain size data given in table 5 reveals that deformation at 4.2×10^{-3} s^{-1} heralds a change in microstructural development. Deformation at the other strain rates, all below 4.2×10^{-3} s^{-1}, resulted in an increase in the mean linear intercept values in the three principal directions. However, although deformation at 4.2×10^{-3} s^{-1} has led to an increase in the grain size in the longitudinal direction (parallel to the tensile axis) the grain sizes in the transverse directions have decreased; this type of behaviour where the change in grain shape reflects the change in specimen shape is consistent with dislocation creep. This together with the observation that m ~ 0.4 at this strain rate suggests that the specimen was indeed deformed in the region where the mode of deformation was in transition from superplastic flow (region II) to dislocation climb controlled creep (region III). An important microstructural characteristic was the appearance of a thin film of β separating some α grain boundaries after deformation. The occurrence intergranular β is a common observation in studies concerning the superplastic deformation of two phase titanium alloys [2,12] and emphasises the important role of the softer β phase in that can act as a deformable layer or mantle accommodating the relative motion between adjacent α grains.

Strain hardening was observed at all strain rates and can be attributed chiefly to concurrent grain growth. However, specimen oxidation, which was evident on all specimens, would contribute to hardening during deformation particularly at low strain rates. Grain elongation and dislocation activity could also result in a degree of hardening and would be of most importance at the highest strain rates.

The texture analyses revealed that static annealing at 850°C prior to deformation led to well defined $\{11\bar{2}0\}<0001>$ and $\{200\}<110>$ type textures in the α and β phases, respectively. A randomisation of texture in the α and β phases occurred as a result of the superplastic deformation at both high and low strain rates within the superplastic range with the degree of randomisation been similar in each case. This reduction of crystallographic texture with superplastic deformation is in agreement with previous studies concerning IMI550 [3]. The most frequently considered mechanisms for superplastic flow all involve grain boundary sliding and grain rotation which is accompanied by an accommodating process such as diffusional flow, for example as proposed by Ashby and Verrall [13], or by a dislocation slip process, for example as proposed by Gifkins [14]. A reduction in texture during superplastic flow would be consistent with these mechanisms.

5.0 Conclusions

1 Alloy modification of IMI550, by replacing 1%Mo with 1%Fe, has reduced the flow stresses required for superplastic flow at 800 and 850°C by more than 20%.
2 Specimens manufactured from the iron modified IMI550 sheet alloy were successfully superplastically deformed at 850°C and strain rates ranging from ~ 2×10^{-5} to 5×10^{-3} s^{-1}.
3 The microstructures after superplastic deformation at all strain rates consisted of fine and generally equiaxed grains of α+β, with some β existing as a thin layer between adjacent α grains. All of the coarser α features had been eliminated.
4 The modified alloy displayed optimum superplasticity at intermediate strain rates of approximately 5×10^{-4} s^{-1} at 850°C. At this strain rate a maximum strain rate sensitivity of 0.58 was determined.
5. The greatest amount of strain enhanced grain growth was observed in specimens deformed at the lowest strain rates. However, this grain growth could not account for the reduction in strain rate sensitivities in region I of the stress-strain rate curve.
6 The greatest grain growth rates occurred in the specimens deformed at intermediate strain rates in region II. This enhanced grain growth could be associated with the dislocation activity in this region, i.e. pipe diffusion, which may in turn benefit superplastic flow.
7 Superplastic deformation resulted in the randomisation of well developed annealing textures in both the α and β phases.
8 The observations are consistent with grain boundary sliding and grain rotation accommodating the superplastic flow.

6. References

1. IMI Titanium 550, Data sheet, IMI Titanium Ltd, PO Box 216, Birmingham 6, UK.
2. J. Ma, R. Kent and C. Hammond, J.Mat.Sci, 1986, (21), 475-487.
3. D.S. McDarmaid, Mat. Sci. and Eng., (70), 1985, 123-129.
4. J.R. Leader, D.F. Neal and C. Hammond, Metall. Trans., (17A), 1986, 93-106.
5. J. Wert and N.E. Paton, Metall. Trans. A, (14A), 1983, 2535-2544.
6. M.L. Meier and A.K. Mukherjee, Sci. Metall., (25), 1991, 1471-1476.
7. N.E. Paton and J.A. Hall, US Patent 4,299,626, Nov. 10, 1981.
8. IMI Titanium Ltd Technical Report, "Production of a titanium alloy with improved SPF at low temperatures", IMI Titanium Ltd, UK.
9. M.L. Meier, D.R. Lesuer and A.K. Mukherjee, Mat. Sci. and Eng., (A136), 1991, 71-78.
10. J. Weertman, J. Appl. Phys., (28), 1957, 362-364.
11. D. Lee and W.A. Backofen, Trans. TMS-AIME, (239), 1967, 1034-1040.
12. M.T. Cope, D.R. Evetts, N. Ridley, Mat. Sci. and Eng., (13) 1987, 455-461.
13. M.F. Ashby and R.A. Verrall, Acta Metall., (21), 1973, 149-163.
14. R.C. Gifkins, J.Mat.Sci., (13), 1978, 1926-1936.

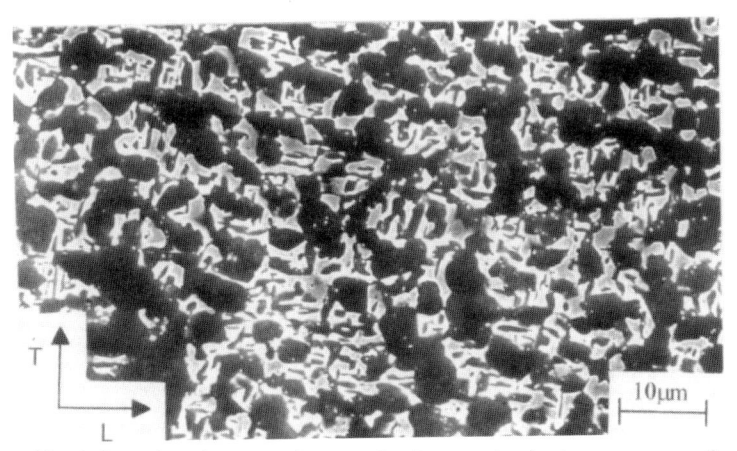

Fig. 1 Scanning electron micrograph of as received microstructure of IMI550-1Mo-1Fe sheet alloy.

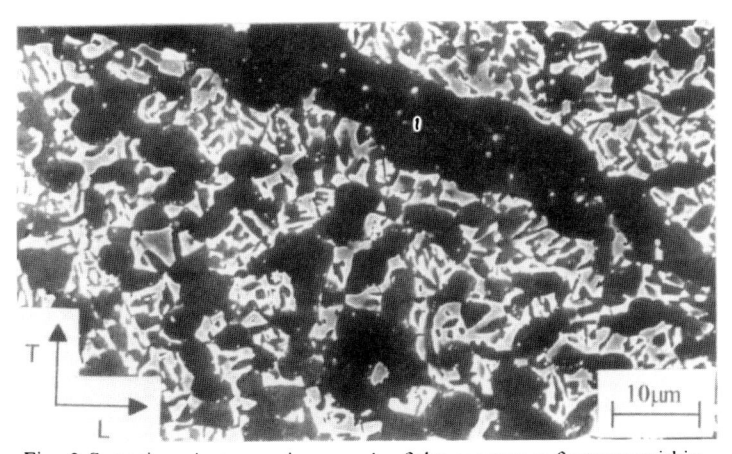

Fig. 2 Scanning electron micrograph of the coarser α features within the as received micostructure of the IMI550-1Mo+1Fe sheet alloy.

170 *Superplasticity: 60 Years after Pearson*

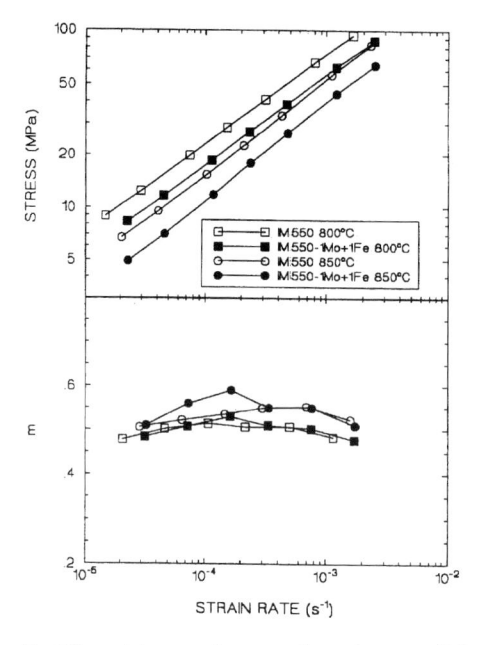

Fig. 3 Stress-strain rate and corresponding strain rate sensitivity data for IMI550 and IMI550-1Mo+1Fe at 800 and 850°C.

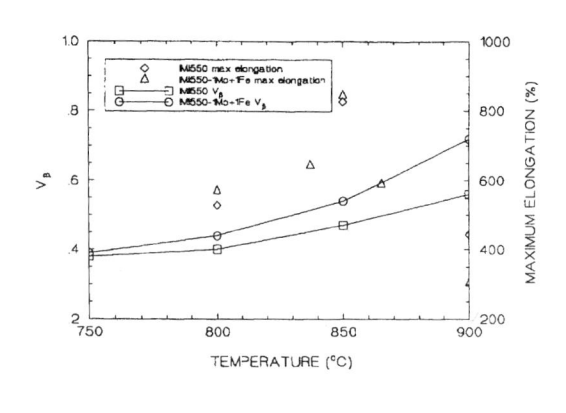

Fig. 4 β Volume fraction and strain to failure data for IMI550 and IMI550-1Mo+1Fe at various temperatures.

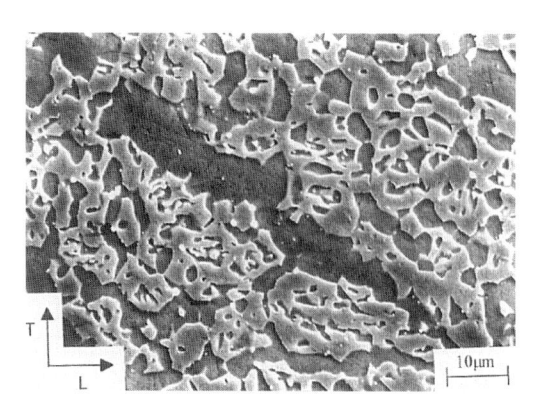

Fig. 5 Scanning electron micrograph of IMI550-1Mo+1Fe sheet after an anneal at 850°C for 30 mins.

Fig. 6 Sheet tensile specimens of IMI550-1Mo+Fe sheet alloy (a) undeformed, (b) deformed to 276% strain by strain rate cycling at 850°C, and (c) deformed to 200% at an initial strain rate of 1.7×10^{-3} s^{-1} and 850°C

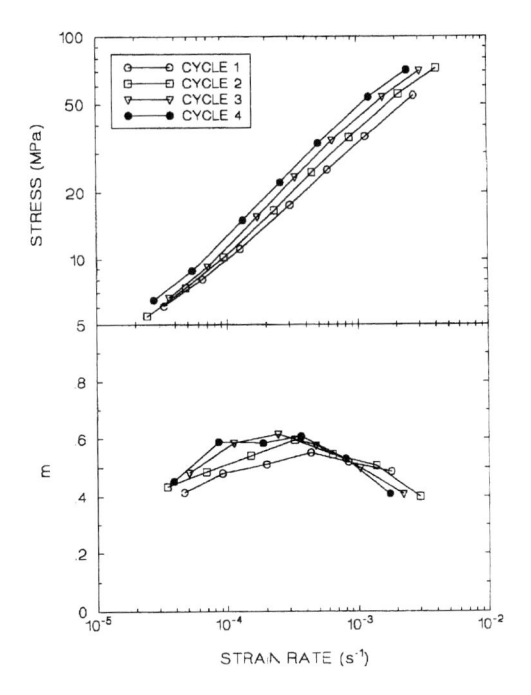

Fig. 7 Stress-strain rate and corresponding strain rate sensitivity data for IMI550-1Mo+1Fe tested at 850°C, only decreasing cycles are plotted.

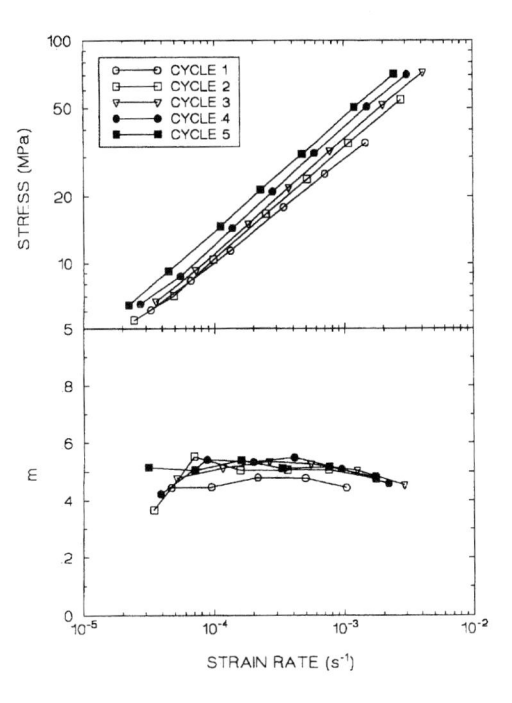

Fig. 8 Stress-strain rate and corresponding strain rate sensitivity data for IMI550-1Mo+1Fe tested at 850°C, only increasing cycles are plotted.

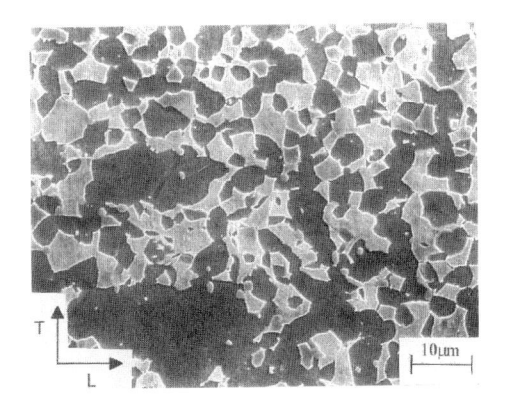

Fig. 9 Scanning electron micrograph of grip section of an IMI550-1Mo-1Fe specimen superplastically deformed by strain rate cycling at 850°C. $D_\alpha = 3.1$ μm, $V_\beta = 0.53$.

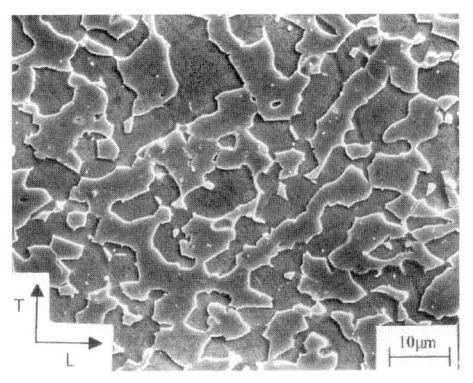

Fig. 10 Scanning electron micrograph of gauge length of an IMI550-1Mo+1Fe specimen superplastically deformed by strain rate cycling at 850°C. $D_\alpha = 3.8$ μm, $V_\beta = 0.54$.

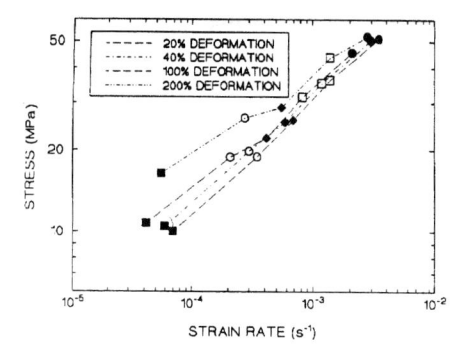

Fig. 11 Stress-strain rate behaviour of specimens deformed at 850°C using the constant cross head speed procedure at initial strain rates of ■ 8.3×10^{-5} s^{-1}, O 4.2×10^{-4} s^{-1}, ◆ 8.3×10^{-4} s^{-1}, □ 1.7×10^{-3} s^{-1} and ● 4.2×10^{-3} s^{-1}.

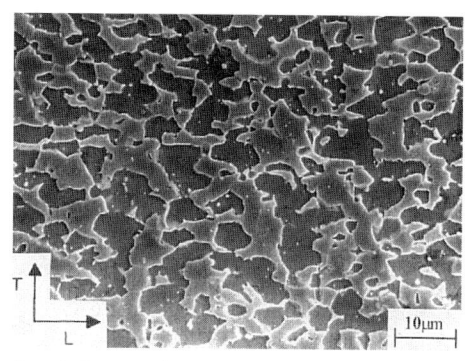

Fig. 12 Scanning electron micrograph of IMI550-1Mo+1Fe superplastically deformed at an initial strain rate of 4.2×10^{-3} s^{-1} and 850°C to 200% strain.

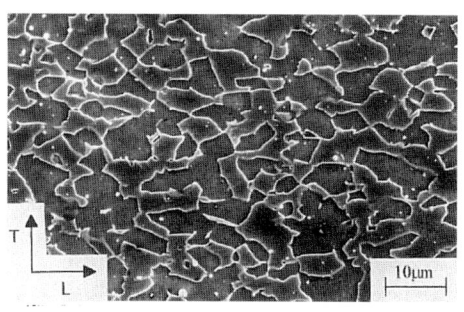

Fig. 13 Scanning electron micrograph of IMI550-1Mo+1Fe superplastically deformed at an initial strain rate of 8.3×10^{-4} s^{-1} and 850°C to 200% strain.

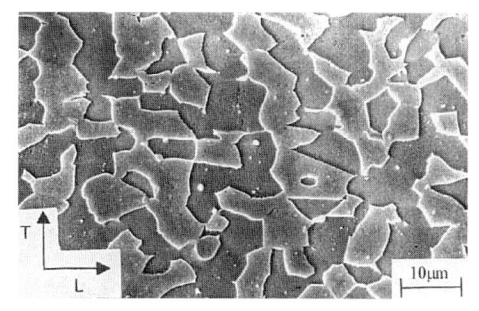

Fig. 14 Scanning electron micrograph of IMI550-1Mo-1Fe superplastically deformed at an initial strain rate of 8.3×10^{-5} s^{-1} and 850°C to 200% strain.

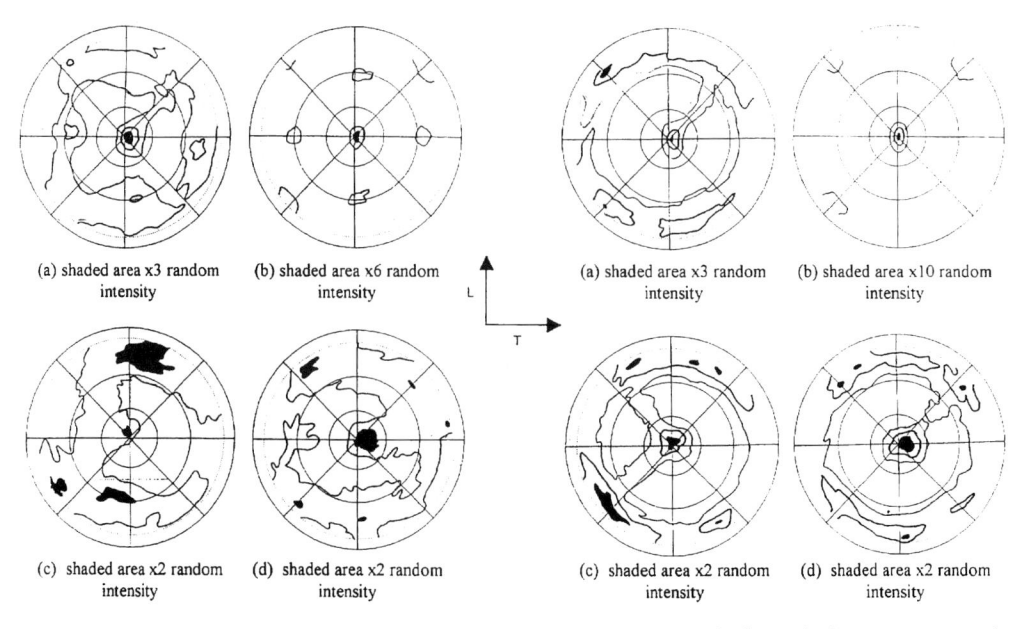

(a) shaded area x3 random intensity
(b) shaded area x6 random intensity
(a) shaded area x3 random intensity
(b) shaded area x10 random intensity

(c) shaded area x2 random intensity
(d) shaded area x2 random intensity
(c) shaded area x2 random intensity
(d) shaded area x2 random intensity

Figure 15. {11$\bar{2}$0} α phase pole figures (a) as received IMI550-1Mo+1Fe sheet, (b) statically annealed at 850°C for 47 minutes, (c) 200% elongation at 1.67×10^{-3} s^{-1} and 850°C, and (d) 200% elongation at 8.33×10^{-5} s^{-1} and 850°C

Figure 16. {200} β phase pole figures (a) as received IMI550-1Mo+1Fe sheet, (b) statically annealed at 850°C for 47 minutes, (c) 200% elongation at 1.67×10^{-3} s^{-1} and 850°C, and (d) 200% elongation at 8.33×10^{-5} s^{-1} and 850°C

Microstructural Evolution of Al-Cu-Zr Alloys During Thermomechanical Processing.

E. Cullen, F.J. Humphreys and N. Ridley
University of Manchester/ U.M.I.S.T.
Material Science Centre, Grosvenor Street, Manchester, M1 7HS

Abstract.

Microstructural analysis and misorientation measurements (via EBSP) have been made on Al-Cu-Zr alloys having a range of Cu and Zr contents. The alloys were given different thermomechanical treatments with prior deformation, temperature and strain as variables.

It was evident that cold/warm working prior to high temperature straining was essential in order to achieve the fine equiaxed grain structure required for superplastic behaviour. Higher Cu and Zr contents and prior cold/warm working both introduced more initial low angle boundaries on static annealing. During concurrent straining the number of high angle boundaries increased with strain whilst the number of low angle boundaries increased initially to a maximum, then declined. The apparent increase in the number of high angle boundaries and simultaneous decrease in the number of low angle boundaries suggests that in-situ strain enhanced recrystallization is important for the development of a fine grain equiaxed structure.

1. Introduction

For an alloy to be superplastic, it must have a uniform, equiaxed fine (<10µm) grain structure which is stable at temperatures > $0.5T_m$ (where T_m is the lowest melting point of its constituents). The required structure is either obtained by static recrystallization prior to superplastic forming (SPF) or it is developed during the initial stages of SPF. Superplastic materials have one of two starting conditions; fully recrystallized or heavily deformed.

Supral, Al-6Cu-0.4Zr, consists of a pseudo single phase microstructure which is made superplastic by the latter production route. The material is essentially a single phase (Al-Cu solid solution) containing some $CuAl_2$ precipitates and a uniform dispersion of very fine (<10nm) particles of $ZrAl_3$. To prevent the introduction of coarse $ZrAl_3$ dispersoids and to achieve a high level of Zr in solid solution, the material is cast at a temperature greater than 780°C and chilled rapidly. Fine $ZrAl_3$ is precipitated out on heating to 360°C, and solutionising at 520°C and subsequent hot rolling takes the majority of the Cu into solid solution. Following break-down rolling, the material is warm/cold rolled to produce a heavily worked structure which evolves to a fine grain structure during the initial stages of SPF.

The purpose of the present work was to investigate the microstructural evolution in Supral type alloys by analysing a range of alloys containing varying amounts of Cu and Zr. The primary aim of the investigation was to ascertain the significance of cold deformation prior to hot compression, and the effects of the compression temperature and compressive strain.

2. Experimental

Four alloys were supplied by Alcan International Ltd. in the form of 10cm diameter DC cast billets. The composition of each alloy is given in Table 1. In the present work these alloys will be designated as listed in Table 2.

Table 1. Chemical compositions of the four alloys.

ALLOY	COMPOSITION (wt%)									
	Si	Fe	Cu	Mn	Mg	Cr	Zn	Ti	B	Zr
D	0.004	<0.001	1.86	<0.001	<0.001	<0.001	0.013	<0.001	<0.001	<0.001
E	0.001	<0.001	3.46	<0.001	<0.001	<0.001	0.013	<0.001	<0.001	<0.001
F	0.006	<0.001	2.01	<0.001	<0.001	<0.001	0.013	<0.001	<0.001	0.29
G	0.021	0.002	3.81	<0.001	<0.001	<0.001	0.014	0.007	<0.001	0.39

Table 2. Designated compositional values.

ALLOY	COMPOSITION (wt%)	
	Cu	Zr
D	2.0	-
E	4.0	-
F	2.0	0.3
G	4.0	0.4

Discs of 10mm and 20mm thickness were sliced from each billet and subjected to a heat treatment of 16hrs at 360°C, air cooled and 2 hrs at 500°C, water quenched. The discs of 20mm in full thickness were cold rolled to a 50% reduction. Compression specimens (10x10x20mm) were removed from both the cold rolled and undeformed discs of each material. Specimens were then uniaxially hot compressed using an Instron 40505 with a three zone split furnace attached, to a nominal strain of 0.5 at 400°C, 450°C and 500°C, respectively, at a strain rate of ~$4 \times 10^{-4} s^{-1}$. To minimize barrelling a high temperature commercial lubricant was used. Optical microscopy at this stage revealed that the ternary alloys, F and G developed a much finer microstructure than the binary alloys D and E; therefore, further experiments involving compression to a range of strains were limited to the two ternary alloys. Each sample took approximately 2 hours to reach the deformation temperature and the average time of the compression test was 30 minutes. In order for a comparison to be made between microstructures which evolved during high temperature straining and those which evolved during annealing in the absence of strain, specimens were heated at the deformation temperature for 2.5 hours. This included the time to heat up the furnace. These specimens also enabled an approximate starting microstructure prior to hot deformation to be established. Compression tests were carried out on alloys F and G using a strain rate of $4 \times 10^{-4} s^{-1}$, to strains of 0.25 and 0.75. After hot deformation all samples were quenched in water within 3 seconds.

A summary of the experimental route for alloys F and G is shown in Figure 1.

Specimens for optical microscopy were cut from the deformed material parallel to the compression direction. They were ground, polished and electropolished in a solution of 10%$HClO_4$ and 90% CH_3OH and then anodized in HFB_4, (7gms Boric acid and 970ml of distilled water). After optical examination, these sections were repolished and electropolished for SEM. Backscattered electrons were used to obtain Kikcuchi patterns (electron backscattered patterns, EBSP). For each sample, the orientations of 100 (sub)grains lying along the rolling direction and 100 (sub)grains normal to the rolling direction were determined and the distance travelled in both directions was recorded. It is important to note that it was only possible to detect misorientations greater than 0.5°, therefore, many subgrains with misorientations less than 0.5° will not have been recorded.

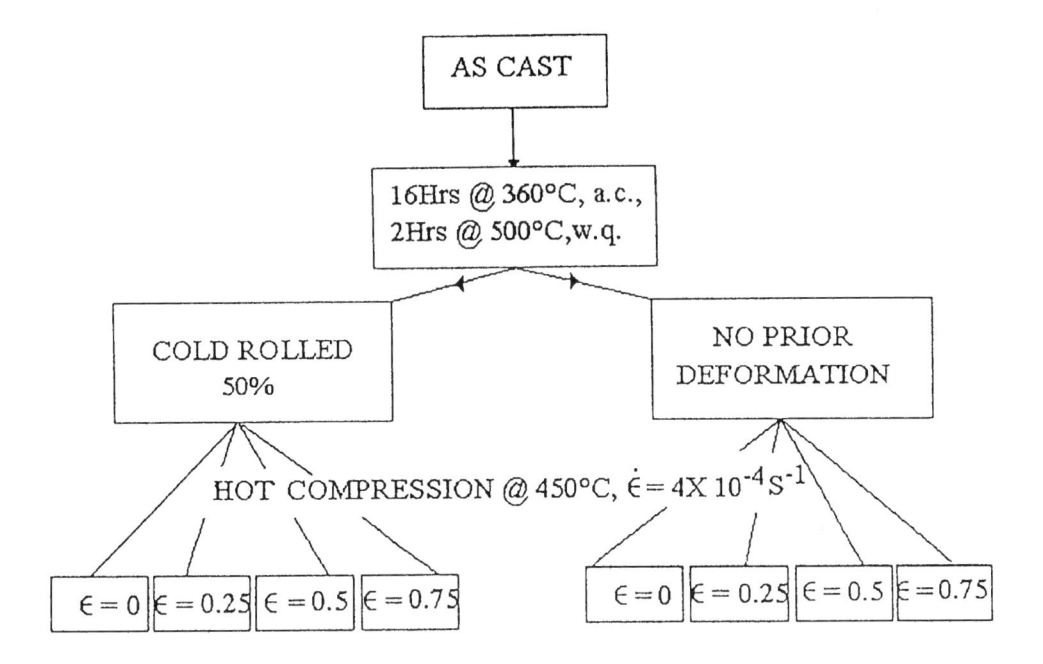

Figure 1. Summary of experimental route.

3. Results

3.1 Effect of composition

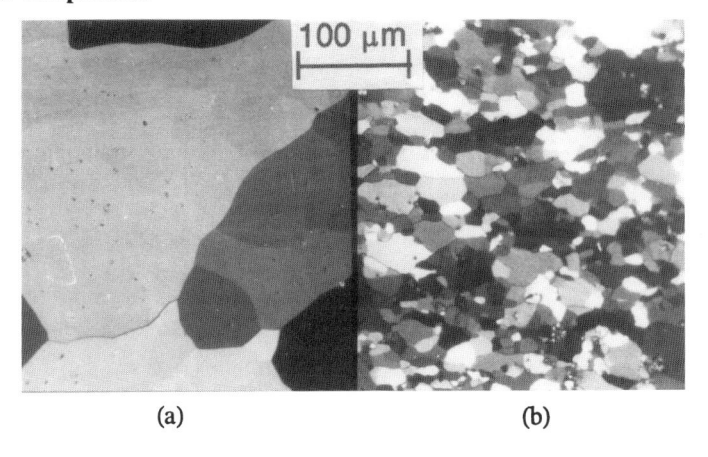

Figure 2. Alloys cold rolled and compressed to a nominal strain of 0.5 at 500°C (a)Al-4Cu and (b) Al-4Cu-0.4Zr

Comparing Figure 2 (a) and (b), it is clear that the addition of Zr to the binary Al-4Cu produces a finer grain structure. A higher Cu and Zr content leads to a larger number of low angle boundaries i.e. subgrain boundaries (figure 3) and also a more refined grain structure as seen in figure 4, with a larger number of high angle boundaries per unit length.

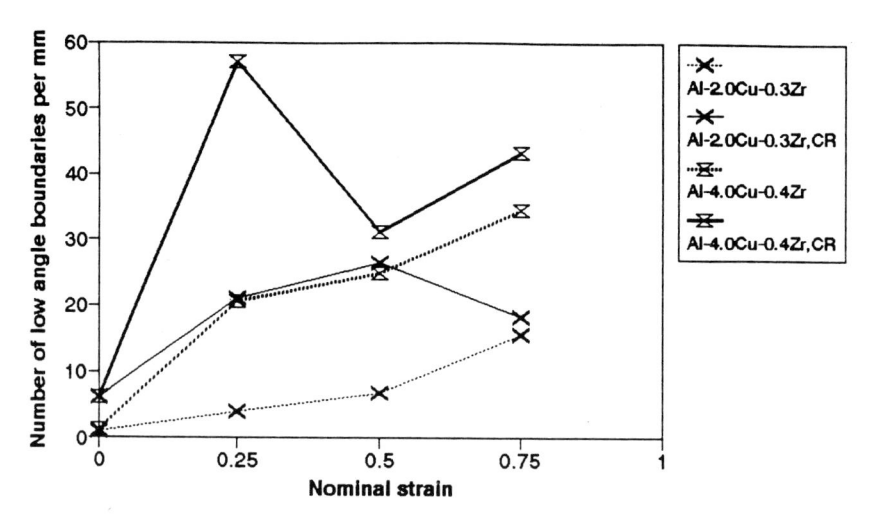

Figure 3. The effect of compressive strain on the number of low angle boundaries ($0.5° < x < 15°$) per unit length.

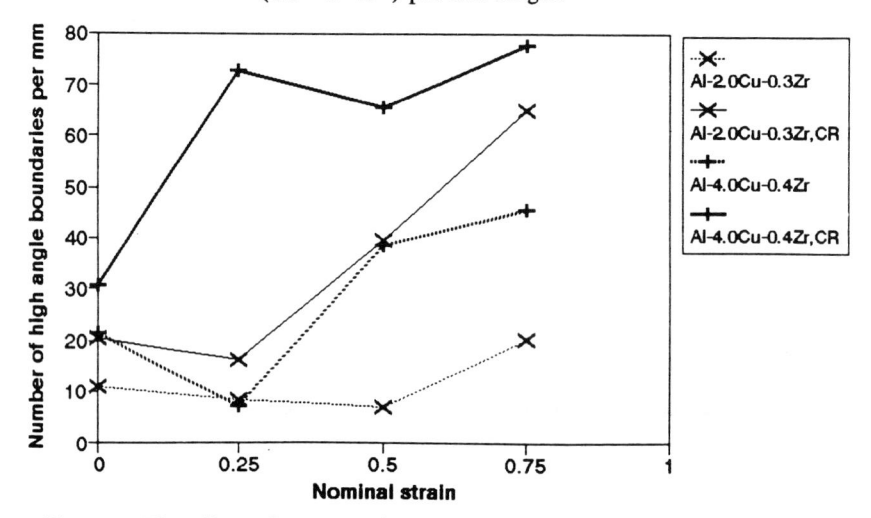

Figure 4. The effect of compressive strain on the number of high angle boundaries ($>15°$) per unit length.

3.2 Effect of temperature

It can be seen from figure 5(a) that hot deformation at 400°C produces flattened, elongated grains. Straining at 500°C (Figure 5b) reveals a larger grain structure which is more equiaxed. This indicates that strain induced boundary migration occurs more readily at 500°C
From the three temperatures used, 450°C was found to give the optimum results producing a more uniform, fine (sub)grain size.

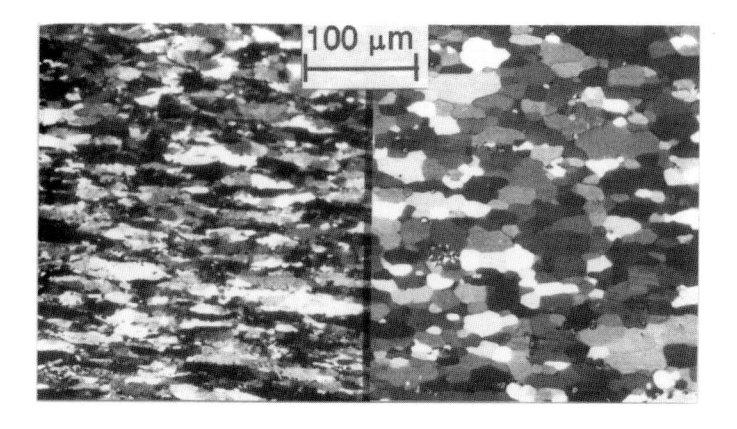

(a) (b)

Figure 5. Al-4Cu-0.4Zr cold rolled, compressed to a nominal strain of
0.75 at a temperature of (a) 400°C and (b) 500°C

3.3 Effect of prior cold work

Optical microscopy revealed that on high temperature compression, Al-4Cu-0.4Zr, without prior cold work, evolved to coarse elongated (sub)grains (figure 6a). In contrast, the same material which had undergone 50% cold work, tends to a smaller, more equiaxed structure (figure 6b). Further examination using EBSP allowed frequency histograms to be obtained (figure 7). The histograms show that cold working prior to hot compression introduces a large number of low angle boundaries. These are still present after static annealing although during hot compression high angle boundaries are formed.

(a) (b)

Figure 6. Al-4Cu-0.4Zr compressed to a nominal strain of 0.5 at 450°C
with (a) no prior deformation and (b) 50% cold rolled

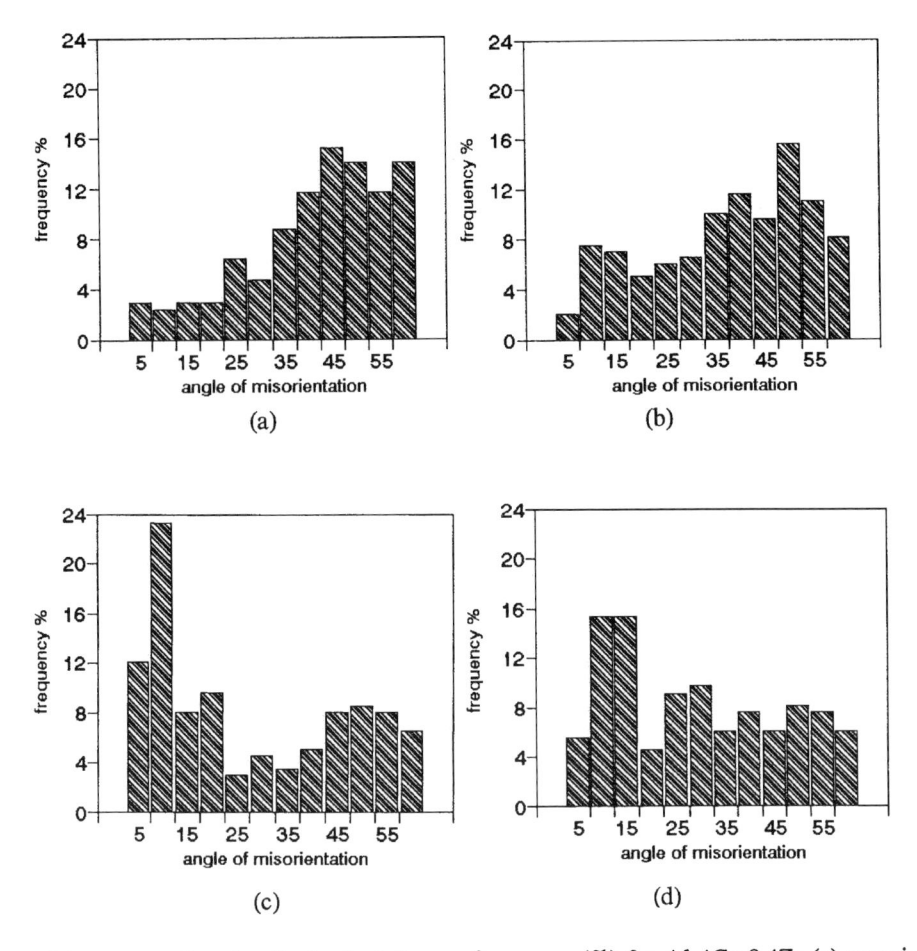

Figure 7. Histograms of angle of misorientation vs frequency (%) for Al-4Cu-0.4Zr (a) no prior deformation, statically annealed at 450°C (b) 50% cold rolled, statically annealed at 450°C, (c) no prior deformation, compressed to a nominal strain of 0.75 at 450°C and (d) 50% cold rolled, compressed to a noimnal strain of 0.75 at 450°C

3.4 Effect of compressive strain

Figure 8. Al-4Cu-0.4Zr, cold rolled 50% and compressed at 450°C to a nominal strain of (a) 0 and (b) 0.75

Optical microscopy reveals that the deformed microstructure is retained on heating a 50% cold rolled specimen of alloy G to a deformation temperature of 450°C. However, straining at 450°C at a strain rate of $4 \times 10^{-4} s^{-1}$ and to a strain of 0.75 produces a fine equiaxed structure that is very uniform. The micrographs, figure 8(a) and (b), together with the measurement of grain boundary misorientations (figure 3 and 4) reveal that the proportions of high angle boundaries increases with increasing strain.

4.0 Discussion.
4.1 Factors affecting microstructural evolution.
During the present investigation, a number of material and processing parameters which are likely to influence the formation of a superplastic microstructure were independently varied.

4.11 Zirconium content
The presence of zirconium in the form of a fine dispersion of $ZrAl_3$ particles can be seen to have a dramatic effect on the development of the microstructure during high temperature straining (fig. 2). In the binary Al-Cu alloy (fig. 2a), very large (146μm) grains, typical of a single phase aluminium alloy develop. However, in the zirconium containing alloy (fig. 2b) the structure is much finer (21μm). The main role of the particles is thought to be the pinning of high and low angle grain boundaries which leads to the prevention of grain or subgrain coarsening during high temperature deformation or annealing. These results are consistent with the earlier work of Watts et al [1] which showed that both copper and zirconium were required if a superplastic microstructure were to be developed.

4.12 Prestrain
Previous work on microstructural evolution in Supral and related alloys has been carried out with a worked microstructure as the starting material. In the present work, by comparing the behaviour of specimens from both the unworked (cast and heat treated) and cold worked material, the role of prior cold (or warm) working on microstructural evolution was investigated. The micrographs of figure 6, and the misorientation results, figures 4 and 7, show that within the range of hot working strains investigated, i.e. less than 0.75 nominal strain, prior deformation is necessary if a fine grained microstructure is to be developed during high temperature straining. The material with no prior straining develops a microstructure which is much coarser (fig 6a) and which has a much higher proportion of low angle grain boundaries (figs 4 and 7). Of course the prior work is only effective if the material does not recrystallize when raised to the hot-working temperature (400°-500°C) and in order to achieve this microstructural stability, it is thought that a dispersion of particles ($ZrAl_3$) is necessary. In order for recrystallization to be inhibited, a balance between pinning and boundary mobility must be retained[2]. Recrystallization is often found to be inhibited when the ratio of volume fraction of particles, F_v, to the mean particle diameter, d, is greater than $0.1 \mu m^{-1}$ [3] and in the Al-4Cu-0.4Zr alloy, F_v/d was found to be $498 \mu m^{-1}$

4.13 Deformation Temperature
The effect of deformation temperature for the Al-4Cu-0.4Zr alloy, given a prior cold reduction of 50% and then deformed to a strain of 0.5 is seen in figures 5a,b and 6b. It may be seen that at the lowest temperature of 400°C, the microstructure remains elongated in the cold working direction. At the optimum temperature of 450°C, a microstructure of small equiaxed grains/subgrains is formed (fig 6b), whereas at 500°C, a much coarser microstructure evolves. The deformation and restoration processes occuring during the high temperature straining are thermally activated. At low temperatures, such processes are slow, and the microstructure retains

many of the features of a deformed material. At the highest temperature, grain boundary mobility is high, and it is clear that substantial dynamic grain/subgrain growth has occured. The optimum microstructure corresponds to the balance between these two factors. The possibility that dynamic grain growth is assisted by the coarsening of the $ZrAl_3$ particles at very high temperatures should also be considered. Higashi and Uno[4] observed that $ZrAl_3$ particles grew more rapidly during hot deforemtion than by static annealing alone. However, the precise role of $ZrAl_3$, such as whether the particles are rendered incoherent as dislocations pass through and if the pinning force diminishes as deformation proceeds, is still unclear.

4.13 High temperature strain
The present work shows that the grain/subgrain shape and the proportion of high angle to low angle boundaries changes progressively with strain. This indicates that the development of a predominantly high angle boundary microstructure from a predominantly low angle boundary microstructure is strain-induced dynamic event, and not merely the static annealing of a cold worked microstructure.

4.20 The mechanism of microstructural evolution
The present work shows that in Al-Cu-Zr alloys which have undergone prior cold working, a fine equiaxed microstructure containing a large proportion of high angle boundaries develops during subsequent hot working at a temperature of 450°C. The investigation of grain and subgrain misorientations (figs 3 and 4) show that in the Al-4Cu-0.4Zr material the total length of high angle boundary increases by a factor of approximately 3, implying the operation of a mechanism whereby high angle boundaries are created during high temperature deformation.

It is reasonable to describe the microstructural evolution in the present work as a form of **dynamic recrystallization.** As the microstructure evolves progressively with strain with no distinguishable nucleation and growth stages, a fuller description would be **continuous dynamic recrystallization.** It should however be noted that this is a phenemenological description which does not imply the operation of any particular micromechanism.

The mechanism of the high temperature deformation is predominantly dislocation glide, with associated dynamic recovery in which dislocations annihilate and also form and are absorbed into low angle grain boundaries. In pure aluminium or binary Al-Cu alloys the mean subgrain misorientation generally reaches a steady state of $< 0.5°$ at low strains, and this steady state is thought to be achieved by the continuing formation and re-formation of the low angle boundaries which have a finite lifetime[5]. The $ZrAl_3$ particles undoubtedly pin the boundaries, and it is likely that this prevents much of the boundary migration necessary for boundary annihilation. Therefore the lifetime of a boundary is greater, and this leads to both a finer substructure and to greater misorientation due to dislocation accumulation by the boundaries. The importance of copper in solid solution in developing this microstructure has already been demonstrated[1], although its precise role is not known. However, it is known that other solutes such as magnesium and zinc in aluminium lead to the development of misorientations during tha high temperature deformation of aluminium, particularly adjacent to grain boundaries[6,7], and it has been suggested[7] that this is achieved by a strain induced subgrain rotation mechanism similar to that which occurs in many minerals.

The stress difference across a grain boundary may cause adjacent grains to slide relative to each other. It is known that boundaries as low as 7° can undergo grain boundary sliding.[8,9] However, the rate of sliding of such low angle boundaries is much slower than for high angle boundaries, and such a mechanism is unlikely to be important in the early stages of the evolution of the microstructure. It is well established that during superplastic deformation, when **grain boundary sliding** becomes a dominant mechanism, then grain rotation does occur. Dislocation and diffusional models for the accomodation of grain boundary sliding are discussed by Pilling and Ridley[10]. However, it is still not clear if such a mechanism contributes to the increase in misorientations found in the present work, as has previously been suggested[8,11,12,13]

As discussed above, the existence of a deformed/recovered substructure is important in the development of the fine-grained microstructure. There are indications, e.g. figure 4, that even in the absence of prior deformation, there is an increase in the high angle boundary content during high temperature straining. Therefore it may well be that the prior cold deformation merely **accelerates** the continuous dynamic recrystallization rather than leading to the operation of a different mechanism.

5.0 Conclusions

1. Cu and Zr are both required to produce a fine grained equiaxed microstructure.
2. Deformation to produce a worked structure prior to hot compression is essential to produce a fine, equiaxed microstructure. It was observed that 50% prior cold work increased the number of high angle boundaries per unit length.
3. The temperature at which SPF is carried out is important. It was found that a deformation temperature of 450°C gave the optimum microstructure.
4. As the compressive strain increased during hot deformation, the number of high angle boundaries increased to a plateau and low angle boundaries increased initially then declined. An initial deformed structure evolved to a uniform, fine equiaxed grain structure during high temperature straining.
5. Although several mechanisms have been proposed, experimental evidence is inadequate and further investigation is required to fully understand the mechanisms of strain enhanced recrystallization.

6.0 References

1. B.M. Watts, M.J.Stowell, B.L. Baikie & D.G.E. Owen, Met. Sci. June 1976, 189-197

2. S.K. Chang, Mat. Sci & Tech., Vol. 8, 1992, 760

3. F.J. Humphreys, Acta Met. 25, 1977, 1323

4. K. Higashi, M. Uno, S. Matsuda, T. Ito & S. Tanimura, Recrystallization '90, Edited by T. Chandra, The Minerals, Metals & Materials Society, 1990

5. H.J. McQueen and J.J. Jonas, Treatise on Material Science and Technology, 6, Edited by Arsenault Academic Press, N. York, 1975, 393

6. K.J. Gardener and R. Grimes, Metal Sci., 13,1979, 216

7. M.D. Drury and F.J. Humphreys, Acta Met., 34, 1986, 2259

8. S.J. Hales & T.R. McNelley, Acta Met., Vol. 36, No. 5, 1988, 1229-1239

9. J. Lui & Dhrubra J. Chakrabarti, Microstructure Evolution & Micromechanisms of Superplasticity in High Strength Al-Zn-Mg-Cu alloys, 4th International Conference on Al Alloys, 1994

10. J. Pilling & N. Ridley, Superplasticity in Crystalline Solids, The Institute of Metals, Camelot Press plc, 1989, p73

11. H. Gudmundsson, D. Brooks & J.A. Wert, Acta Metall.Mater., Vol. 39, No. 1, 1991, 19-35

12. L. Qing, H.Xiaoxu, Y. Meu & Y. Jin Fang, Acta Metall.Mater., Vol. 40, No. 7, 1992, 1755-1762

13. M.T. Lyttle & J. A. Wert, J.O.M.S, 29, 1994, 3442-3350

Acknowledgements

The work was carried out with the support of an EPSRC Case award (E.C.) in collaboration with Alcan International Ltd. The authors would like to thank Dr. G. Mahon and Dr. M. Stowell for valuable discussions.

Macro and Microtexture development during the Superplastic deformation of AA8090

P.L. Blackwell and P.S. Bate
IRC in Materials, University of Birmingham, B15 2TT, U.K.

Abstract

A comparison is made between the results obtained from X-ray macrotexture measurements and microtexture measurements obtained using Electron Backscatter Diffraction (EBSD) on a series of sections of AA8090 (Al-Li-Cu-Mg-Zr). The AA8090 had been processed such that structural development took place by what has been termed 'strain-assisted continuous recrystallisation'. A series of tests were carried out under optimum superplastic conditions to examine the microstructural development as a function of strain with the intention of gaining an enhanced understanding of the deformation mechanism of this material. Macrotexture measurements made previously using X-ray diffraction had revealed the overall behaviour of the crystallites within the polycrystalline array during superplastic deformation, but gave no detail regarding local changes in texture. EBSD was used here to complement the X-ray data by providing information regarding local texture changes on a grain to grain basis.

1. Introduction.

The present Conference on the subject of Superplasticity was organised partly in celebration of the 60th anniversary since one of the seminal papers on the subject was published by Pearson [1]. Since that time much research and development has taken place in order to try to understand the factors that control and influence the phenomenon. Part of the motivation for this work has been that through understanding the mechanisms that control superplastic deformation then it may be possible to design improved alloys and to optimise existing alloys, and processes, that already find commercial application.

The material under investigation for this project was an Aluminium - Lithium alloy : AA8090. This was produced by Alcan Plate Ltd. at their Kitts Green plant. The alloy AA8090 contains approximately 2.5 wt.% Li, the presence of which reduces the density of the alloy by about 10% as compared to AA2014 against which it was designed to compete. Applications for AA8090 have been mainly in the field of transport, primarily aerospace - interested readers are referred to volume two of the proceedings of the 6th International Aluminium -Lithium Conference [2].

Some of the earliest work on the effect of superplastic deformation on crystallographic texture was carried out by Matsuki, Morita, Yamada and Murakami [3]. They studied texture evolution in a Al-Zn-Mg-Zr alloy which had been processed to give a recrystallised grain size of about 10μm. Testing was carried out at a number of strain rates and it was found that at strain rates which generated high strain rate sensitivity exponents (i.e. 0.5), there was a tendency for the preexisting texture to become progressively reduced in intensity with increasing strain. This type of effect has also been seen in Al-Li alloys [4] which, like Matsuki et al's material, are pseudo single phase (the grain size being stabilised by the presence of a fine dispersion of $AlZr_3$), and Ti-6Al-4V [5] which is a dual phase material. This tendency for the texture to become more random following deformation under optimum superplastic conditions indicates that grain rotation must take place, with the grains rotating independently of each other. This is usually assumed to occur by a process of grain boundary sliding i.e. by some type of viscous shear taking place at the grain boundary or in a region close to the boundary [6].

Regarding superplasticity in aluminium alloys, two broad approaches have been adopted with respect to initial microstructures. The first approach has been to commence deformation with a fully recrystallised grain structure, this would typically contain a predominance of high angle boundaries which, supposedly, should facilitate grain boundary sliding. The second method is to begin with an unrecrystallised semi-recovered microstructure which contains many low-angle boundaries and which may not initially be superplastic. In the latter case a process termed "strain assisted continuous recrystallisation" takes place [7]. At low strains the material has been reported to exhibit a relatively high work hardening rate, this stabilises the material against plastic instability and thus prevents premature necking at low strains. As straining continues the work hardening rate falls but this is compensated for by a rising strain rate sensitivity [8]. Microstructurally, the average boundary misorientation has been found to increase with increasing superplastic strain and a point is reached at which there exist sufficient high angle boundaries to support the imposed strain rate via grain boundary sliding [9]. The material is then superplastic in the accepted sense.

2. Experimental Procedure.

The alloy used for this work had the following composition : Al-2.44Li-1.19Cu-0.67Mg-0.12Zr (wt.%). This had been rolled from 12.5mm plate to 2mm sheet and had been processed such that it contained a non-recrystallised, recovered microstructure with an average subgrain size of approximately 3µm. Testpieces were taken from the sheet and tested in uniaxial tension at a temperature of 800K and a strain rate of $10^{-3}s^{-1}$. Further details on the process route used and the test procedures adopted were given in [4]. This combination of temperature and strain rate is within the so-called Region 2 for this material. Samples were tested to various strains and sections were then taken and prepared for examination using Electron Backscatter Diffraction (EBSD). The preparation for EBSD involved grinding and polishing, finishing using colloidal silica, followed by a light etch in Kellers reagent.

EBSD is a technique that has become more widely available over the last few years, it is carried out using a suitably equipped scanning electron microscope and provides orientation information on a grain to grain basis and may thus be used to study texture changes within discrete areas of material microstructures. A review of the technique is given in [10].

Previous research using X-ray diffraction had shown that this material exhibited a texture gradient from the surface to the centre of the sheet [4]. EBSD was therefore carried out (in the longitudinal-short transverse direction) both at the sheet centre and at a point 0.15mm in from the surface. Photographs were taken from the regions which had been examined and the grain structures were converted into a digital form so that the spatial information regarding grain shape and position could be combined with the orientation data to produce 'orientation images' of the microstructure. These are microstructural images which contain orientation information. In-house software was used to produce projections showing the distributions of the <100> directions and to calculate the misorientation distributions.

3. Results.

Figure 1 shows the distribution of <100> directions obtained, using EBSD, from the centre and edge positions. In each case approximately 100 grains was examined. The upper two figures were obtained from the start material and, as noted above, there was a marked difference in the textures present at these two positions. The central region was found to contain the sharper texture which lay close to {110} <112>, this is often referred to as the "brass orientation". At the surface the texture approximated to (001) [1$\bar{1}$0]. The effect of superplastic strain is shown in Figure 1(c) and (d), which were obtained following deformation to a strain of 0.72. It will be seen that in the outer layers of the material the initial texture had virtually disappeared, while in the central region there was still some evidence of the initial texture being retained, albeit much less pronounced.

The results from the analyses of the misorientation distribution are shown in Figure 2 and Figure 3. Here the X axis shows ω, the misorientation angle (in degrees), while the Y axis shows $L/L_{tot}.d\omega$, where L is the length of grain boundary that lies within a particular misorientation angular range; $d\omega$. L_{tot} is the total length of grain boundary measured in the area examined for each specimen.

Figure 2(a) illustrates the initial distribution existing at the edge region of the sheet, the smooth curve superimposed on this figure shows how the distribution would appear for a randomly textured material. With increasing strain (Figure 2(b) and (c)) the number of higher angle

boundaries increases and the overall distribution approaches that for a randomly oriented material.
A slightly different situation occurred at the centre of the sheet. The initial spread of misorientations in the undeformed material is given in Figure 3(a). Here there was a marked peak at low misorientations, with very few boundaries of intermediate angles, and a second, more modest, peak at misorientation angles above about 50°. Figure 3(b) shows the situation at a strain of 0.5 and Figure 3(c) the distribution at a strain of 0.9. The continuous lines in Figures 3(a) and (b) show the distributions calculated from results previously obtained using X-ray diffraction [4]. These show the same general trends although clearly there were differences in detail, which will be commented on in the next section.
Superplastic deformation often produces enhanced rates of grain growth [11]. It was possible to measure the grain sizes from the scanning electron micrographs, the result being shown in Figure 4. Measurements carried out on the static grain growth rates for this material indicated that the grain size was essentially stable at this temperature, the grain growth measured was therefore entirely dynamic. The second curve on this figure indicates the grain growth that would be required in order that grain growth alone were sufficient to balance the strain being imposed on the testpiece. As can be seen the result obtained was in excess of this over the range of strains used.
Figures 5 shows orientation images illustrating the orientation distribution present at the central region of the sheet at strains of 0, 0.5 and 0.72. In these figures the different shades represent levels of misorientation from some reference texture component, in this case {110} <112>. In the undeformed material it was apparent that the (sub-) grains existed as colonies having similar orientations which were elongated in the rolling direction. The {110} <112> texture consists of two components : (110) [1$\bar{1}$2] and (101) [1$\bar{2}$1] which are related by a 60° rotation. Both components were represented here as illustrated, with one component appearing light and the other one dark. At a strain of 0.5 (fig. 5(b)) it was apparent that the banded structure had begun to break up though it was still very evident. By a strain of 0.72 (fig. 5(c)) the break up of the structure had proceeded a stage further but was still by no means complete.

4. Discussion.

The texture components revealed by the EBSD investigation agree well with previous work carried out using X-ray diffraction (XRD). Each method has it's strengths and weaknesses. Depending on the grain size of the material being investigated, XRD may sample data from many hundreds of grains simultaneously and is fairly efficient in terms of the time required to obtain the basic orientation data. Conventional XRD does not however provide any spatial information e.g. regarding any tendencies for grains to exist in colonies, or for grains to rotate as groups - as some theories for superplasticity have suggested. In this instance it appeared that, using EBSD, a sample of 100 grains was sufficient to provide overall textures similar to those seen using XRD. The (001) [1$\bar{1}$0] texture seen at the surface is sometimes referred to as the 'shear' texture and is associated with shearing effects which occur between the sheet and the rolls during rolling. The {110} <112> texture seen in the central regions is often developed in metals or alloys having a face centred cubic crystal structure and fairly low stacking fault energies. It is unlikely that the presence of the levels of lithium and copper etc. in the present alloy would, in themselves, be sufficient to reduce the stacking fault energy of aluminium enough to generate such a strong {110} <112> texture. A number of proposals have been made to try to explain why such a texture is generated [12], but none of these are entirely satisfactory.
The common feature of both misorientation distributions was that initially there existed a predominance of boundaries having relatively low misorientations. This was predictable given that the processing route was controlled to suppress recrystallisation. The misorientation spread in the centre of the sheet was bimodal. This was reflected in the microstructure in this region which as shown in Figure 5(a), consisted of grains elongated in the rolling direction, separated by high angle boundaries, which were divided into colonies of subgrains by low angle boundaries oriented normal to the rolling direction. The break up of the banded structure reflected what had been observed previously using optical microscopy and polarised light [4]. The fact that the banding tended to persist to strains as high as 0.72 (fig. 5 (c)) suggested that a grain

switching type mechanism was unlikely; if such a mechanism were occurring then the banding should have shown evidence of disappearing in a more uniform manner.

It was interesting to note that by a strain of 0.9 both edge and centre misorientation populations were close to those that would be expected for a material having a random texture. It would appear that the rates at which the two regions misorientation populations approached this random situation were not dissimilar, which is somewhat surprising, given a grain boundary sliding hypothesis and the rather different starting points in terms of the original distributions.

The continuous curves shown in Figure 3 were obtained previously using X-ray diffraction. These show similar trends to those obtained using EBSD although, as was observed in the Results, there were differences in detail. As has been discussed, the X-ray data does not make any spatial differentiation at the level of the basic microstructural unit, be it a grain or subgrain. The method used to obtain the X-ray misorientation functions was detailed in [4], it basically depended on a statistical analysis which assumed that there existed, firstly, no correlation between the spatial position and orientation of grains and, secondly, that there was no preferred spatial distribution of grain boundaries. In this instance neither of these assumptions was entirely true, in fact in the central region of the sheet both were violated, this was particularly true in the undeformed material. In Figure 3(b) it will be noted that the two distributions were more closely comparable, this can be attributed to the fact that by a strain of 0.75 the increasing randomisation of the texture was making the above assumptions more reasonable.

Figure 4 illustrates an interesting point. Many models which attempt to describe the relative movement of grains during superplastic deformation have little regard for the effects of strain enhanced grain growth. What Figure 4 shows is that in this instance there was no need to invoke mechanisms such as grain translation or grain emergence to explain the microstructural behaviour. The grain growth which took place was such that it more than balanced the elongation which occurred during tensile testing. This does not necessarily prove that grain boundary sliding did not occur in some form, since clearly there must have existed some mechanism by which the grains were able to rotate with respect to each other in a random way. Grain boundary sliding would provide such a mechanism, but the question arises as to what it is that drives the grains to rotate as they do. The fact that the grains rotate in a non-cooperative way suggests that the applied stress does not, directly, provide the driving force for the rotation. It is possible, since a grain will be surrounded by boundaries having different misorientations, and given that sliding rates are thought to be influenced by this [13], that this could provide a net torque on the grain as the assembly of grains of which it is a part is forced to change shape.

Another possibility is that a limited form of slip takes place, perhaps just on one or two slip planes, this could certainly generate grain rotation. Evidence for the occurrence of such limited slip has recently been found in a superplastic Al-Mg alloy [14]. It is often said that slip plays little or no part in superplastic deformation, except perhaps as an accommodation mechanism to facilitate boundary sliding. It may however be that in fact some process, such as grain boundary sliding, functions as an accommodation mechanism for such a restricted type of slip. To produce a random grain rotation however the slip would have to be inhomogeneous. Blackwell & Bate [4], following the ideas of Raj & Ghosh [15], suggested that if inhomogeneous slip tended to occur in the larger grains only, with diffusional flow in the remainder then this could provide a possible mechanism. Li et al [14] used an *in-situ* quenching technique, followed by ageing under load, to preserve the dislocation structures existing during superplastic deformation. It was observed that there was a tendency, within the range of temperatures and strain rates at which the rate sensitivity exponent exceeded 0.5, for dislocations to appear in the larger section grains only. Often, in grains containing dislocations, it was found that the dislocations all had the same Burgers vector i.e. slip was indeed inhomogeneous. The basic criteria were therefore fulfilled and it may be that such a process could explain, in part at least, the occurrence of textural randomisation following superplastic deformation.

5. Conclusions.

1. It has been shown that there was a good agreement between the results generated by EBSD and those found previously using a mixture of X-ray diffraction and optical microscopy.

2. A sample size of 100 grains for EBSD orientation analysis proved sufficient to generate results for overall texture that were closely comparable to those obtained using X-ray diffraction.

3. Given that EBSD has the capability to relate individual grain orientations and boundary misorientations to the microstructural topology, then the technique is a powerful tool for investigating microstructural evolution following superplastic deformation.

References.

[1] C.E. Pearson, J. Inst. Metals, Vol. 54, 1934, 111-124.
[2] Proceedings of the Sixth International Aluminium-Lithium Conference, 1991, Garmish-Partenkirchen (FRG), Eds. M. Peters & P-J. Winkler, Pub. DGM Informationsgesellschaft mbH, 1992, 1259-1394.
[3] K. Matsuki, H. Morita, M. Yamada & Y. Murakami, Met. Sci. J., Vol. 11, 1977, 156-163.
[4] P.L. Blackwell & P.S. Bate, Met. Trans. A., Vol. 24, 1993, 1085-1093.
[5] P.G. Partridge, A.W. Bowen, C.D. Ingelbrecht & D.S. McDarmaid, Int. Conf. on Superplasticity, Eds. B. Baudelet & M. Suery, Grenoble, Sept. 1985, Editions du C.N.R.S., Paris, 1985, 10.1-10.4.
[6] R.C. Gifkins, Superplastic Forming of Structural Alloys, Eds. N.E. Paton & C.H. Hamilton, AIME, New York, 1982, 4-26.
[7] R.A. Ricks & P-J. Winkler, Proceedings of the Sixth International Aluminium-Lithium Conference, 1991, Garmish-Partenkirchen (FRG), Eds. M. Peters & P-J. Winkler, Pub. DGM Informationsgesellschaft mbH, 1992, 1035-1046.
[8] B.A. Ash & C.H. Hamilton, Scripta Metall., Vol. 22, 1988, 277-282.
[9] A.J. Shakesheff, D.S. McDarmaid & P.J. Gregson, Technical Report Number 91033, Defence Research Agency, RAE, Farnborough, HMSO London, 1991.
[10] V. Randle, Microtexture Determination and its applications, Institute of Materials, 1992.
[11] D.S. Wilkinson & C.H. Caceres, Acta Metall., Vol. 32, 9, 1984, 1335-1345.
[12] A.W. Bowen, Mats. Science & Technology, Vol. 6, 1990, 1058-1071.
[13] F. Weinberg, Trans. TMS-AIME, Vol. 212, 1958, 808-817.
[14] F. Li, W. T. Roberts & P.S. Bate, Scripta Metall. et Mat., Vol. 29, 1993, 875-880.
[15] R. Raj & A.K. Ghosh, Acta Metall., Vol. 29, 1981, 283-292.

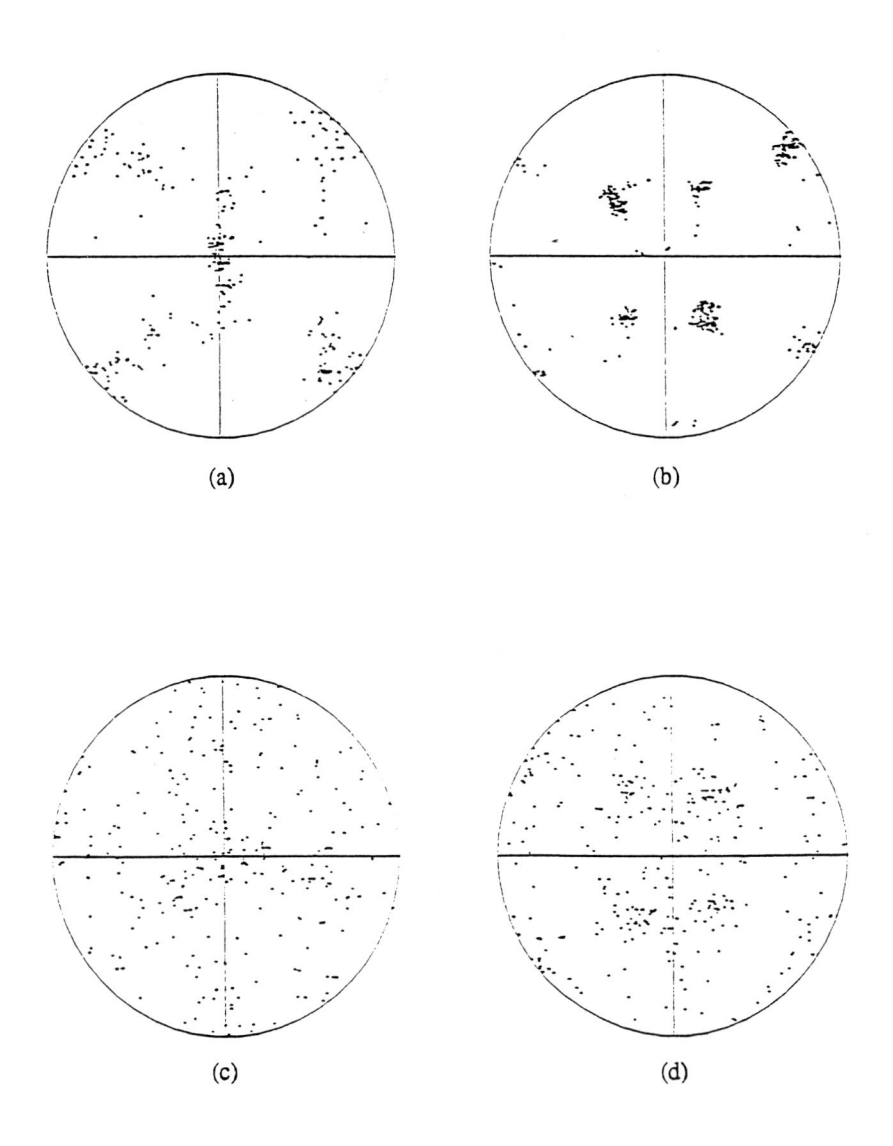

Figure 1. The distribution of <100> directions determined by EBSD. Shown as stereographic projections, they are analogous to 200 pole figures measured using XRD. Figure 1(a) and (b) were obtained from the undeformed sheet in the edge and centre regions respectively. Figures (c) and (d) show the result following deformation to a strain of 0.72, again at the edge and centre locations respectively.

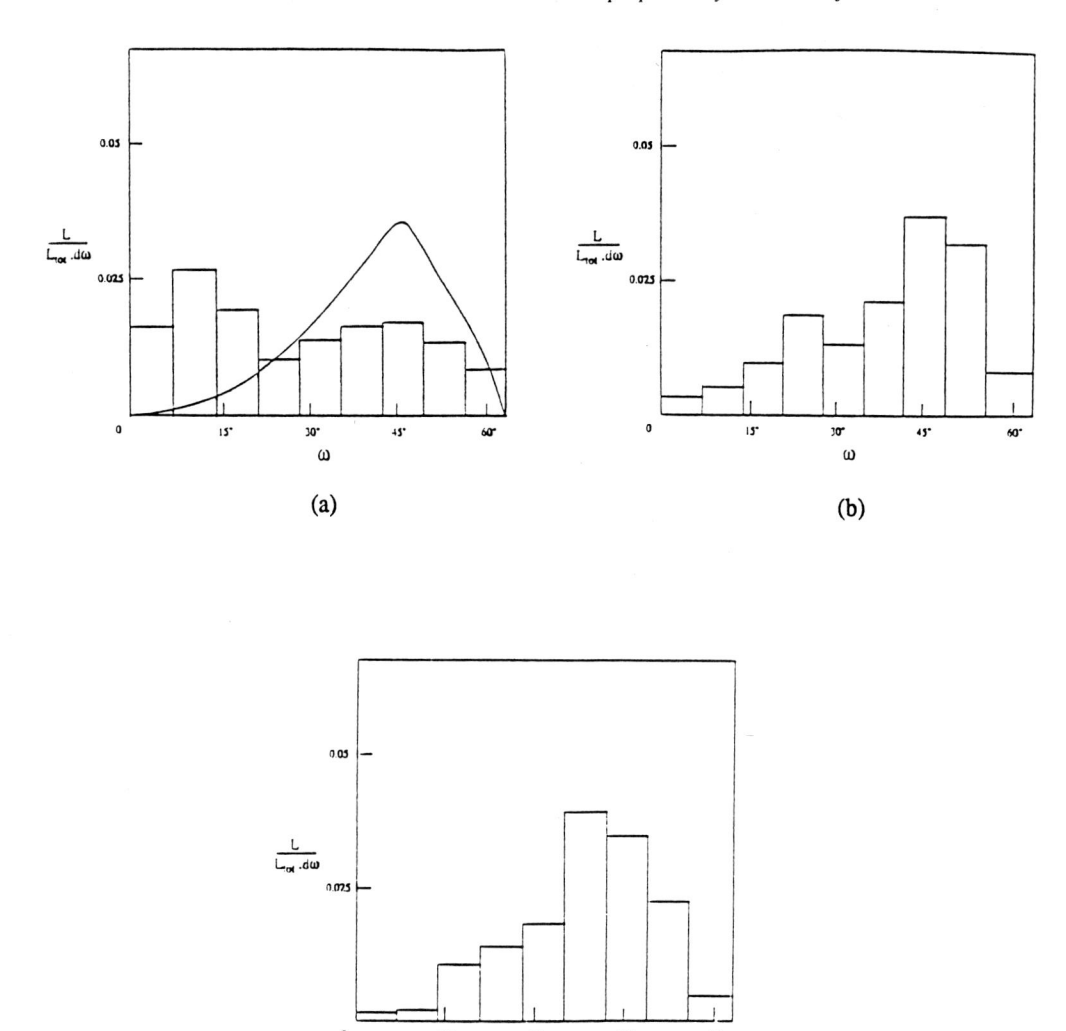

Figure 2. Misorientation distributions from the edge region of the sheet: (a) shows the original distribution at $\varepsilon = 0$. Superimposed on this is the distribution that would be obtained if two samples having random textures were compared and the relative level of misorientation determined as a function of ω. Parts (b) and (c) of this figure show the results at strains of 0.5 and 0.9 respectively.

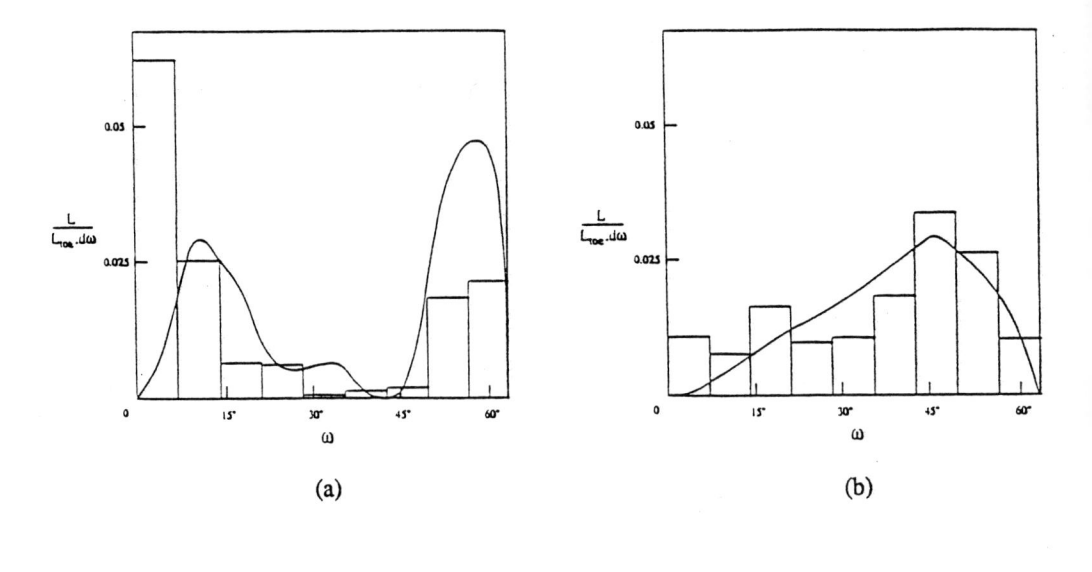

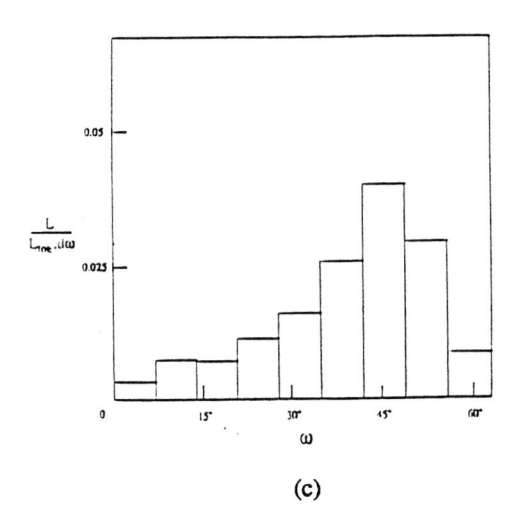

Figure 3. Misorientation distribution at the centre of the sheet: at (a) $\varepsilon = 0$, (b) $\varepsilon = 0.75$ and (c) $\varepsilon = 0.9$. The continuous curves in 3(a) and 3(b) were the results obtained previously using XRD [4].

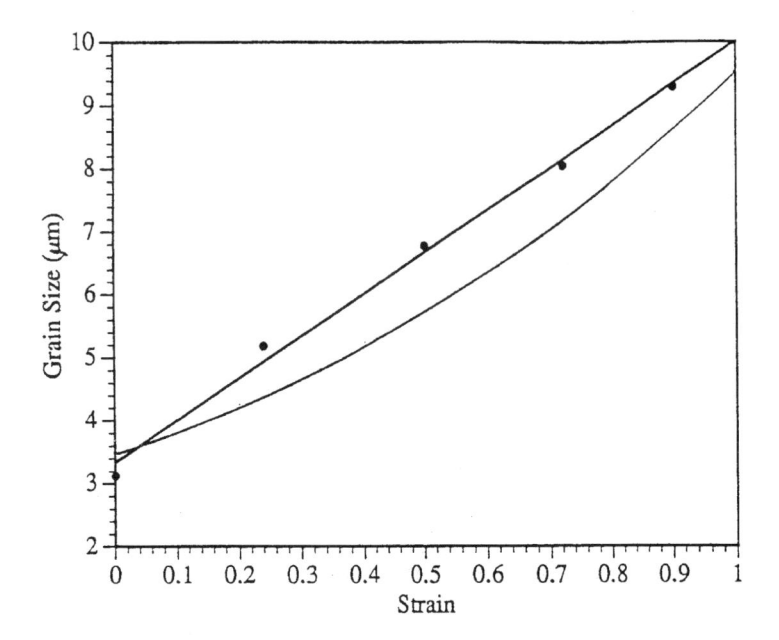

Figure 4. Grain growth as a function of strain. The upper line shows the experimental result, while the lower line is a plot of : $d_i = d_o.e^\varepsilon$, where d_i is the current grain size at strain ε, and d_o the initial grain size (taken as 3.5µm). The lower line shows the grain size required to balance the strain by grain growth alone.

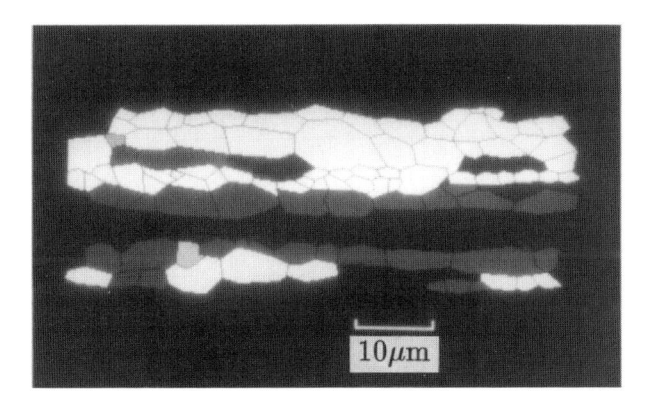

Figure 5(a)

Figure 5. Orientation images taken from the central region of the sheet: (a) at $\varepsilon = 0$, (b) at $\varepsilon = 0.5$, and (c) at $\varepsilon = 0.72$. Here the different colours reflect the levels of misorientation with respect to one of the two components of the brass texture detailed in the Results. The colour scale was as follows:
Dark Grey for $\omega \leq 20°$
Mid Grey for $20° > \omega < 50°$
White for $\omega \geq 50°$

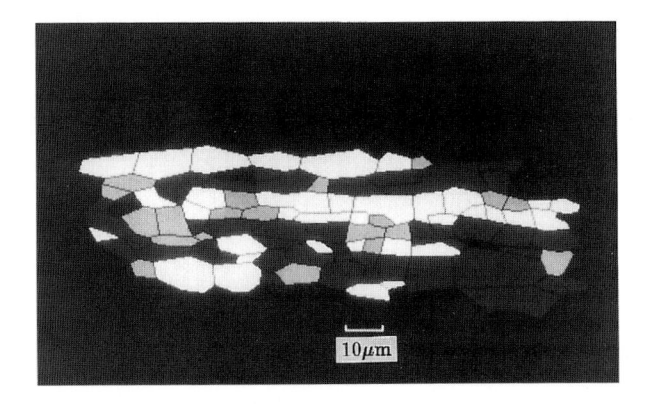

Figure 5(b)

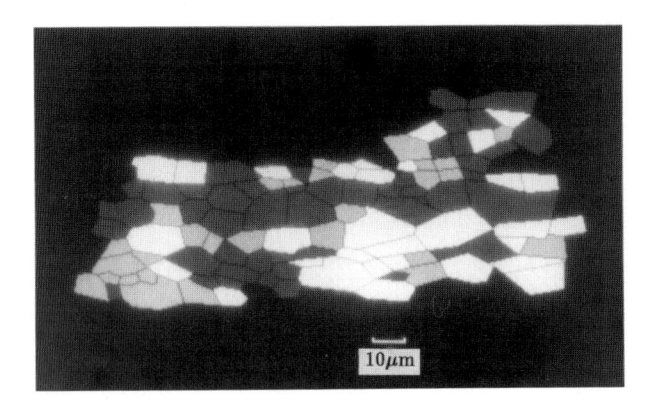

Figure 5(c)

SUPERPLASTICITY IN ALUMINIUM ALLOYS IN RECRYSTALLIZED AND UNRECRYSTALLIZED CONDITIONS

A.A.Alalykin, I.I.Novikov, V.K.Portnoy, V.I.Pavlov
Moscow Steel and Alloys Institute
Moscow, Russia

ABSTRACT

The unrecrystallized and recrystallized alloys exhibit equal values of the strain rate sensitivity and have similar flow stress-strain rate curves. But analyses of the tensile tests data reveal different flow behavior of these two groups of superplastic alloys. Flow hardening is typical for superplastic deformation of the recrystallized alloys. The uncrystallized alloys exhibit flow softening during superplastic deformation. These alloys have negative flow hardening coefficient. The flow hardening is caused by grain coarsening which takes place during deformation of recrystallized alloys. The flow softening of alloys deformed in unrecrystallized condition is a result of the development of a continuous dynamic recrystallization. The evidence of a sacrificing effect of flow softening on deformability at superplastic forming when strain proceeds not uniformly is demonstrated.

INTRODUCTION

Before deformation the superplastic (SP) aluminium alloys may be in a recrystallized (R) condition with the equiaxed grains, or have an unrecrystallized (UR) structure with the elongated heavily deformed grains with the high dislocation density [1,2,3]. The equiaxed grain structure is common for SP materials. Fine grains are formed by the static recrystallization of a heavily deformed alloys with a matrix containing relatively coarse particles of the second phase. These materials are recrystallized at heating for SP forming (such alloys as the Al-5Zn-5Ca) or a special heat treatment-annealing with the rapid heating is needed to produce a fine grain structure (such alloys as 7475) [2,3].
UR structure is typical for SP materials containing the second phase in the form of ultrafine precipitates. These precipitates stabilize the substructure, suppress the development of the static recrystallization at heating for SP forming (such alloy as Supral and 1570). The equiaxed grain structure is formed after SP deformation to strains about hundreds per cent [2]. Thus it is obvious to categorize all SP aluminium alloys according to the structure type before the start of SP deformation. These two groups are alloys in R and UR conditions. Materials of both types exhibit high SP properties: the total elongation is up to 1000%, the strain rate sensitivity index is about 0.5. But the structural changes during SP flow are different for R and UR alloys. The main microstructural change for R condition is the grain coarsening enhanced by SP deformation. SP deformation of UR alloys cause the development of the equiaxed grain structure and after only a significant strain the grain growth prevails [4]. In spite of the enormous data available upon the structure and rheological properties of the aluminium alloys the principle characteristics of SP flow in R

and UR conditions are not analyzed. To reveal and discuss the main features of a rheological behavior of SP aluminium alloys is the purpose of this article.

EXPERIMENTAL

SP sheets of the aluminium alloys Al-5Zn-5Ca, Neopral and 1570 were examined. The Neopral alloy was produced from a semicontinuously casted ingot according to [5]. The technology includes homogenization, hot and cold rolling. The cold rolled sheets of the Neopral alloy were recrystallized at heating for SP forming. The Al-5Zn-5Ca and 1570 sheets were received in the cold rolled condition. At heating of the Al-5Zn-5Ca alloy the fine grain structure of the intermediate type was formed [6]. The 1570 alloy after heating to SP temperature conserved UR condition [4]. To determine the stress and strain rate sensitivity versus strain rate dependence the tensile tests with step increasing strain rate were carried out at the temperature range of 300-500°C. The strain rate sensitivity index was calculated as a slope of the best fit curve through the logarithm stress against the logarithm strain rate data using a glide cubic spline. The tensile tests with the constant strain rate were used for obtaining flow curves. The flow hardening was analised according to approach developed by Jonas [7]. The microstructure parameters were evaluated by the quantitative metallographic methods.

RESULTS AND DISCUSSIONS

Quantitative parameters of the structure for the investigated sheets are presented in table 1. The Al-5Zn-5Ca alloy contains a significant amount of the second phase. The particle size of the phase is about an order of magnitude of the grain size in this alloy. At heating the recrystallization of the matrix produced a fine grain of an intermediate between the matrix and microduplex type. The volume fraction of the second phase in the Neopral alloy is significantly lower than in Al-5Zn-5Ca alloy. The precipitate size is about of 0.1 m and the matrix type is developed by the recrystallization taking place at heating to the SP temperature. The 1570 alloy contains ultrafine dispersoids having a coherent interface with the matrix. These dispersoids stabilize the structure of the cold rolled materials so hard that at heating to SP temperature the static recrystallization is suppressed. Before the start of deformation the alloy possesses a polygonized structure with the cell size of about 1 μm. It may be concluded that these materials are typical representatives for SP aluminium alloys.

Table 1

Sheet microstructure parameters

Alloy	System	Structure type	Second phase volume fraction,%	particle size, μm	Grain size,
Al-5Zn-5Ca (08050)	Al-Zn-Ca	intermediate	2.8	0.8	2-3
Neopral	Al-Mg-Cu-Mn	matrix	7-8	0.1	6
1570	Al-Mg-RE	matrix	0.8	0.005-0.02	-

All alloys were tested in the temperature range of 400-600°C with the step increasing strain rate. The maximum value of the strain rate sensitivity of the Al-5Zn-5Ca alloy flow stress is nearly constant in this temperature range and it is equal approximately to 0.4. The value is constant in a broad range of the

strain rate. The stress dependence against the strain rate is like a straight line. The Al-5Zn-5Ca alloy exhibits high total elongations in a wide range of the strain rates and temperatures [6].
For the Neopral alloy a sigmoidal plot of the logarithm stress versus the logarithm strain rate is observed. This shape is typical for SP materials. As the temperature raises the strain rate sensitivity increases. The m index is enlarged from 0.37 at 410°C to 0.7 at 530°C and approach approximately 1 when the temperature increasing proceeds. This alloy gives the maximum tensile elongation values in only a narrow region of the high temperature where m exceeds 0.5.
The 1570 alloy has also a sigmoidal mode of the stress strain curve at different temperatures (Fig.1a,b). But opposite to the Neopral alloy the 1570 m index does not vary with temperature. Its maximum values lie within the range of 0.4-0.5 as at the optimum temperatures (450-470°C) where the total elongations of 500-700% have been achieved so as at the lower ones (380-420°C) where the elongations do not exceed 300%. For the 1570 alloy the high elongation are observed in a wide temperature range.

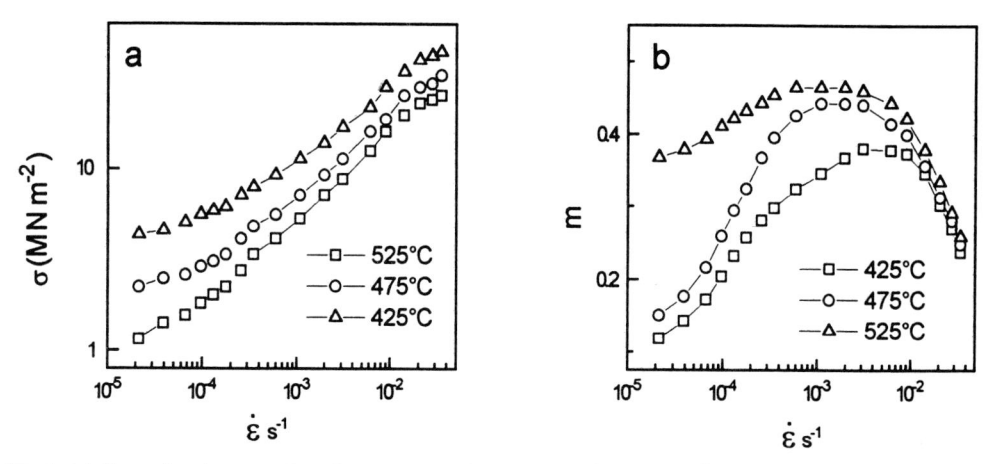

Fig.1 (a) Stress/strain rate plots from step strain rate tests for 1570 alloy at various SP deformation temperatures and (b) the corresponding plots of m value.

The activation energy of SP deformation was calculated for the investigated alloys. The activation energy for the SP flow of the Al-5Zn-5Ca and the 1570 alloys is equal 170 kJ/mol, the value usually obtained for the lattice diffusion of aluminium. The activation energy of the Neopral alloy is significantly higher. It lies in the range of 350-400 kJ for the optimum SP temperature range the activation energy of SP flow approaches the values of the activation energy of the lattice diffusion. The activation energy values obtained in these experiments do not accord with the predictions of SP theories developed for classical SP alloys the activation energy of which is about a half that of the lattice diffusion. Such a high activation energy values point out that SP flow of these alloys can have a special mechanism.
The tensile tests with constant strain rate at the optimum temperature revealed the difference in the flow behavior for these alloys. The recrystallized Al-5Zn-5Ca and Neopral alloys exhibit flow hardening at straining. The dependence of flow stress and flow hardening rate versus strain at a various strain rates is

shown on fig.2 (left) a and b respectively. At low strain rates linear proportion of the flow stress against strain is observed. At the high strain rates the flow stress increases, gets its maximum and then falls. The flow softening occurs only at the maximum strain rates and is caused probably by the flow localization. The flow hardening rate is decreasing with strain and remains positive at strain rate range of SPD. For the optimum SP strain rate the flow hardening rate values is about 1. These magnitudes correspond to the stable flow. The strain rate sensitivity index fall down progressively with a strain.
The 1570 alloy hardens only at initial stages of straining (fig.2 right). The flow hardening interval is very short and after the tension to the strains of 0.03-0.1 the flow softening takes place. At the high strain rate there is a

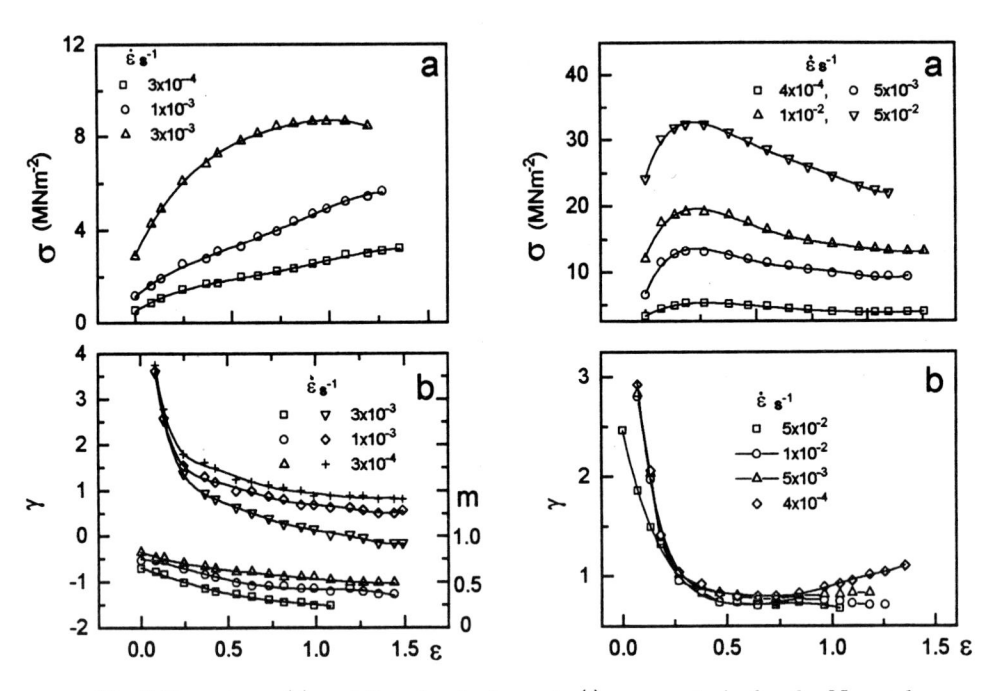

Fig.2 True stress (σ) and flow hardening rate (γ) versus strain for the Neopral alloy at 500°C (left) and for the 1570 alloy at 475°C (right).

small plateau with a constant flow stress. The stress decrease becomes monotonic when the maximum stress is bypassed. The flow hardening rate falls of to the negative values passes over the minimum then slightly increases holding negative sign. This means that flow curves of the alloy are concave upwards and according to the considerations of [7] this is the reason for high elongations observed on this alloy.
The positive sign of flow rate at SP deformation of the alloys in R condition is a result of the grain growth. The results of measurements of grain sizes in longitudinal and and lateral directions of the Neopral alloy samples carried out at different strains are presented on Fig. 3. At annealing grain structure of Neopral alloy is stable. SP deformation induces intensive anisotropic grain growth resulting in developing elongated grains structure. Usually SP alloys preserve equiaxed structure or equiaxed grains develop during straining. For example strain stimulated grain growth in Al-5Zn-5Ca alloy does not effect distinct grain elongation. The nature of such behavior of Neopral alloy and

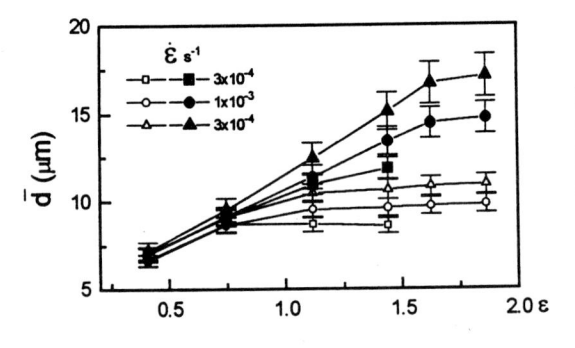

Fig.3 Grain growth kinetics for Neopral alloy at 530°C and different strain rates. Shaded points - logitudinal direction; open points - transverse direction.

several other alloys at SP deformation has been discussed in [8]. Nevertheless one can consider the flow hardening to be a result of grain growth during SP deformation of the alloys in R condition.
To understand the nature and behavior of UR alloy 1570 is not so easy because of the lack of information on this subject. Flow softening on the initial stage of deformation was observed on brasses, which had a banded grain structure before straining [9]. The flow softening rate `d mode depended on the orientation of the sample to the rolling direction. These data were treated as a result of the lamellar break-up during the deformation. After the deformation exceeds strains of 0.2-0.3 the flow softening has been arrested and converted to the flow hardening. The change in the sign occurs when the equiaxed structure has been formed. The 1570 alloy gives flow softening behavior in all the investigated range at least up to 1. The flow softening is a result of the continuous dynamic recrystallization. The matrix grain size remains constant while softening proceeds. Thus no grain occurs which may be a possible explanation for the softening. So we suggest that flow softening reveals a substructure is enormously stable due to the presence of ultrafine dispersoids of the second phase during the static annealing at SP temperatures. The deformation evokes a rearrangement of dislocations which causes the softening process. Relatively large particles as in Neopral can not suppress the recrystallization at heating and the grain growth at straining, thus the flow hardening occurs. It is interesting to compare the Neopral and 1570 alloys behavior because these alloys almost coincide with in the matrix composition, but are different in respect to the particle size.
The different behavior of the alloys in UR and R condition at SP flow does not effect the tensile elongation. All the investigated alloys give the high total elongation exceeding 500-700%. But a complicated strain state deformation mode can reveal the effect of the rheological properties on the plasticity and fracture as well as on the wall thickness distribution in parts produced by SP forming. For example we present the data on fracture from the tensile with bend test of the Al-5Zn-5Ca and 1570 alloys. To demonstrate the evidence of this we show the specimens subjected simultaneously to the tension bend. The Al-5Zn-5Ca alloy demonstrates high plasticity in this test. The failure of specimen locates in the middle of the tensile zone (fig.4a) and this complicated strain state does not deteriorate deformability. The 1570 alloy specimen rupture occurs at the bend zone where at the beginning of deformation the bend strain is prevail (fig.4b). This discrepancy between the tensile with bend test results for these alloys can not be accounted for by considering the strain rate sensitivity and total elongations obtained by tensile tests because these parameters almost coincide with the both alloys. At the initial stages of the deformation the bend strains prevail in a narrow zone. In the materials has the negative flow hardening rate this zone becomes soft and the strain rate in this zone increases progressively. This localized deformation yields rupture of specimens along the bent region. Bending of the Al-5Zn-5Ca alloy on the contrary yields increasing

Fig.4. Specimens of the Al-5Zn-5Ca alloy (a) and the 1570 alloys (b) after the tensile with bend to failure testing at 500°C and 475°C respectively

of its hardness. The bent zone is strengthened and deformed at the lower strain rate than the tensile zones. The fracture is not tied up with bending and occurs after the significant tensile strains. Another effect which is worth to allow for flow softening: the cavity in the corner of the superplastically formed 1570 alloy box probably develops due to the flow softening of this material.

CONCLUSION

Thus it has been established that the aluminium base alloys with UR structure as the 1570 alloy would give a negative sign of the flow hardening rate at SP temperature and strain rate range. This behavior is contrasted to R alloys which are strengthening when flow. The effects of flow softening on SP forming process has been demonstrated.

REFERENCES

1. I.I.Novikov, V.K.Portnoy Super plastizitat von Legierungen (Leipzig, VEB Deitche Verlag fur Grundstoffindustrie, 1984).
2. D.J.LLoyd, D.M.Moore. 'Aluminium alloy design for superplasticity', in: Superplastic forming of structural alloys, ed. N.E.Paton and C.H.,Hamilton (New York, NY: The Metallurgical Society, 1982), 147-172.
3. C.H.Hamilton, C.C.Bampton, N.E.Paton. "Superplasticity in high strength aluminium alloys", in: Superplastic forming of structural alloys. ed. N.E.Paton and C.H.Hamilton (New York, NY: Tehe Metallurgical Society, 1982), 173-190.
4. A.M.Diskin, A.A.Alalykin. "Superplasticity of duraluminium and magnalium with unrecrystallized initial structure", Tsvetnye Metally, 1987, No 5, 84-87.
5. Watanabe et al. "Development of new superplastic aluminium alloy Neopral". Bulletin of the Japan Institute of Metals. 24 (1985), 313-315.
6. V.K.Portnoy et al. "AZ5K5 alloy for superplastic forming" Tsvetnye Metally, 1984, No 9, 72-74.
7. J.J.Jonas. "Implications of flow hardening and flow softening during superplastic forming", in: Superplastic forming of structural alloys, ed.N.E.Paton and C.H.Hamilton (New York, NY: The Metallurgical Society, 1982), 57-68.
8. V.S. Levcenko, V.K. Portnoy, I.I.Novikov. ' Unusual low grain boundary sliding in aluminum alloy with classical features of micrograin superplasticity', proc. int.conf. Superplasticity in Advanced Materials, Japan 1991, p39.
9. E.Sato, J.K.Karibayashi, R.Horiuchi. "Superplastic induced grain growth in microduplex and second phase dispersed alloys", in: Superplasticity and superplastic forming. The Minerals, Metals & Materials Society, 1988), 115-119.

Superplastic Forming: Modelling

Finite Element Modeling of Superplastic Forming with Precise Dies

Reza Sadeghi, San Diego State University, San Diego, CA
Zachary Pursell, San Diego State University, San Diego, CA

Abstract

Finite Element Modeling of superplastic forming with precise die models is developed, using the analysis procedures of the MARC-Mentat system by MARC Analysis Research Corporation. The incremental approach based on Rigid Plastic formulation incorporates the experimental superplastic material law as proposed by C.H.Hamilton, which includes both static and dynamic grain growth terms. Frictional contact conditions yield constraint relations which, together with equilibrium equations, form the basis for derivation of the consistent tangent matrix.

I. Introduction

The simulation of superplastic forming process (SPF) has become a requirement in all industries currently using this process. In addition to aerospace, the automobile industry has recently become an active participant in SPF fabrication. More and more, multiple parts manufactured by traditional methods are being replaced by one superplastically formed part. As a result, the shapes that are fabricated using SFP have become much more complex and demand more accurate computer modeling.

For modeling contact between rigid to deformable bodies (sheet and die), a number of contact algorithms have been developed over the years [References 1-5]. Most of those algorithms focus on enforcing contact constraints. It is assumed that the possible contact points on the die surfaces (rigid surfaces) are already known, or may be obtained by considering that the contact regions are composed of straight segments or patches, and the normal vector is constant over subdivided segments or patches. However, due to the discontinuity of the first derivative between patches, the accuracy of solution techniques, whether using Lagrange multipliers, penalty functions or solver constraints (to enforce the constraint in the solution of incremental-iteration nonlinear analysis), depends on the number of discretization of the contact surface. Piecewise polynomial functions or bezier curves might be applied to areas of severe discontinuity, such as corners or singular points. However, those approaches are not useful for highly nonlinear contact conditions.

The general stages for contact algorithms are: checking for contact, enforcing the contact constraint, checking separation, and checking penetration. Each stage involves the geometric problem of calculating the normal vector at the contact point. The sudden change of the normal vector (due to the piecewise linear discretization) or the use of inaccurate normal vector (due to large deformation) will result in unexpected contact and separation. The consequence is a reduction of loading step size during analysis, unnecessary iterations, or incorrect results. Inaccurate calculation of friction force is a common phenomenon due to the use of an inaccurate normal vector. Nonsymmetrical behavior may be observed for symmetrical loading and geometry unless extremely fine segments or patches are used.

An improved contact algorithm has been developed and applied to rigid analytical surfaces (commonly called master segments). The surface can be represented analytically by either quadratic surfaces or nonuniform rational B-spline surfaces (NURBS). Simple quadratic surfaces include: spheres, cylinders, cones, ellipsoids, etc. Some free-form surfaces and surfaces of revolution may also be described as analytical surfaces. Generally, all of these surfaces can be described by NURBS. NURBS have the advantage of continuity of the normal vector along the surfaces, and the flexibility to model multiple complex surfaces with a single mathematical description. The major improvement of the new contact algorithm is that it takes advantage of the characteristics of NURBS to locate the contact points, calculates the normal, tangency, and the second derivatives (if necessary) at the point on the surface.

During the first stage of contact (checking for contact), a fast searching technique was developed to find the minimum distance between the contacting node and the NURB surface in one to five iterations with an accuracy of sine (0.1 degrees). Once an accurate contact point is located, the new incremental displacement is calculated to ensure that the contacting node slides on the surfaces every iteration until separation occurs or a converged solution is obtained. Separation is checked at every iteration except the first one, so that unnecessary iterations are avoided. The friction calculation includes the effect of changes in the direction of normal vector from iteration to iteration in order to obtain an accurate contribution. These improvements have been implemented in the contact capabilities of MARC. A wide variety of tests have been made from simple quadratic surfaces to three-dimensional sculptured surfaces. Applications include: single sheet as well multi sheet SPF parts. The number of iterations to obtain the stable solution for incremental nonlinear analysis has been reduced by 10 to 50 percent, and better results are obtained, especially when friction contributions are included. The computation costs associated with contact algorithms are: (1) locating the point on NURBS, and (2) calculating the incremental displacement for every iteration. Comparing the reduction in the number of iterations needed to obtain a converged solution, the time spent locating the contact point on the NURBS is trivial. The accuracy and the computational improvements are significant.

II. Contact Searching

The first stage for general contact algorithms is to detect the contact. The procedure involves a significant amount of searching and comparison. Usually, the potential contact regions are composed of straight segments or patches and the normal vector is calculated based on the discretized surface. The discontinuity of the first derivative exists and the unexpected detection of contact may be obtained. However, the nonuniform rational B-spline surface (NURBS) provides a single precise mathematical form capable of representing the common analytical shape -- circle, conic curve, free-form curves, surfaces of revolution, and sculptured surfaces -- used in computer-aided design. It also can be expressed in simple mathematical form to model complex multiple surfaces with advantage of continuity in first and second derivative. Those analytical forms are expressed in four-dimensional homogeneous coordinate space by:

$$P(u,v) = \frac{\sum_{i=1}^{n+1} \sum_{j=1}^{m+1} B_{i,j} h_{i,j} N_{i,k}(u) M_{j,l}(v)}{\sum_{i=1}^{n+1} \sum_{j=1}^{m+1} h_{i,j} N_{i,k}(u) M_{j,l}(v)} \quad (1a)$$

For curves, it simplifies to:

$$P(u) = \frac{\sum_{i=1}^{n+1} B_i h_i N_{i,k}(u)}{\sum_{i=1}^{n+1} h_i N_{i,k}(u)} \qquad (1b)$$

where the B are 4D homogeneous defining polygon vertices, $N_{i,k}$ and $M_{j,l}$ are non uniform rational B-spline basis functions, $h_{i,j}$ is homogeneous coordinates [Reference 6]. For given parameters u and v in local system, the location (x, y, z) in three-dimensional space, first derivative and second derivative (if required), are calculated through Equation 1. Given a point with x, y and z Cartesian coordinates, there is no explicit mathematical form to find out the parameters u and v in the local space, so they are obtained using an iterative procedure.

The goal of the searching algorithm is to locate the point on the NURB to which the nodal point will contact. That is to find the point, P, on the NURB which has a minimum distance to the given node, O. There are several iterative schemes used to find the parameters u and v for 3D NURBS surface, or parameter u for 2D NURBS curve of the contact point which allow the prescription of required accuracy. A binary searching algorithm, for example, is one iterative scheme used to locate the point on the NURBS. For given parameters u1, and u2, initially 0 and 1, respectively, the average value of u1 and u2 yields an estimated parameter, t. Substituting the parameters t into Equation 1, the coordinates and normal vector of point P on the NURBS are obtained. One then determines if the line along the normal direction at the point P will pass through the potential contact node O. This procedure is repeated until the required accuracy is reached and the contact point on the NURBS is found. The procedure is accurate, but too many iterations are required and the computational costs are too large.

New searching techniques for locating the contact point on the NURBS surface have been developed in order to obtain the minimum distance between the given node, O, and the contact point, P, using a few iterations as possible. The criterion used to determine the contact point on the NURBS surface is that the line along the normal direction to the NURBS at the contact point should pass through the given node O. In terms of mathematical forms:

$$n1 \cdot n2 = \pm 1 \qquad (2)$$

where n1 is unit normal vector, obtained from Equation 1, of contact point P on the NURBS surface, and n2 is the unit vector between the point P and the given node O. When the condition in Equation 2 is satisfied, the distance between point P and node O is calculated. If the distance falls within the contact tolerance, contact is detected. From Figure 1, the following equation is satisfied to locate the estimated point P on the NURBS:

$$[OP1 + t (P2 - P1)] \bullet (P2 - P1) = 0 \tag{3}$$

The new estimated parameter t in Equation 3 is obtained from the parameters u1 and u2, then normal vector and coordinates at the point p on the NURBS is obtained through Equation 1. Equation 2 is used to check if the estimated t is accurate. If it fails, choose parameters u1 and t as new estimated u1 and u2, respectively, if t1 is less than average value of u1 and u2. Or choose t and u2 if t is greater than average value of u1 and u2. If t is less than zero or greater than one or the number of iteration is larger than 50, the searching procedure terminates and contact between the node and the surface does not occur.

Usually, ;the initial values of u1 and u2 are zero and one, respectively. However, a bounding boxes strategy is used to rule out the possibility of potential contact. The original NURBS are internally subdivided into many pieces. If the given potential contact node is inside the bounding box which is made of the subdivided parameters (segments), the searching procedure described above proceeds. The number of subdivisions will not affect the detection of contact nor the accuracy. The average number of iterations to detect contact is between two and five with an accuracy of sine (0.1 degree).

III. Calculation of Contact Constraint

When a node comes into contact with the NURBS surface, the displacement of the contact point is calculated in order to ensure that the contact node slides on the surface during iterations. That is, searching for the new location of the contact node on the NURBS is repeated every iteration through the similar scheme of contact searching. the contact point at the new location on the NURBS also satisfies the criterion in Equation 2. The difference of the contact point at the new position and at the original position forms the incremental displacement and the projection of this displacement vector along the normal direction of NURBS surface at the new location is calculated. This magnitude is enforced to be the contact constraint. Either Lagrangian multipliers, penalty approach, or solver constraints are used to enforce the contact constraints. It is not in the scope of this paper to discuss the ways of imposing contact constraints which have been extensively documented in the finite element literature [Reference 2-4].

IV. Improved Contact Strategies

Once a normal reaction force of a contact node turns tensile in nature, the node is considered free and separates from the surface and the solution in the increment has to be repeated. Normally, the checking of separation, which is the third stage of contact algorithm, is made after the solution is convergent. However, in the improved contact strategy, the checking occurs in every iteration except the first one. Because the discontinuity of normal vector does not exist for smooth surfaces, the normal reaction forces in nature will not change sign abruptly as long as all of the constraint conditions remain the same. If actual separation does occur, many unnecessary iterations are avoided.

Friction effects are generally dependent on the relative velocity and the normal stress or force. For accuracy and stability, incorporating friction requires a contribution to the load vector and the stiffness matrix. The use of the piecewise linear approach results in an accurate calculation of the friction behavior due to the discontinuity of the first derivative which is used to calculate the relative sliding direction. When an accurate normal vector based upon the NURBS is used, the relative sliding velocity is calculated based upon the current normal direction. Hence, the change of sliding velocity is accounted for from iteration to iteration. An accurate calculation of the contribution to the stiffness matrix can be made. This results in a more accurate solution.

When modeling a geometry using piecewise linear segments, two situations occur when a node slides from one segment to the next. For either concave or convex surfaces, the node may simply slide over the transition if the associated angles are small. If the angle is large, the node may separate in the convex case or stick in the concave case. The results of the discretization are discontinuous results, increased computational costs and potential instabilities.

V. Summary

When NURBS are used to represent die surfaces with the new contact strategies, the following improvements have been made:

1. Continuity of the first derivative along the surface shape, allowing a more accurate description of the geometry.

2. Improved accuracy of the solution (especially for friction condition).

3. Number of iterations required to obtain the solution is reduced for complex geometric model.

4. Instability near corners or singular points have been avoided.

References

1. MARC reference manual, Vol. F, "Development of Contact Algorithms for a General Purpose Finite Element Program", MARC Analysis Research Corporation, Palo Alto, California U.S.A., 1994.

2. David J. Benson and J.O. Hallquist. "A Single Surface Contact Algorithm for the Post-Buckling Analysis of Shell Structures", Computer Methods Appl. Mech. Engrg., (78), 1990, pp. 141-163.

3. J.O. Hallquist etc. "Sliding Interfaces with Contact-Impact in Large-Scale Lagrangian Computations", Computer Methods Appl. Mech. Engrg., (51), 1985, pp. 107-137.

4. N. Kikuchi and J.T. Oden. "Contact Problems in Elasticity: A Study of Variational Inequalities and Finite Element Methods for a Class of Contact Problems in Elasticity", SIAM Studies, Vol. 8 (SIAM 1986).

5. Ted Belytschko and Mark O. Neal. "Contact-Impact by the Pinball Algorithm with Penalty and Lagrangian Methods", Int. J. Numer. Meths. Engrg., Vol. 31, 1991, pp. 1413-1418.

6. David F. Rogers and J. Alan Adams, "Mathematical Elements for Computer Graphics", 2nd Edition, 1990.

7. C. Howard Hamilton, "5083 Al Alloy Constitutive Parameters for Superplastic Deformation", Washington State University, August 4, 1993. In Press.

8. C. Howard Hamilton, "Designing Optimized Deformation Paths for Superplastic Ti-6Al-4V", Washington State University, Fall 1992. In Press.

ASPECTS OF THE NUMERICAL SIMULATION OF SPF INCLUDING MATERIAL PARAMETER EVALUATION

R.D. WOOD and J. BONET

Institute for Numerical Methods in Engineering, University of Wales, Swansea, U.K.

ABSTRACT

The paper reviews the techniques used for the finite element analysis of superplastic forming processes using the incremental flow formulation. Additionaly, some preliminary investigations are presented that explore the possibility of using inverse methods for the determination of the material parameters in the SP constitutive equations. Finally, a realistic application of the use of FE simulation is presented.

INTRODUCTION

The manufacture of a superplastically formed component is a complex, difficult and often time consuming exercise. Traditionally the manufacturing process engineer relies on a great deal of skill and experience of the SP material and how it is likely to behave during the forming process. This skill is directed towards achieving a part that has the required final thickness distribution and, in the case of diffusion bonding, to establishing the layout of the cells in order to produce the desired disposition of the webs. To be succesful the engineer must attempt, one way or another, to determine initial blank thickness distribution and predict the final thickness distribution and the forming pressure cycle that generates optimum forming.

The numerical analyst working in the field of SPF simulation has precisely the same aims. The finite element (FE) method, being the most succesful and versatile simulation tool, is capable, in principle, of predicting all of the items discussed above. It is suggested that the finite element analysis of SPF [1,2,3,4,5] can substantially enhance the skill of the process engineer and indeed enable the feasability of innovative designs to be examined and thereby reduce the number of expensive trial and error tests.

This paper will focus on some recent developments in the FE anlysis of SPF in an effort to promote the integration of FE analysis into the design cycle.

Firstly, the constitutive equations that are commonly used in conjunction with FE analysis will be reviewed. An algorithmic constitutive equation will then be introduced to enable the SPF analysis to be carried out using an existing finite deformation displacement based formulation. This is known as the *incremental flow formulation*.

Secondly a reiteration of the authors energy dissipation method of determining the pressure cycle is very briefly presented [5]. This draws attention to the possibility that the pressure cycle for optimum forming could be based on the average rate of energy dissipation per unit volume of forming material rather than on a strict adherence to a strain rate criterion.

Finally some very preliminary investigations are presented that explore the possibility of using inverse methods [12,13] for the determination of the material parameters in the SP constitutive equations. Constitutive equations are crucially important in a numerical analysis and with the advance numerical solution techniques now available the constitutive equations are the Achilles heel of the problem. Tensile tests carried out in order to evaluate material parameters make assumptions regarding the stress and strain rate distribution that may not be warranted (see Dunne and Katramados in this publication). This may lead to erroneous parameters that result in an inability for the FE simulation to replicate the test. In this paper a simple inverse method is presented that uses a genetic algorithm in order explore if material parameters could be obtained from a numerical simulation. A non finite element numerical model was employed for expendiency in this pilot study.

MATERIAL DESCRIPTION

The successful finite element analysis of the forming of even very simple components requires an accurate constitutive description of the alloy. The simplest type of equations relate the flow stress to the strain rate by means of two material parameters K and m to give the well-known equation,

$$\sigma = K \dot{\varepsilon}^m \tag{1}$$

For titanium alloy Ti-6Al-4V this equation can be considerably improved by including the evolution of the grain size g by letting K and m vary with g [1]. This necessitates the evaluation of the changes of grain size with time. For this purpose an experimentally based equation such as,

$$\frac{g}{g_0} = \left(\frac{t}{600}\right)^{N(\dot{\varepsilon})} \quad ; \quad N(\dot{\varepsilon}) = 1.8(\dot{\varepsilon} + 0.00005)^{0.237} \tag{2}$$

has been successfully used [1,6].

Recently combined power laws have been employed for the constitutive modelling of superplastic alloys [7,8]. These give strain rate as a function of flow stress as,

$$\dot{\varepsilon} = \Phi(\sigma, g) = \frac{K_{II} D}{T g^p} \left(\frac{\sigma - \sigma_0}{G}\right)^{1/M} + \frac{K_{III}}{T} \left(\frac{\sigma}{G}\right)^n \tag{3}$$

where σ is the flow stress; $\dot{\varepsilon}$ is the strain rate, σ_0 is threshold stress, G is the shear modulus of the material; T is the temperature; g is the grain size of the material; K_{II}, K_{III}, M, n and p are material parameters; and $D = D_L + (\pi D_B \delta)/g$ is the appropiate diffusivity with lattice and grain boundary components D_L and D_B repectively. The grain size evolution is now obtained by adding the static and dynamic grain growth components as,

$$\dot{g} = \dot{g}_s + \dot{g}_d \ ; \quad \dot{g}_s = \frac{B}{q}(g_0^q + Bt)^{\frac{1}{q}-1} \ ; \quad \dot{g}_d = \frac{\lambda \dot{\varepsilon} g}{g^q} \qquad (4)$$

where B, q and λ are material parameters.

Equation (3) gives the strain rate in terms of the flow stress. The reverse equation Φ^{-1} giving the stress in terms of strain rate cannot be obtained analytically and a numerical Newton-Raphson procedure has to be employed to obtain σ for each given value of $\dot{\varepsilon}$.

Finally, the extension of the above equations to three dimensional situations is achieved by re-writing the above equations in terms of Von Mises equivalent strain and stress measures, $\dot{\bar{\varepsilon}}$ and $\bar{\sigma}$ respectively, and by defining the viscoplastic potential $\psi(\dot{\bar{\varepsilon}})$ as [9],

$$\bar{\sigma} = \frac{d\psi}{d\dot{\bar{\varepsilon}}} \ ; \quad \psi(\dot{\bar{\varepsilon}}) = \int \Phi^{-1}(\dot{\bar{\varepsilon}}, g) \, d\dot{\bar{\varepsilon}} \qquad (5)$$

DISCRETIZATION

A discrete sequence of sheet postions is used to simulate the continuous forming process. Figure 1 shows two consecutive stages at times t and $t + \Delta t$. The geometry of the sheet at each position is discretized using an arbitrary number of finite elements in terms of the coordinates of the mesh nodes $a = 1, 2, ..., n$ and the shape functions N_a. For instance, the geometry at time t is approximated as,

$$\boldsymbol{X} = \sum_{a=1}^{n} N_a \boldsymbol{X}_a \qquad (6)$$

where $\boldsymbol{X}_a$ are the nodal positions at time t. Similarly, at time $t + \Delta t$ the new geometry is given in terms of the current nodal positions $\boldsymbol{x}_a$ as,

$$\boldsymbol{x} = \sum_{a=1}^{n} N_a \boldsymbol{x}_a \qquad (7)$$

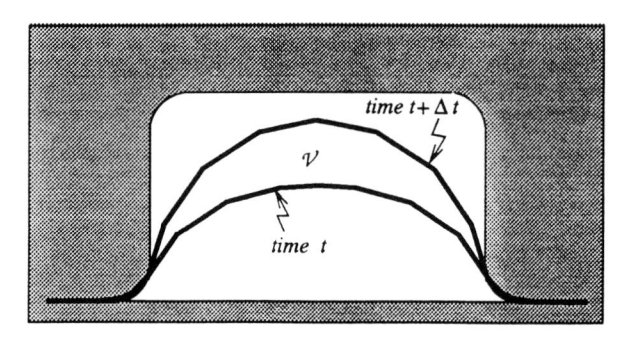

FIGURE 1 Consecutive discrete forming stages

The derivative of the new geometry with respect to the previous particle positions defines the *deformation gradient* tensor F as,

$$F = \frac{\partial x}{\partial X} = \sum_{a=1}^{n} x_a \otimes \nabla_0 N_a \, ; \qquad \nabla_0 N_a = \frac{\partial N_a}{\partial X} \qquad (8)$$

An approximation to the average Von Mises equivalent strain rate during the motion from time t to $t + \Delta t$ can be evaluated from the deformation gradient tensor as [10,11],

$$\dot{\bar{\varepsilon}} = \frac{1}{3\Delta t^2}(J^{-2/3} F : F - 3) \, ; \qquad J = \det F \qquad (9)$$

The above value is used as the variable in the viscoplastic potential defined in equation (5), thereby extending the material description given above to 3-dimensional discrete processes. This results in an algorithmic constitutive equation.

VARIATIONAL EQUATIONS

The basis of the finite element equations is the minimization of a potential energy functional. Assuming that the sheet geometry at time t is known, this functional depends on the new position at time $t + \Delta t$ as,

$$\Pi[x] = \Pi_{\text{dis}}[x] + \Pi_{\text{vol}}[x] + \Pi_{\text{ext}}[x] + \Pi_{\text{bon}}[x] \qquad (10)$$

where the distortional energy component Π_{dis} is given by the integration of the viscoplastic potential over the volume of material V as,

$$\Pi_{\text{dis}}[x] = \int_V \psi(\dot{\bar{\varepsilon}}) \, dV \qquad (11)$$

The volumetric component Π_{vol} enforces the incompressibility of the deformation by requiring the Jacobian J to equal 1 using a penalty parameter λ_V as,

$$\Pi_{\text{vol}}[x] = \int_V \lambda_V (J - 1)^2 \, dV \qquad (12)$$

Note that this term is only required if 3-dimensional elements are used. In the case of membrane elements, the incompressibility condition can be easily enforced by simply adjusting the thickness of the element. The external component Π_{ext} reflects the potential energy stored in an amount of gas $\mathcal{V}$ at a forming pressure p and is given as,

$$\Pi_{\text{ext}}[x] = p\mathcal{V} \qquad (13)$$

Finally, $\Pi_{\text{bon}}[x]$ enforces the condition that the gap g between two bonding parts of the sheet, or between the sheet and the die, should vanish by means of a penalty number λ_B as,

$$\Pi_{\text{bon}}[x] = \int_{A_B} \lambda_B \, g \cdot g \, dA \qquad (14)$$

where A_B denotes the area of material in contact.

EQUILIBRIUM EQUATIONS

The equilibrium position at time $t + \Delta t$ is obtained by minimizing the above functional with respect to the nodal coordinates. This process leads to the following equilibrium equations at each mesh node a,

$$T_a + V_a + B_a - pF_a = 0 \qquad (15)$$

where, assuming that simple elements with a single quadrature point are used, the deviatoric internal equivalent forces T_a are given as,

$$T_a = \frac{\partial \Pi_{\text{dis}}}{\partial x_a} = P : \int_V \nabla_0 N_a \, dV \; ; \qquad P = \frac{\partial \psi}{\partial F} \qquad (16)$$

The volumetric equivalent forces V_a, which are used to enforce incompressibility in the case of 3-dimensional elements, emerge as,

$$V_a = \frac{\partial \Pi_{\text{vol}}}{\partial x_a} = \lambda_V \left(\frac{v - V}{v} \right) \int_v \nabla N_a \, dv \qquad (17)$$

where, v and V are the current and previous volumes of a typical element. The diffusion bonding or contact forces B_a are,

$$B_a = \frac{\partial \Pi_{\text{bon}}}{\partial x_a} = \lambda_B \int_{A_B} g N_a \, dA \qquad (18)$$

Finally, the external pressure forces are evaluated as,

$$F_a = \frac{\partial \Pi_{\text{ext}}}{\partial x_a} = \int_{A_p} n N_a \, dA \qquad (19)$$

where A_p is the surface, with normal unit vector n, where the gas pressure is applied.

SOLUTION PROCEDURE

The above equilibrium equations at a given node a are assembled in the standard finite element manner to yield a set of global nonlinear equilibrium equations in terms of the global vector of new nodal positions **x** and the global residual **R** as,

$$R(x) = T + V + B - pF = 0 \qquad (20)$$

In this equation the applied pressure can be prescribed or evaluated by controlling the rate of energy dissipated in the mesh [5]. This energy rate $\dot{E}$ is identical to the rate of work done by the pressure which s given as,

$$\dot{E} = p\dot{V}VV \; ; \qquad \dot{V} = \int_{A_p} n \cdot v \, dA \qquad (21)$$

In an incremental context the rate of change of $\mathcal{V}$ is approximated as,

$$\dot{\mathcal{V}} \approx \frac{\mathcal{V}_{t+\Delta t} - \mathcal{V}_t}{\Delta t} \qquad (22)$$

Given a target strain rate $\dot{\bar{\varepsilon}}_0$ or stress $\bar{\sigma}_0$ it is possible to obtain the equivalent target energy rate $\dot{E}_0$ and the pressure required to achieve this target becomes,

$$p = \frac{\dot{E}_0 \Delta t}{\mathcal{V}_{t+\Delta t} - \mathcal{V}_t} \qquad (23)$$

The above nonlinear equations (20) are solved by means of a Newton-Raphson iterative procedure whereby, given an initial guess of the solution $\mathbf{x}_0$, an iterative increment $\mathbf{u}$ in the k-th solution is evaluated by soving a linear set of equations in terms of the tangent matrix $\mathbf{K}$ as,

$$\mathbf{Ku} = -\mathbf{R}(\mathbf{x}_k) \; ; \quad \mathbf{x}_{k+1} = \mathbf{x}_k + \mathbf{u} \; ; \quad \mathbf{K} = -\frac{\partial \mathbf{R}}{\partial \mathbf{x}} \qquad (24)$$

Full expressions for the tangent matrix $\mathbf{K}$ are given in references [10–11]. The above Newton-Raphson iteration will quickly converge to the new equilibrium position at time $t + \Delta t$ and the whole process is then repeated by taking further steps until full contact with the die is reached.

MATERIAL PARAMETER EVALUATION

Obtaining constitutive equation parameters is an inverse problem whereby some form of curve fitting using, say, a least squares method, is employed to ensure that the parameters represent the experimental data. A fundamental assumption contained in this approach is that the physical behaviour of the test specimen is sufficiently well understood for the essential micro constitutive quantities such as stress and strain rate to be obtained from macro measurements such as cross-head force, displacement and velocity.

It is conjectured that a better technique would be to abandon the physical assumptions regarding the micro behaviour and rely on the numerical (perhaps FE) simulation to determine the behaviour and thereby assist in the material parameter determination. Such an approach has been investigated by Gaurus et al. in [12] with respect to torsion tests and a viscoplastic material.

Essentially such an inverse procedure involves using the non-linear numerical simulation with estimated material parameters and then systematically modifying these parameters until some objective function representing the difference between the experimental and numerical results is minimized. Whilst this may be computationally time consuming it does not require dubious physical assumptions regarding the micro behaviour of the test specimen .

Minimizing the objective function representing the error between the experimental and numerical results usually involves a sensitivity analysis which requires either an analytical or numerical evaluation of the response of the numerical model to changes in the material parameters. In situations where this is difficult a genetic algorithm (GA) [14] may provide a suitable alternative optimisation method. GA's are computationally costly but reasonably robust and can generally search a large parameter space without getting constrained to a local spurious solution. Whilst GA's involve a degree of random selection the process is not entirely random, but rather is a guided move through the parameter space based on three key operations, namely, *reproduction, crossover* and *mutation*. These operations are associated with a survival of the fittest members of a parameter population to find the optimum parameters. In order to focus ideas and provide a (very) preliminary evaluation of the use of a GA to solve the inverse SP material parameter evaluation a simple (non FE) numerical model based on a uniaxial specimen will be employed.

Simple numerical model

The constitutive model for Ti-6Al-4V will be based on the experiments reported by Ghosh and Hamilton [6] and given in equations (1-2). In what follows the behaviour of the uniaxial specimen determined using these equations will be considered to be the *experiment* and the GA is used to find the parmeters ($p = p_1, p_2, p_3$) that determine N and thus the grain growth from which K and m can be found. Consequently in the GA,

$$N = p_1(\dot{\varepsilon} + p_2)^{p_3} \tag{25}$$

The relationship between the grain size and K and m is given in Table 1.

Ti-6Al-4V Alloy		
g (μm)	log K (Mpa)	m
6.4	3.72	0.865
9.0	3.72	0.793
11.5	3.51	0.675
20.0	2.90	0.450

Table 1 Values of log K and m for different grain sizes.

It is simple to show that for a uniaxial model,

$$\frac{dx}{dt} = \left(\frac{F}{KV}\right)^{\frac{1}{m}} x^{(\frac{1}{m}+1)} \tag{26}$$

where F is the force, V the constant material volume, x the length and t the time. For a choice of parameters (p_1, p_2, p_3) equations (1-2) and (26) provide a means of establishing the extension time behaviour which in the general case would be provided by the FE analysis of the actual specimen.

The genetic algorithm

Coding the parameters: the first step in the GA is to encode the parameters (p_1, p_2, p_3). For ease of explanation these will be coded using 3 strings (genes) of 5 binary digits concatenated to form a, so called, *chromosome* as,

$$11001, 10010, 00110$$

Initially a population of such chromosomes is randomly established.

Decoding the chromosome: Each 5 digit string can represent numbers 0 to 31 which are eventually scaled to lie between selected limits for **p**.

Fitness evaluation: For each member of the population having parameters $(p_1, p_2, p_3)_i$, the numerical simulation of equation (26) enables an objective fitness function f_i, based on a least squares error to be evaluated as,

$$f_i(p_1, p_2, p_3) = C - \int_{x_0}^{x_t} (t_{exp} - t_p)^2 \, dx \qquad (27)$$

where the integral is evaluated numerically (or in the real case using an FE analysis), C is a suitably chosen large number so that a maximum f_i corresponds to the minimum least square error.

Reproduction, crossover and mutation: The population of chromosomes can be considered as a population of *parents*. Pairs of parents are chosen that have the highest fitness function values $f_i(p)$ and a crossover of genetic (binary digit) information takes place between the two parent chromosomes to reproduce and form a new population of *children*. A single crossover site is selected at random and digits are interchanged as,

```
parent 1:    11001100|1000110
parent 2:    01010111|1010111


child 1:     11001100|1010111
child 2:     01010111|1000110
```

Finally a small probability of mutation is introduced whereby a digit may randomly be exchanged between the children in order to maintain some population diversity, particularly during the later stages of the optimisation.

The new population is again accessed by the numerical model and a new set of fitness values f_i calculated, hopefully showing that the parameters **p** are better able to represent the 'experimental' data. Eventually the average parameter values of the parameters converge and an optimum is achieved. Since the procedure is based on a random generation of the initial population and of subsequent crossover sites the final converged result may not always be the same. Indeed experience with this simple case reveals the parameters are not unique but are nevertheless generally similar for different initial populations. For more information on GA's see the excellent text by Goldberg [14].

GA - example: the data for the model is: $x = 100$, $V = 2000$, $F = 110$ and simultaneously 140 for the evaluation of f_i. The total chromosome length was 30 digits (3×10) and the decoding set limits on the parameters as,

$$1.0 \leq p_1 \leq 2.0$$
$$0.0 \leq p_2 \leq 0.0001$$
$$0.2 \leq p_3 \leq 1.0$$

Typical results were obtained with a population size of 40 with the fitness being scaled as $f_i^{1.5}$ after 10 generations in order to promote the fittest parents in the selection process. The probability of mutation was 0.002. Figures 2 show typical graphs of the convergence of the average value of the parameters with the generation number. The converged values were $p_1 = 1.8076$, $p_2 = 0.00005645$, $p_3 = 0.2383$. Figure 3 shows the uniaxial SPF extension time behaviour using these parameters, for a force of 110 and 140. The curves are practically identical to the 'experimental' values obtained using $p_1 = 1.8$, $p_2 = 0.00005$, $p_3 = 0.237$.

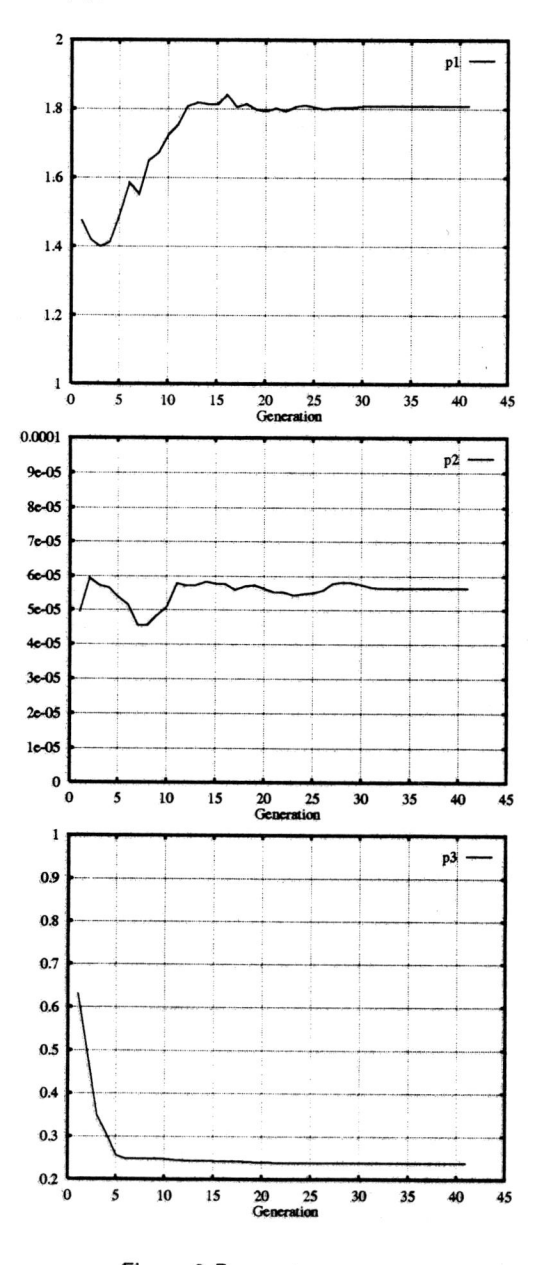

Figure 2 Parameter convergence

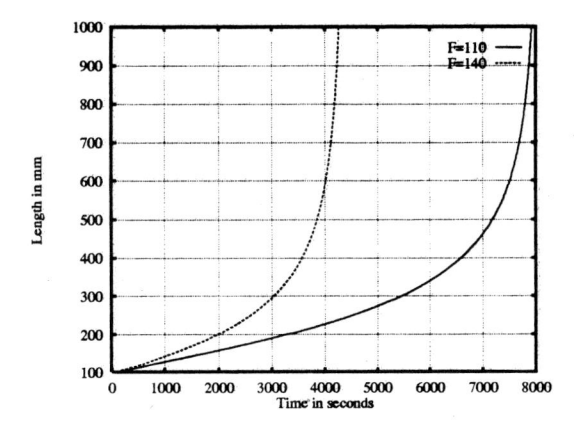

Figure 3 Elongation history

APPLICATION

This example has been included in order to illustrate the ability of the finite element method described above to simulate the forming of realistic components. A 5 mm thick blank of dimensions 120×120 mm^2 is formed into a double square box of compartments heights of 30 and 60 mm. The geometry of the die is depicted in figure 4.

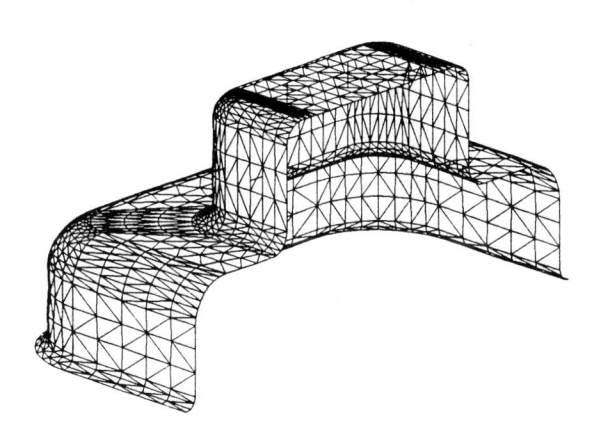

Figure 4 Double box die

The material is Ti-6Al-4V and the initial grain size is 6μm. The obvious symmmetry of the problem enables only a quarter of the problem to be modelled. The computational blank mesh consists of 4218 tri-linear hexahedral elements and 6582 mesh nodes. Two layers of elements have been taken through thickness of the sheet. The deformed finite element meshes at various times are shown in Figure 5. The complete run took 156 timesteps involving 280 Newton-Raphson iterations.

Time = 3100 seconds

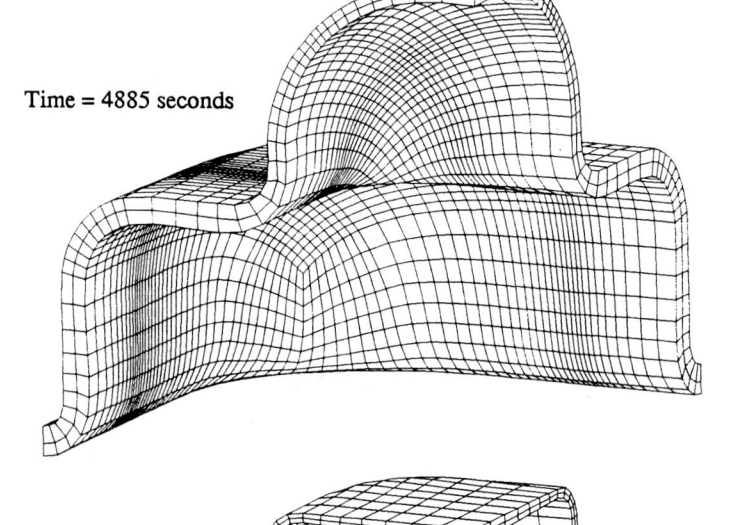

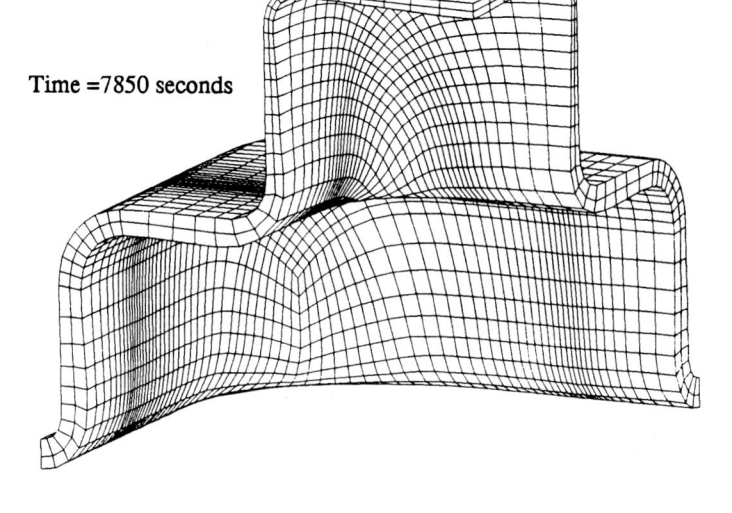

Time = 4885 seconds

Time = 7850 seconds

Figure 5 Double box shapes at various times

ACKNOWLEDGEMENTS

The authors gratefully acknowledge the financial support provided by British Aerospace, Airbus ltd. and the permision given to publish the developments reported.

REFERENCES

1- J.H. Argyris and I. St. Doltsinis, *A primer on superplasticity in natural formulation*, Compt. Meths. Appl. Mech. Engrg., **30**, 1984, pp. 83–132.

2- R.D. Wood and J. Bonet, *Superplastic forming simulation using the finite element method — A review*, Journal of Metals, Materials and Processes 4(3), 1993, pp. 229–257.

3- J. Bonet, R.D. Wood and A.H.S. Wargadipura, *Numerical simulation of the superplastic forming of thin sheet components using the finite element method*, Int. J. Num. Meth. Engrg. 30, 1990, pp. 1719–1737.

4- J. Bonet, A.H.S. Wargadipura and R.D. Wood, *A pressure cycle control algorithm for superplastic forming*, Comm. in Appl. Num. Meth. **5**(2), 1989.

5- J. Bonet, R.D. Wood and R. Collins, *Pressure control algorithms for the numerical simulation of superplastic forming*, Int. J. Mech. Sci., **36** (4), 1994, pp. 297–309.

6- A.K. Ghosh and C.H. Hamilton, *Mechanical behaviour and hardening characteristics of a superplatic Ti-6Al-4V alloy*, Metall. Trans. A, **10A**, 1979.

7- C.H. Hamilton, *Simulation of static and deformation enhanced grain growth effects on superplastic ductility*, Metall. Trans. A **20A**, 1989.

8- C.H. Hamilton, H.M. Zbib, C.H. Johnson and S.K. Richter, *Microstructural coarsening and its effects on localization of flow in superplastic deformation*, in E. Hori et al. (Eds.) *Superplasticity in Advanced Materials*, The Japan Society for Research on Superplasticity, 1991.

9- J.L. Chenot, M. Bellet, *The viscoplastic approach for the finite element modelling of metal forming processes*, in *Numerical Modelling of Material Deformation Processes: Research, Development and Applications*, Springer-Verlag, Berlin, in Hartley et al. (eds), 1992, pp. 179–224.

10- J. Bonet, *The incremental flow formulation for the numerical analysis of plane stress and thin sheet forming processes*, Comp. Meth. Appl. Mech. Engrg., **114**, 1994.

11- J. Bonet and P. Bhargava, *The incremental flow formulation for the numerical analysis of 3-dimensional viscous deformation processes: continuum formulation and computational aspects*, to appear in Comp. Meth. Appl. Mech. Engrg., 1995.

12- A. Gavrus, E. Massoni & J.L. Chenot, *Computer aided rheology for nonlinear large strain thermo-viscoplastic behaviour formulated as an inverse problem*, in *Inverse Problems in Engineering Mechanics*, But et al. (eds.), Balkema, 1994.

13- H.D. But, M. Tanaka et al., *Inverse Problems in Engineering Mechanics*, Balkema, 1994.

14- D.E. Goldberg, *Genetic Algorithms in Search, Optimization and Machine Learning*, Addison-Wesley, 1989.

NUMERICAL SIMULATION OF
SUPERPLASTIC FORMING/DIFFUSION BONDING PROCESSES

O. Baldó, J. Díaz
Construcciones Aeronáuticas, S.A., Av. John Lennon s/n, 28906 Getafe, Spain

F. Martínez, F.J. Beltrán, A. Mateos, G.Rico
Principia, S.A., Velázquez 94, 28006 Madrid, Spain

Abstract

The paper reports on the results of four numerical simulations of superplastic forming processes carried out for Construcciones Aeronáuticas, S.A. (CASA) by Principia, S.A. in order to show the capabilities of general purpose finite element code ABAQUS/Standard in this field. Each problem corresponds to a real manufactured component for which experimental data are available.

The first simulation deals with the forming of two simple cylindrical Al-Li cups. Several axisymmetric computations were made with different assumptions for the friction between the metal sheet and the die. The second analysis corresponds to the fabrication of Ti-6Al-4V parts. The third problem is aimed at the simulation of the forming of an Al-Li component with general three dimensional geometry. Finally the last analysis shows the simulation of the forming of Ti-6Al-4V aircraft leading edge.

In all cases the main point of interest is the final geometry of the formed parts, mainly the distribution of thicknesses, that the program is able to predict.

Numerical results are compared with experimental measurements and, from the comparison, conclusions are extracted about modeling assumptions and the overall suitability of the program for this kind of applications.

1. Introduction

In order to increase the range of applicability of superplastic forming (SPF) within the aerospace industry, one of the main problems faced by engineers is the difficulty in predicting the thicknesses of the formed parts.

The process produces a very high level of strain in the material. Strains on the order of several hundred per cent are not uncommon. As a result, dramatic reductions of the original plate thicknesses usually take place. Except for very simple die geometries the distribution of thicknesses which results from a given process is difficult to predict using simplified procedures. This is specially true when due account is taken of the influence of certain factors, such as the friction between plate and die, which are not easy to control.

A common way to tackle this problem is to carry out an extensive experimental program. However, the high cost of physical testing and the number of tests required to determine the optimum pressure cycle and the resulting thickness distribution clearly justify the use of a numerical simulation tool. The role of numerical simulation is to reduce the time and costs needed to design a particular forming process. In this context, four problems were set up by CASA engineers in order to assess the capabilities of the general purpose finite element code ABAQUS/Standard (HKS, 1993); the numerical predictions were to be compared with results already obtained after considerable experimental work.

The rest of this paper is dedicated to the description of these four problems, the way in which they were modeled with ABAQUS/Standard and the comparison of analytical and experimental results. The experimental results were not communicated to the modellers prior to conducting the analyses; the calculations are therefore true predictions, as opposed to "postdictions".

2. Simulation I: Forming of two "top hat" Al-Li 8090

2.1 Description

The first case deals with the forming of two "top hat" Al-Li 8090 parts. Figure 1 shows the corresponding die geometry. The formed parts are basically two 370mm diameter circular cylinders with a rim, resembling "hats". They have a very simple geometry, with axial symmetry. The difference between them is the depth of the hat, which is 71mm in one case (50% top hat) and 114mm in the other (100% top hat).

The process starts from a 2.0mm plate which is clamped at a radius of 213mm with respect to the axis of revolution of the die by means of a sealing lip.

The forming takes place at a constant temperature of 530°C. At this temperature the creep of the alloy has been experimentally determined to follow a potential law of the type:

$$\sigma = A \dot{\varepsilon}^m \tag{1}$$

where: σ = effective stress
$\dot{\varepsilon}$ = effective strain rate
A, m = material constants

The optimum strain rate for the material at the temperature of the process is approximately $10^{-3} s^{-1}$. The actual pressure cycles used to form the components were not given as part of the data for the simulation. The determination of an optimum pressure cycle for each of the components was part of the problem posed to Principia.

2.2 Finite element modelling

Taking advantage of the axial symmetry of the problem, 2D finite element models were used to simulate the forming process.

The material behavior was represented using a creep law of the type:

$$\dot{\varepsilon} = A \sigma^n t^m \tag{2}$$

with m=0 and A and n adjusted to the given data.

The two forming process were simulated using a range of different values for the friction coefficient between the superplastic alloy and the die.

The plate was assumed to be fully restrained at the position of the sealing lips. This boundary condition was considered adequate since the study of local effects near the sealing lips was outside the scope of the benchmark.

As an average, the resulting models had about 900 degrees of freedom.

The ABAQUS capabilities for automatic amplitude control were used in order to obtain a pressure cycle compatible with the optimum strain rate given for the material.

Each computer run took a few hours of CPU time in an Silicon Graphics Indigo MIPS 3000 workstation.

2.3 Results

Figure 2 shows the evolution of the deformed shape during the simulation of the forming of the 100% top hat.

Figure 3 compares the analytical and experimental thickness profiles along sections of both components. The solid lines correspond to the experimental profiles and the shaded areas represent ABAQUS predictions for a particular value of the friction coefficient taking into account the initial tolerance for plate thickness (± 0.1mm). It can be seen that a fairly good correspondence exists between analysis and experiment except for the area near the boundaries of the finite element model, where the discretization cannot represent local effects accurately.

As could be expected, the agreement between experiments and computations is better for the 50% top hat, in which deformation levels are smaller.

In figure 4 the optimum pressure cycle computed by ABAQUS is compared with the pressure cycles used in the experiment. Significant differences are apparent both in respect of the total time and the maximum pressure. The optimization algorithm used by ABAQUS produced shorter cycles with higher maximum pressures. Nevertheless, it is well known that the influence of the pressure cycle on the final predictions for thickness distribution is small; this was verified in all cases.

3. Simulation II: Forming of a Ti-6Al-4V aircraft component

3.1 Description

The second numerical simulation has a more of qualitative nature. The goal in this case was to test whether the program was able to predict the occurrence of a specific geometrical effect during the manufacturing process of a certain Ti-6Al-4V alloy component. This effect was unforeseen at the time of designing the process but was observed during prototype testing. Its nature will be discussed in the section on results.

The superplastic forming of this component is preceded by the diffusion bonding (DB) of a package of four 0.8mm plates. A picture of the actual part can be seen in figure 5. Figure 6 presents a schematic view of the idealized two dimensional process which was the object of the numerical simulation.

The process has two main stages. The first one is the diffusion bonding of the plate package. During this phase the plates are heated up to 925°C and subjected to a very high pressure, typically 2MPa, for a long period of time, around 2 hours. The plates become perfectly bonded everywhere except in the areas previously impregnated with an inhibiting chemical ("stop-off" chemical).

The simulation of this phase is of some interest because it helps in predicting potential defects (pores) in the bonded package.

After the first stage of diffusion bonding of the plates, the superplastic forming takes place. This is accomplished by injecting gas between the plates in the region which remained unbonded due to the presence of the stop-off material.

At the process temperature the behavior of the alloy can be represented by a law similar to (1). As in the previous problem, the pressure cycle for the second phase of the process was not supplied as part of the data. The pressure cycle for the superplastic forming had to be determined with the knowledge that the optimum strain rate for the alloy at the temperature of the process was $2 \times 10^{-4} s^{-1}$.

3.2 Finite element modelling

The assumption of plane-strain conditions was considered acceptable. Friction was taken into account at all contacts. The bond in the areas where the stop-off material was not present was modelled. The edges of the model were considered clamped, without attempting to capture local effects in these areas. Once again, the automatic pressure control capability within ABAQUS was used in order to determine the optimum pressure cycle.

Figure 7 shows a view of the finite element model, which has over 4500 degrees of freedom. The computer run took approximately two days of CPU time using the same workstation as in the first benchmark.

5.3 Results

Figures 8 to 10 summarize the results of the simulation. Figure 8 shows the configuration of the plates at various stages during the diffusion bonding phase and figure 9 gives a sequence of deformed shapes in the superplastic forming of the product. The simulation was interrupted when a buckling instability appeared in the straight portion of the component. At this moment the computed geometry of the corrugated area was that shown in figure 10 together with the experimentally observed geometry.

As can be seen in figure 8, a small pore is left after bonding of the outer plates. For a more rigorous analysis, the mesh should probably be refined in the area.

It was mentioned earlier that the ability to predict a specific effect provided the motivation for conducting this analysis. The observation was that, from the beginning of the superplastic forming phase, the edge of the bonded region penetrates into the cavity of the dies (figure 9). The amount of penetration increases until the plates establish contact with the dies. From this moment onwards the penetration reduces but cannot be totally eliminated; consequently, the component is not formed properly.

The comparison with experimentally obtained shapes (figure 10) demonstrates that the simulation is able to capture well the actual phenomenon.

4. Simulation III. Forming of an Al-Li 8090 product

4.1 Description

The third problem proposed was of a three dimensional nature. The goal was to simulate the superplastic forming of an Al-Li 8090 product and to obtain a map of the resulting thicknesses.

Figure 11 shows the geometry of the die. It covers an area of $932 \times 622 mm^2$ and its maximum depth is 60mm. The process starts from a 2.0mm plate and takes place at a temperature of 530°C. The material behavior is that already described in the first simulation.

As in previous cases, the questions posed included the determination of a pressure cycle compatible with the range of strain rates required for superplastic behavior.

4.2 Finite element modelling

Since the die has two planes of symmetry, only a quarter of the plate and die was included in the finite element model.

The contact with the die was simulated by means of rigid surface contact elements. The geometry of the die was represented using a Bezier surface defined by 3400 triangular patches. A friction coefficient was assigned to the contact.

The plate was clamped along the edges of the die and symmetry conditions were imposed at the other boundaries of the plate.

The model geometry is shown in figure 12. The mesh had a total of 10500 degrees of freedom. It should be said that, when the problem was solved as described above, the computer time requirements were extremely high: several weeks of CPU time in the same workstation mentioned earlier. The reason for this inordinate time requirements was not clear. However, the analysts have the feeling that the contact logic with Bezier surfaces may strongly benefit from some optimization.

4.3 Results

The results of the simulation are summarized in figures 13 to 15. Figure 13 shows the final deformed shape. In figure 14 two experimentally obtained thickness profiles along orthogonal directions are compared with ABAQUS predictions. The solid lines correspond to experimental results and the shaded areas represent analytical predictions, taking into account the tolerance for initial plate thickness (±0.1mm). It can be seen that the predictions are quite good, always within the margins of tolerance.

Note that, in this three dimensional case, it is extremely difficult to make "manual" or even intuitive predictions on what will be the shape of the final thickness distribution because it is highly dependent on the sequence in which the different areas of the plate come into contact with the die.

Finally, figure 15 compares the pressure cycle computed by ABAQUS with the actual pressure cycle used to form the component. Both cycles have the same shape and approximately the same maximum pressure.

5. Simulation IV: Forming of a Ti-6Al-4V leading edge

5.1 Description

The fourth problem proposed was also three dimensional. The goal was to simulate the forming of a Ti-6Al-4V aircraft leading edge with the aim of comparing the final geometric configuration and the thickness distribution with the experimental results. The forming, as the second simulation explained previously, has two stages.

In the first stage two different plates are subjected to a diffusion bonding process. The upper plate is 0.8mm of thickness while the lower is 4.4 and 3.0mm of thickness as is depicted in figure 16. The plates are previously heated up to 925°C. The bonding is accomplished applying a pressure of 2.1MPa upwards during nearly three hours. In this case the bonding takes place at the same time the plates are formed against an upper die. At the end of this stage the two plates are perfectly bonded except in those areas previously impregnated with an inhibiting chemical ("stop-off" chemical). The aim of this previous forming is to accomplish a corrugated longeron within the leading edge.

The second stage is properly the superplastic forming process. This is achieved applying a gas pressure between the two plates in the area where remains unbonded as a result of the first phase. The pressure cycle is not **known.It** is obtained as a consequence of the numerical simulation keeping the strain rate of the alloy within an optimum range of $2 \ 10^{-4} \ s^{-1}$.

5.2 Finite element modelling

Only a section between two parallel planes was included due to the symmetry of model. The contact with the two dies was modelled by means of rigid surface contact elements. The plate was clamped along the edges of the dies and symmetry conditions were imposed at two parallel planes.

The mesh has a total of 17000 degrees of freedom. The actual process took around three hours, while the simulation required four days of CPU time in the same workstation mentioned in previous sections.

5.3 Results

The results of the simulation are summarized in figure 17. This figure depicts the final deformed shape at the end of each stage. In the same figure it is represented the comparison between the experimentally obtained thickness profiles along the nose and the longeron and the numerical predictions. In this case the initial plate thickness tolerance is ±0.2mm. It can be seen that the predictions are again quite good.

6. Conclusions

The results of the four simulation problems presented here are very satisfactory and indicate that ABAQUS can be a very useful tool in the first stages of the design of a forming process. Such a tool allows making accurate predictions of the thickness reductions, timely detecting unforeseen phenomena and, generally, better understanding the mechanisms of a prototype process.

As in other fields, numerical simulation here can lead to substantial savings in the time needed to develop a particular fabrication sequence, reducing to a minimum the number of physical tests. The latter is specially true in the case of DB/SPF combined processes, where a large number of attempts may be necessary to define an appropriate package of plates with proper distribution of stop-off material.

On the other hand, it should be noted that the computer times involved, specially in three dimensional cases, may be significant. The finite element models used in the some of the simulations were not optimized in this sense and some reductions on the required CPU time could have been achieved with additional work. In any case, computer resources is a factor to bear in mind when dealing with this kind of simulations.

7. References

Hibbitt, Karlsson & Sorensen, Inc. (1993). "ABAQUS User's Manual". Version 5.3.

Figure 1.   Simulation I. Geometry of "top hat" parts

Figure 2.   Simulation I. Sequence of deformed shapes in 100% top hat simulation

Figure 3.   Simulation I. Comparison between analytical & experimental thicknesses

Figure 4.   Simulation I. Comparison between analytical & experimental pressure cycles

Figure 5.   Simulation II. Geometry of real component

Figure 6.   Simulation II. Idealized process

Figure 7.   Simulation II. Finite element model

Figure 8.   Simulation II. Deformed shapes in DB simulation

Figure 9.   Simulation II. Deformed shapes in SPF simulation

Figure 10.   Simulation II. Comparison with experimental deformed shapes

Figure 11.   Simulation III. Geometry of die

Figure 12.   Simulation III. Finite element model

Figure 13.   Simulation III. Deformed shapes

Figure 14.   Simulation III. Profiles of thickness

Figure 15.   Simulation III. Pressure cycles

Figure 16.   Simulation IV. Geometry of dies

Figure 17.   Simulation IV. Comparison between analytical & experimental thicknesses

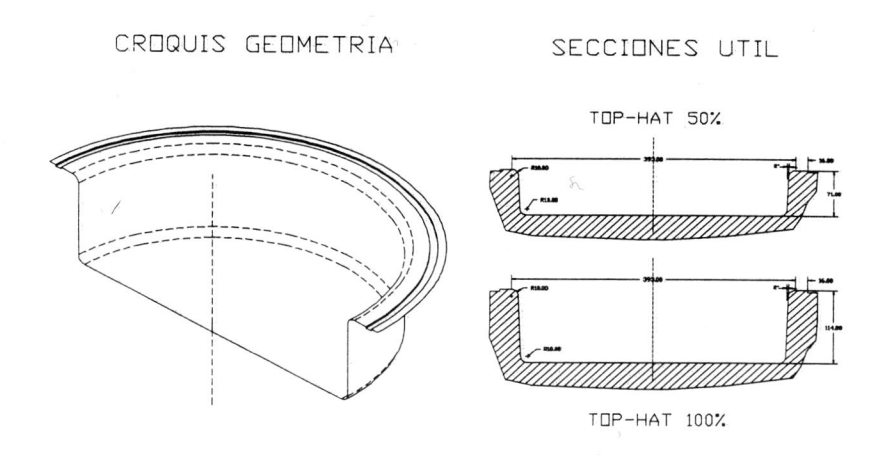

Figure 1. Simulation I. Geometry of "top hat" parts

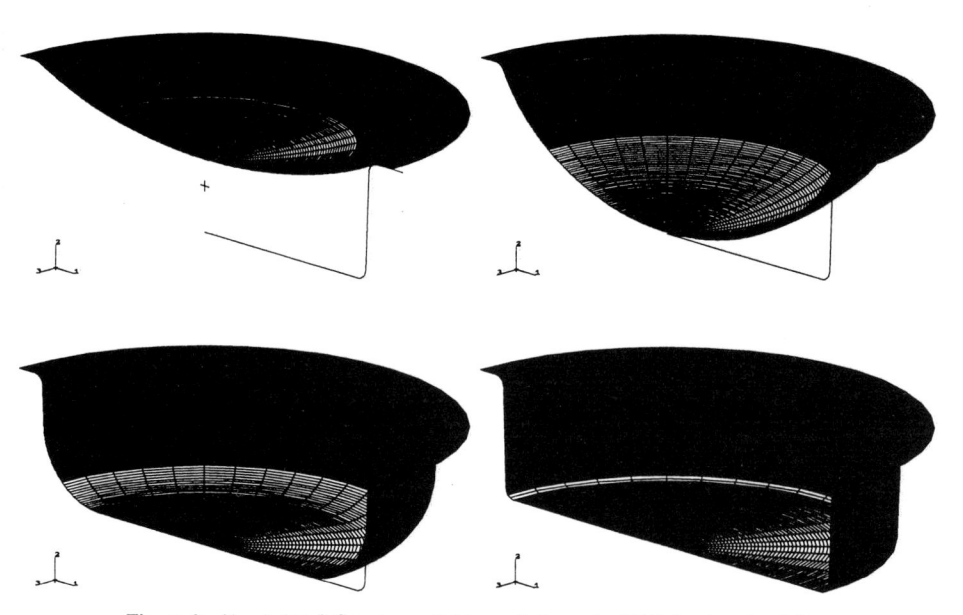

Figure 2. Simulation I. Sequence of deformed shapes in 100% top hat simulation

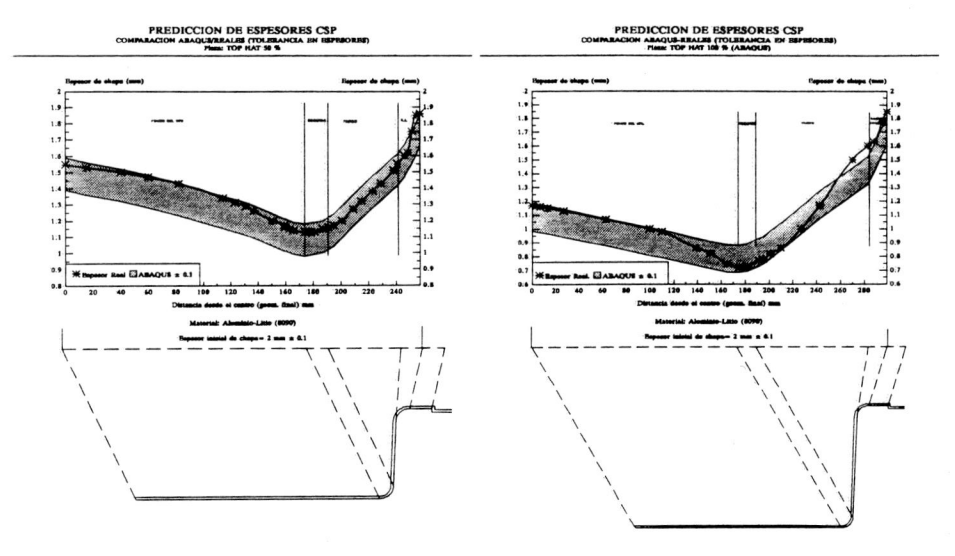

Figure 3. Simulation I. Comparison between analytical & experimental thicknesses

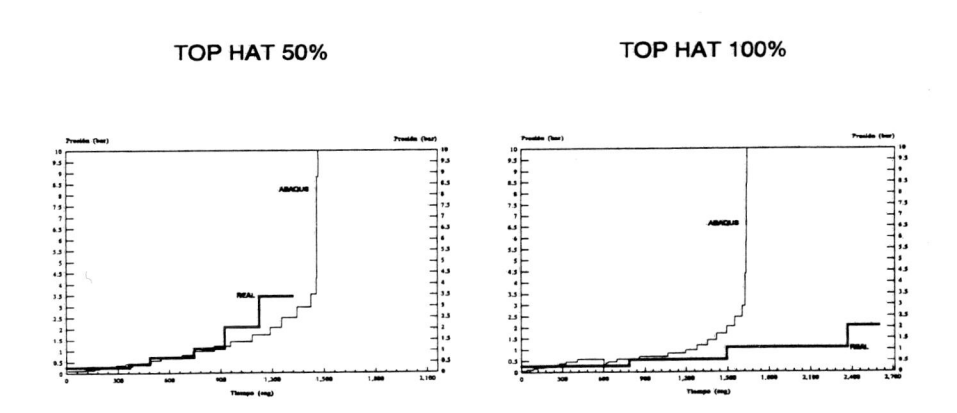

Figure 4. Simulation I. Comparison between analytical & experimental pressure cycles

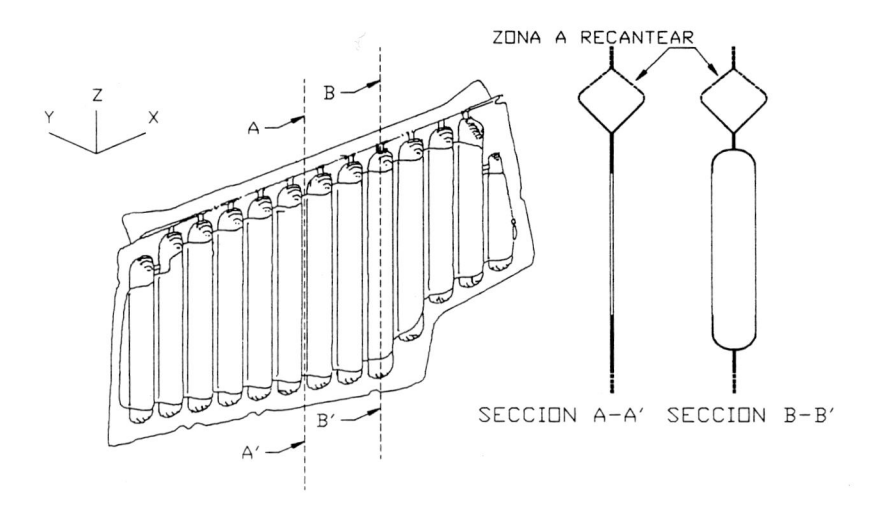

Figure 5. Simulation II. Geometry of real component

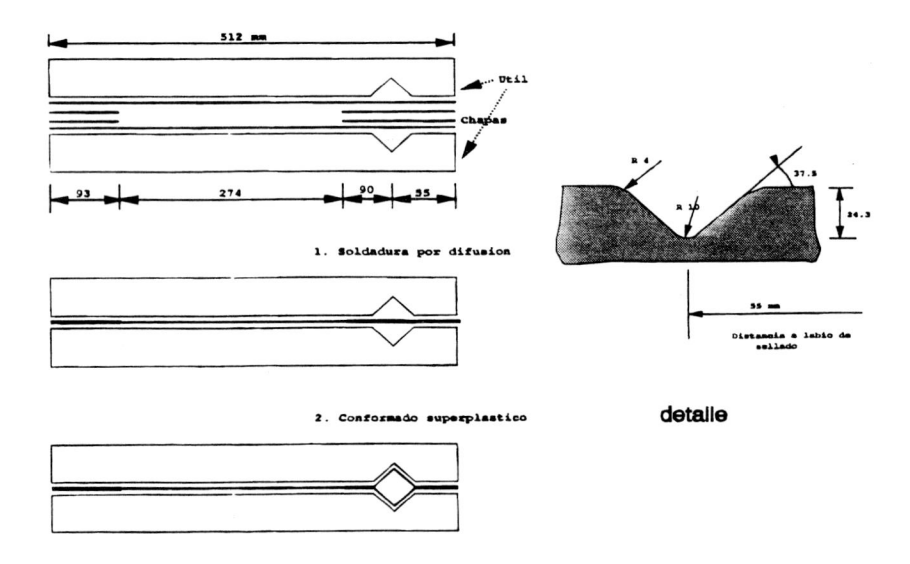

Figure 6. Simulation II. Idealized process

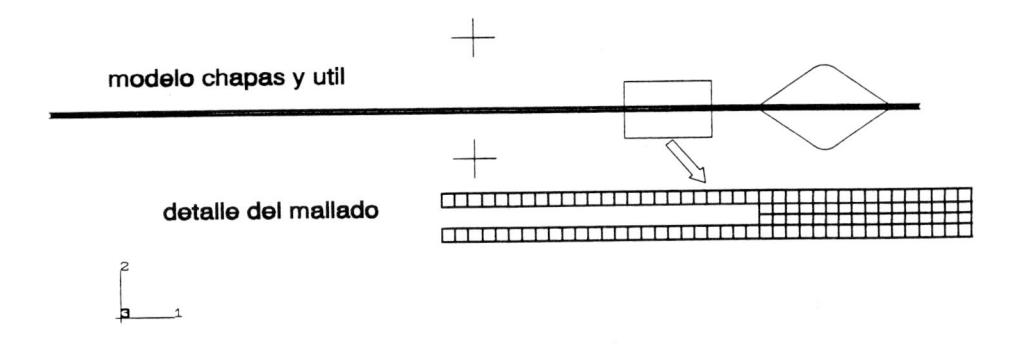

Figure 7. Simulation II. Finite element model

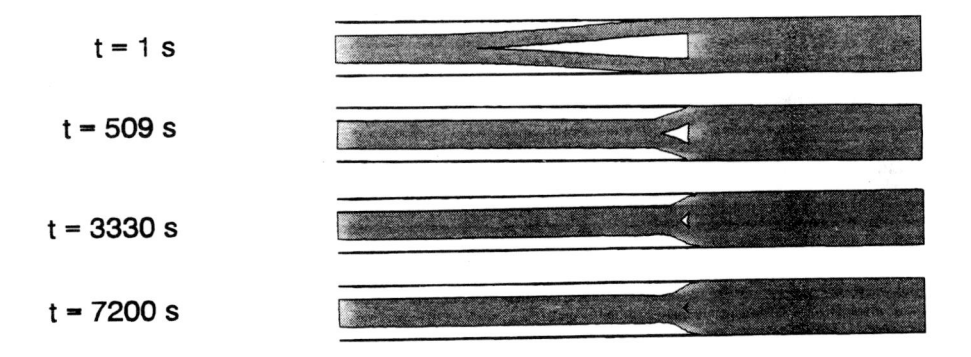

t = 1 s

t = 509 s

t = 3330 s

t = 7200 s

Figure 8. Simulation II. Deformed shapes in DB simulation

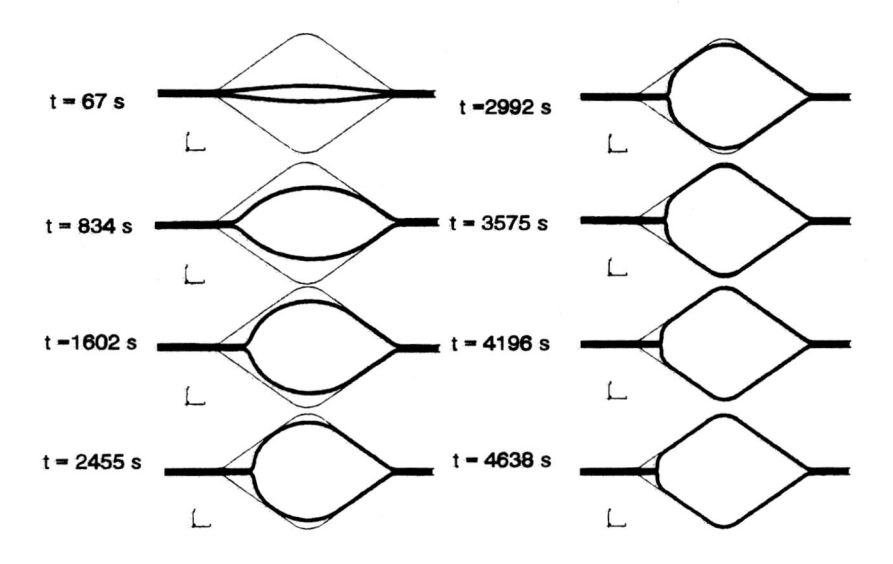

Figure 9. Simulation II. Deformed shapes in SPF simulation

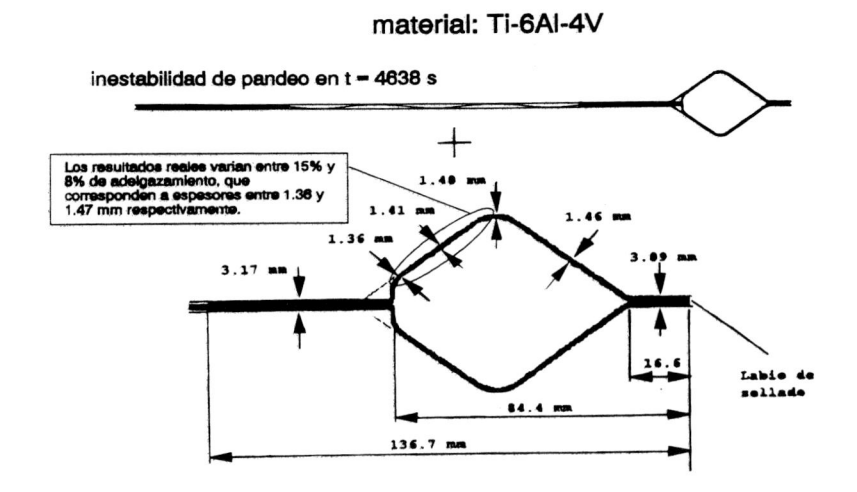

material: Ti-6Al-4V

Figure 10. Simulation II. Comparison with experimental deformed shapes

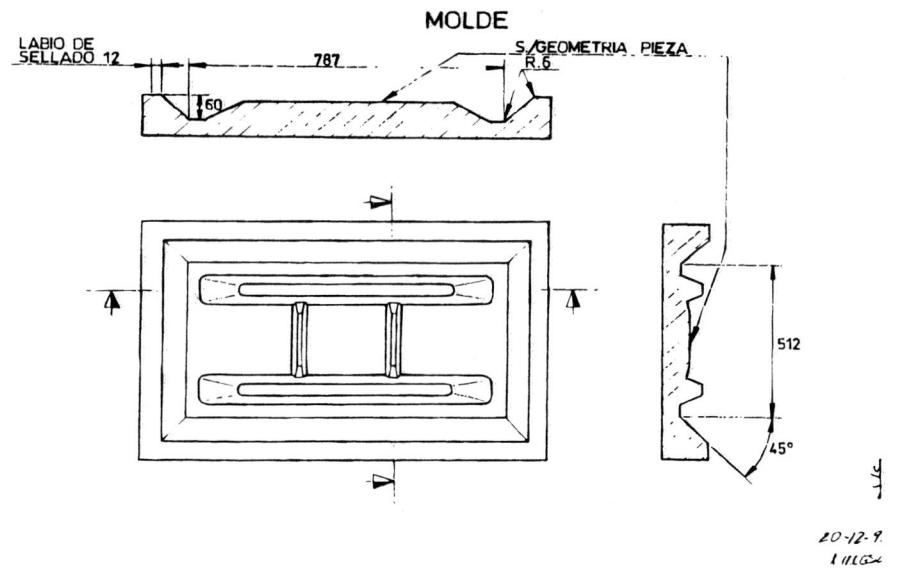

Figure 11. Simulation III. Geometry of die

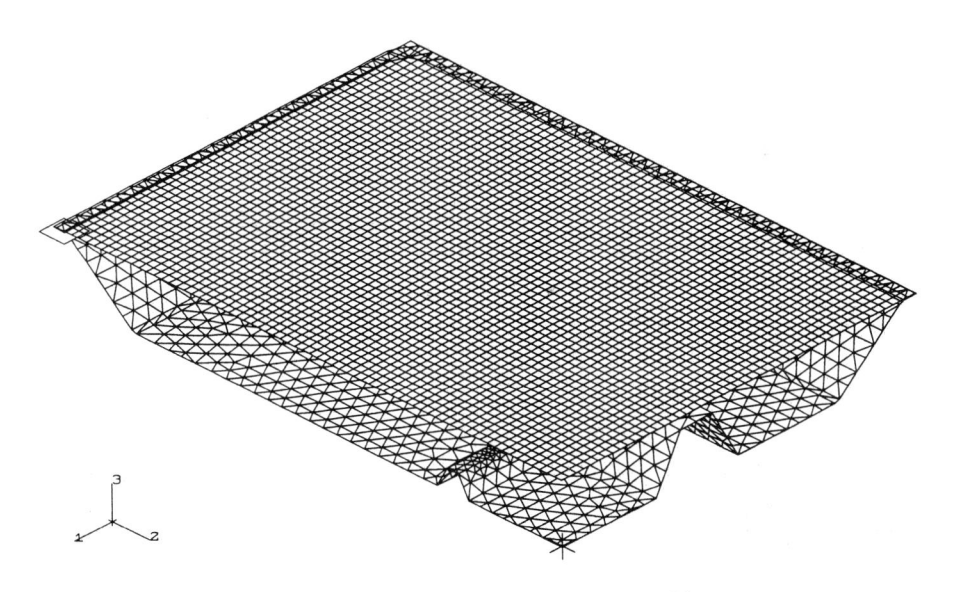

Figure 12. Simulation III. Finite element model

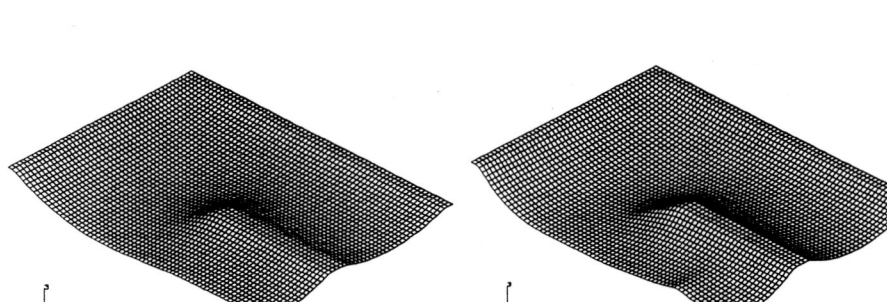

Figure 13. Simulation III. Deformed shapes

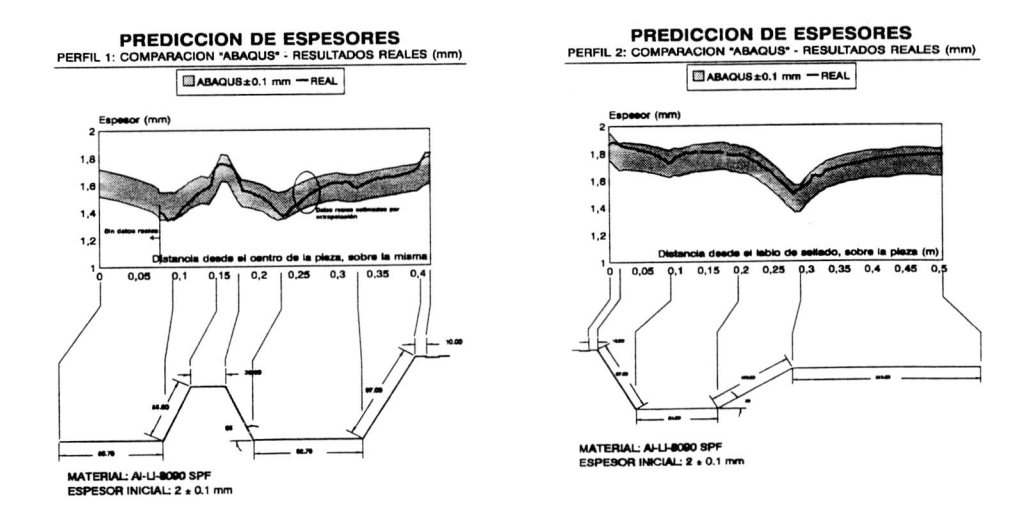

Figure 14. Simulation III. Profiles of thickness

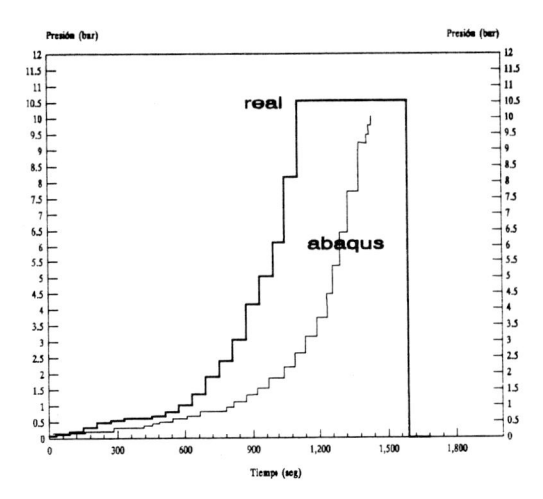

Figure 15. Simulation III. Pressure cycles

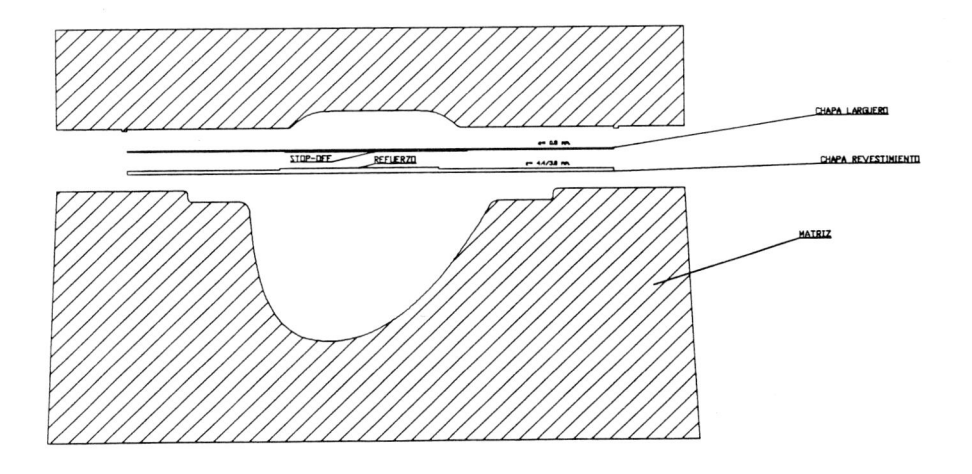

Figure 16. Simulation IV. Geometry of dies

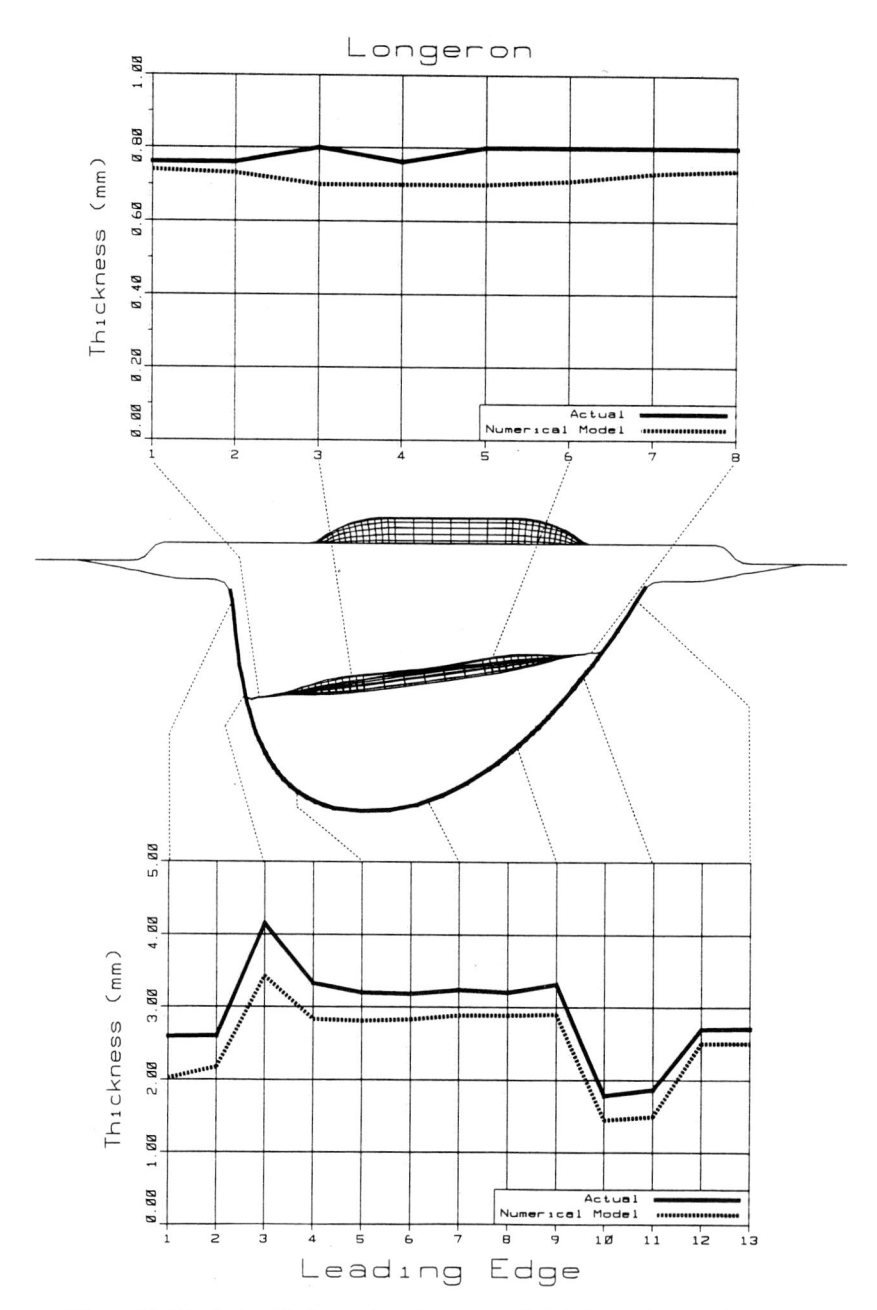

Figure 17. Simulation IV. Comparison Between analytical & experimental thicknesses

Simulation of Uni-axial Superplasticity Specimen Testing in Ti-6Al-4V

F. P. E. Dunne, I. Katramados

Department of Mechanical Engineering, UMIST, Sackville st, PO Box 88, Manchester, M60 1QD, UK

Abstract

Multiaxial constitutive equations have been given for the superplastic deformation of a titanium alloy. The equations include a description of grain size kinetics, and their coupling with the deformation, hence enabling the influence of microstructure on superplastic deformation to be modelled.

The equations have been implemented in a general purpose non-linear finite element solver, and the model employed to investigate the suitability of a typical uniaxial test specimen for the generation of superplastic materials data. While the specimen design enables the establishment of uniform gauge length stress and strain fields, the strain rates obtained are not necessarily those that are desired, and may lead to the determination of stresses that are up to ±20% in error. A modified displacement loading history has been developed to reduce the error.

1. Introduction

Superplastic forming is an important manufacturing process that is employed, for example, for the production of fan blades for aero-engines. An important feature of the process is its ability to enable very large strains to be achieved in the absence of necking, cavitation, and failure. It is therefore used in a range of applications requiring the forming of metal sheet at large strains, under tensile loading.

In order to facilitate superplastic behaviour in a material, it is necessary to prescribe and control the regimes of strain, strain rate and temperature applied [1]. The regimes of strain rate and temperature required for ideal superplastic behaviour may be rather narrowly defined. An additional feature that is recognised to be important in influencing superplastic material behaviour is the grain size [2]. Usually, a small (<10μm), equi-axed grain size is required facilitating grain boundary sliding, which is thought to be the predominant deformation mechanism of superplasticity occurring simultaneously with grain accommodation through diffusion controlled processes [3]. However, in practice, it is not usually possible to obtain the desired completely uniform grain size, and quite considerable variations can be seen in the commercial grade titanium alloy Ti–6Al–4V [2], for example.

During the superplastic forming of engineering components, because of the geometrical and loading conditions, it is not possible to maintain uniform plastic strain rates in the deforming material. The resulting stress fields are therefore also non-uniform. Regions of the material suffering the highest levels of strain may also be subject to local thinning, in the out-of-plane direction, and necking leading possibly to rupture. Occurrences such as these are not permitted in components to be employed, for example, in critical, high performance applications. In order to enable the design of superplastic forming processes that avoid the problems discussed above, it is necessary to establish constitutive equations for superplastic behaviour which, when implemented in appropriate finite element software, enable the simulation of superplastic forming.

1.1 Materials Data for Superplasticity in Ti–6Al–4V

Because of its importance as a structural alloy in aero-engine components, and because of the availability of materials data in the literature, the titanium alloy Ti–6Al–4V is chosen for consideration in the present paper. Ghosh and Hamilton [4] have carried out uniaxial tensile tests on specimens cut from Ti–6Al–4V sheet. The specimen employed is shown in figure 1. The ends of the specimen are gripped between two plates under load, imposed by means of pins, which pass through the specimen ends. The hole through which a pin passes is shown, for example, centred at D in figure 1. The axial load through the specimen is imposed partly through the pin, and partly through the gripping plates discussed above. Poor gripping may lead to the majority of the load being transmitted by the pin, the consequences of which are discussed later. In an attempt to ensure that the gauge length of the specimen undergoes constant strain rate loading, variable displacement control was employed by Ghosh and Hamilton which was calculated on the basis that only the specimen gauge length deforms; that is, the specimen ends are assumed to be rigid. The effectiveness of this approach, and validity of the assumption, are also addressed in a later section.

A sample of the experimental data obtained by Ghosh and Hamilton [4] is shown by the symbols in figures 2(a) and (b). The experimental true stress versus true strain data in figure 2(a) for the three strain rates $5 \cdot 0 \times 10^{-5} \text{ s}^{-1}$, $2 \cdot 0 \times 10^{-4} \text{ s}^{-1}$, and $1 \cdot 0 \times 10^{-3} \text{ s}^{-1}$ show that a pronounced strain rate effect exists for this material at 927°C. By carrying out microstructural examination, Ghosh and Hamilton were also able to identify the occurrence of significant grain growth, as shown in figure 2(b) by the symbols.

1.2 Constitutive Equations for the Superplastic Behaviour of Ti–6Al–4V

Uniaxial constitutive equations, and computational techniques for their determination, have been established by Zhou and Dunne [5] for the superplastic behaviour of Ti–6Al–4V. For the temperature and strain rate range considered, the equations reflect the likelihood of the predominant deformation mechanisms occurring being those of grain boundary sliding together with grain accommodation by dislocation creep. The processes assumed to lead to the material hardening seen are grain growth and the hardening processes associated with dislocation creep. The mechanisms identified by the techniques discussed in [5] are in accord with those discussed by Ghosh and Hamilton [6] and Pan and Cocks [3].

The constitutive equations have been shown to provide a good representation of the uniaxial behaviour of Ti–6Al–4V observed experimentally. In particular the equations successfully predict the evolution of grain size during superplastic deformation. The strain hardening observed to occur through grain growth [4] is also predicted correctly by a direct coupling of the grain size with the strain rate equation, and the true stress-true strain behaviour is therefore correctly represented over the broad range of strain rate considered. It is for the reasons given above that the constitutive equations of Zhou and Dunne [5] are employed in the present work, and are discussed further in a subsequent section.

1.3 Scope of the Present Paper

In the following section, constitutive equations are discussed for the titanium alloy, which are implemented in an appropriate finite element solver enabling solution of initial value - boundary value problems. A typical uniaxial superplasticity test specimen is then examined using the constitutive equations for Ti–6Al–4V implemented in the finite element solver. The suitability of the specimen design for the generation of materials data for superplasticity is evaluated.

2. Constitutive Equations

An elastic-viscoplastic material model is employed for the titanium alloy considered. The importance of rate effects, and time-dependent processes that occur for the material over the range of temperature and strain rate necessitate the use of such a model. The mechanistic background to the development of the constitutive equations is discussed by Zhou and Dunne [5], but the equations are given here in multiaxial form as

$$\underset{\sim}{D}_p = \frac{3\alpha}{2d^\mu} \operatorname{Sinh} \beta \left[J(\underset{\sim}{\sigma} - \underset{\sim}{x}) - k \right] \frac{\underset{\sim}{\sigma}' - \underset{\sim}{x}'}{J(\underset{\sim}{\sigma} - \underset{\sim}{x})} \qquad (1)$$

$$\overset{\nabla}{\underset{\sim}{x}} = \frac{2}{3} C \underset{\sim}{D}_p - \gamma \, \underset{\sim}{x} \dot{p} \qquad (2)$$

$$\dot{d} = \left(\alpha' + \beta \dot{p} \right) / d^{\gamma'} \qquad (3)$$

$$\overset{\nabla}{\underset{\sim}{\sigma}} = E \left(\underset{\sim}{D} - \underset{\sim}{D}_p \right) \qquad (4)$$

and where

$$\dot{p} = \left(\frac{2}{3}\underset{\sim}{D}_p : \underset{\sim}{D}_p\right)^{\frac{1}{2}} \tag{5}$$

$$J\left(\underset{\sim}{\sigma} - \underset{\sim}{x}\right) = \left[\frac{2}{3}\left(\underset{\sim}{\sigma}' - \underset{\sim}{x}'\right) : \left(\underset{\sim}{\sigma}' - \underset{\sim}{x}'\right)\right]^{\frac{1}{2}} \tag{6}$$

$\underset{\sim}{D}_p$ is the rate of plastic deformation, d the grain size, $\underset{\sim}{\sigma}$ the Cauchy stress (which may be considered since the elastic strains are infinitesimally small, making the Kirchhoff stress almost identical to the Cauchy stress), $\underset{\sim}{x}$ an internal variable to describe hardening associated with dislocation creep and grain accommodation, and $\underset{\sim}{D}$ the rate of total deformation. $\underset{\sim}{E}$ is the elasticity tensor, and $\alpha,\beta,\kappa,\mu,C,\alpha',\beta'$ and γ' are material dependent parameters. The Jaumman rate of stress $\overset{\triangledown}{\underset{\sim}{\sigma}}$ is calculated from

$$\dot{\underset{\sim}{\sigma}} = \overset{\triangledown}{\underset{\sim}{\sigma}} + \underset{\sim}{W}\underset{\sim}{\sigma} + \underset{\sim}{\sigma}\underset{\sim}{W}^T \tag{7}$$

in which $\underset{\sim}{W}$ is the spin tensor. The constitutive equations have been implemented in the general purpose non-linear finite element solver ABAQUS [6] by means of a user-defined subroutine, employing an explicit time integration scheme.

2.1 Model Validation

In order to validate the model, the uniaxial, plane stress behaviour of Ti–6Al–4V was simulated under conditions of constant strain rate loading at a temperature of 927°C. The strain rates considered are those employed in the experiments of Ghosh and Hamilton [4], and by Zhou and Dunne [5] in the independent numerical integration of the constitutive equations (1) to (6).

A simple four element mesh was considered, employing eight-noded, plane stress, quadrilateral elements, and the results of the calculations are shown in figures 2(a) and (b). The symbols shown in the figures represent the experimentally obtained data of Ghosh and Hamilton [4], and the solid and broken lines are the predictions obtained by two methods: firstly, the uniaxial form of the constitutive equations given in (1) to (6) were integrated numerically and secondly, predictions were obtained using ABAQUS as discussed above. The broken line is used only to indicate that the experimental data corresponding to this strain rate was not employed in the fitting process to obtain the material parameters [5]. The finite element solutions obtained using the constitutive equations can be seen to be identical to those obtained from the independent numerical integration scheme, which in turn, can be seen to compare well with the experimentally obtained data. Having validated the implementation of the constitutive equations into ABAQUS, the model is employed to evaluate the suitability of a test specimen design for the generation of superplasticity materials data. This is discussed in the following section.

3. Evaluation Of Test Specimen Design

The test specimen considered in the present work is that used by Ghosh and Hamilton [4] for the generation of the superplasticity materials data shown in figures 2(a) and (b). The specimen is shown in figure 1.

It was mentioned in the introduction that the axial loading can be transmitted to the specimen in two ways. The specimen ends are each gripped by means of two plates which are loaded in the direction normal to the specimen plane. The axial load is therefore transmitted to the specimen by shear at the interface of the loading plates with the specimen faces. However, because the uniaxial strains are large, the specimen consequently thins, and the frictional constraint, by which the shear force is transmitted may drop, therefore enabling some sliding of the loading plates relative to the specimen ends. As a result of this, some of the axial load carried by the specimen may be transmitted through the pins, which pass through holes in the specimen ends, shown in figure 1. Testing under these conditions leads to elongation of the pin holes, which is observed in experiments.

The exact conditions of loading vary from test to test, and are therefore difficult to model. Consequently, two extreme conditions are considered so that the bounds of the resulting specimen behaviour may be established. One extreme assumes that the gripping is perfect so that no load is transmitted through the pin. This can be modelled by the imposition of displacement control on an appropriate horizontal boundary of the specimen which is chosen here as line CD in figure 1. The other extreme is that all the axial load is transferred through the pin, and that no load is transferred through the grips. This can be modelled by imposing displacement control on a rigid pin located in the hole in the end of the specimen. Through symmetry conditions, only a quadrant of the specimen can be considered, and for the two cases outlined above, the two regions modelled are shown in figure 1. The boundary conditions imposed in each case are shown in figures 3(a) and (b). For the case of ideal gripping, the hole is not considered in the model, as shown in figure 3(a). The prescribed displacement is imposed along the top boundary CD. In the case of the load being imposed entirely through the pin, the boundary conditions are shown in figure 3(b). In this case, the hole in the specimen end is included, and the prescribed displacement is imposed by means of a rigid pin. The interacting surfaces between the specimen and the pin are assumed to be frictionless.

3.1 Prescribed Displacement

The prescribed displacement imposed is that specified by Ghosh and Hamilton [4] in which the displacement rate was chosen in such a way that constant true strain rate could be maintained in the specimen gauge length. The assumption made by Ghosh and Hamilton is that the specimen ends, because of their significantly increased width, are almost rigid in comparison to the gauge length material. The prescribed displacement, d, therefore varies with time in the following way:

$$d = l_o \left(e^{\dot{\varepsilon} t} - 1 \right) \tag{8}$$

in which l_o is half of the initial specimen gauge length, $\dot{\varepsilon}$ the required strain rate, and t the time. The results of the test simulations for each set of loading conditions are discussed in the following section.

3.2 Results of Test Simulations

For the case of each of the loading conditions shown in figure 3, a strain rate of $2 \cdot 0 \times 10^{-4}$ s^{-1} was selected, and the specimens deformed upto a gauge length strain of approximately 1·1. Results, at this strain level, for the conditions shown in figure 3(a) are shown in figure 4.

The axial stresses (yy–stresses) generated at a gauge length strain of 1·1 are shown in figure 4(a). The original, undeformed configurations of the specimens are shown by the broken black lines in the figures. The results show for this loading case that the stresses generated in the gauge length region are uniform, as desired. In addition, comparison of the stress achieved with that shown in figure 2(a) shows reasonable agreement, but that the stresses are approximately 20% higher, indicating that the stress-strain data obtained from tests carried out using this specimen are good approximations of the true material uniaxial, stress-strain behaviour. While some non-uniformity of stress can be seen in the region of the change of section of the specimen, the stresses in the important region of the gauge length at the specimen mid-point are uniform. The differences in stress observed are discussed later.

Figure 4(b) shows, for the same specimen in the same configuration, the predicted grain size. Again, while a non-uniform field of grain sizes exists in the specimen, the grain size in the important region of the gauge length at the specimen mid-point, can be seen to be uniform. For the purposes of this analysis, the initial grain size was assumed to be uniform throughout the specimen, and to have a value of $6 \cdot 4 \mu m$, which was measured by Ghosh and Hamilton [4]. The final grain size in the gauge length region can be seen to be approximately $10 \cdot 0 \mu m$, which compares reasonably well with that shown in figure 2(b), and calculated on the basis of true uniaxial loading conditions.

The results obtained based on analysis of the specimen loading conditions shown in figure 3(b) are shown in figure 5. Figure 5(a) shows the predicted axial (yy) strain field obtained for a gauge length strain of approximately 1·0. The strain in the region of the gauge length of the specimen can be seen to be uniform. The deformed configuration of the specimen shows that considerable elongation of the originally circular hole has taken place, resulting in strains less than that desired in the gauge length region at any given time. This, in effect, results in strain rates in the gauge length region that are less than those desired and specified by the prescribed displacement control. Specimen gauge length strain rates are addressed in the next section. Elongation of the pin holes in the direction of loading has been observed in experiments on a considerable number of occasions. Figure 5(b) shows, for the same specimen configuration, the resulting grain sizes at the end of the deformation, again predicted on the basis of an initially uniform grain size of $6·4\,\mu m$. The final grain size in the gauge length region can be seen to be close to that obtained on the basis of the previous loading conditions.

3.3 Specimen Gauge Length Strain Rates

It is important to establish that the strain rates imposed on the gauge length material are in fact those that are required, and specified by controlling the variation of the specimen grip displacement with time. This is the case because of the very strong rate dependency of the titanium alloy considered, which shows significant changes in stress level generated for a given level of strain, for comparatively small changes in strain rate. For example, figure 2(a) shows that for a given strain level, increasing the strain rate from $2·0 \times 10^{-4}\,s^{-1}$ to $1·0 \times 10^{-3}\,s^{-1}$, that is a factor of increase of 5, leads to an increase in the stress of 100%.

Strain rates were calculated for the specimen gauge length mid-point, on the basis of the loading conditions shown in figures 3(a) and (b), and are plotted against the axial strain at the same point over the entire loading history of the specimen. The results are shown in figure 6, in which the calculated strain rates have been normalised by the desired constant strain rate, labelled the nominal strain rate, in the figure. Hence, a normalised strain rate of 1 indicates that the strain rate in the gauge length is equal to that desired. The results therefore show that a considerable variation in strain rate occurs. The strain rate varies from a minimum of approximately 0·5, to a maximum of 1·3 of the nominal strain rate.

It is anticipated that early in the loading history, the gauge length strain rates are likely to be less than the nominal strain rates, since it is assumed by Ghosh and Hamilton in choosing the loading conditions that there is no strain in the specimen ends. The effective gauge length of the specimen is therefore somewhat larger than assumed leading to reduced normalised strain rates. As the deformation continues, however, the normalised strain rate can be seen to increase for both sets of loading conditions, and at a strain of approximately 0·75, it exceeds unity; that is, the gauge length strain rate has exceeded that originally specified as the constant strain rate. Because the loading is displacement controlled, this is only possible with a reduction in the effective gauge length. Examination of figures 5(a) and 5(b) reveals how this takes place.

As the deformation proceeds in the gauge length, the average grain size increases, leading to material hardening. However, a region of the specimen can be seen in figure 5(b) in which a smaller grain size exists, permitting further deformation to take place preferentially, and at a higher rate than in the gauge length. This ultimately enables deformation to take place preferentially within an effective gauge length which is smaller than the true gauge length, therefore leading to nominal strain rates greater than 1·0.

Figure 6 also demonstrates that for the loading conditions shown in figure 3(b), that is, for the case in which load is transferred through the pin, considerably larger variations in strain rate arise than for the conditions shown in figure 3(a), as would be expected.

3.4 Discussion of Results

The results of the analyses demonstrate that uniform states of strain, stress, and grain size are generated within the gauge length region near the mid-point of the specimen. The strain, stress and grain size measurements obtained experimentally from such a specimen design can therefore be considered to be representative for uniaxial loading. The uniformity of the fields was not influenced by the two loading types considered therefore indicating that in this context, the method of loading is not a critical feature. This is not necessarily the case, however, when considering gauge length strain rates, and hence stress levels, which can be influenced significantly by the loading conditions.

The results of the analyses have shown that the specimen gauge length strain rate is not constant during a test, but varies within approximately ± 35% of the desired strain rate, even with compensated prescribed displacement control. For most applications, it is unlikely that such a variation in strain rate would be of major concern, since its influence on stress level may not be all that significant. However, the implication of the result is that for the material considered, the stress levels determined from a test may be in error by as much as ± 20%.

In order to reduce the variation of gauge length strain rate obtained in a test, an effective gauge length has been introduced, which is assumed to vary with deformation. The prescribed displacement loading given in equation (8) is therefore re-written as

$$d = l_e \left(e^{\dot{\varepsilon} t} - 1 \right) \qquad (9)$$

in which l_e, the effective gauge length, is written as a function of a specified specimen strain, ε_s, as

$$l_e = l_0 / (a + b\varepsilon_s) \qquad (10)$$

in which a and b are constants. For the pinned specimen, the specimen strain is chosen to be the loading direction strain at a point adjacent to the pin, and a and b take the values 0.451 and 0.424 respectively. The resulting gauge length strain rates obtained are shown in figure 6, in which a considerable improvement over those discussed earlier can be seen. The appropriate selection of the effective gauge length, l_e, through computational studies, can therefore provide a corrected prescribed displacement to enable a more constant gauge length strain rate to be achieved.

4. Conclusions

Multiaxial constitutive equations for superplasticity have been implemented in a finite element solver, and the model employed to investigate the suitability of a uniaxial test piece for the generation of materials data. While the specimen design enables the establishment of stress and strain fields that are uniform in the gauge length region, the strain rates obtained in this region are not necessarily those that are desired, and may lead to the measurement of stresses that are upto ±20% in error.

5. References

1. J. Pilling, N. Ridley, Superplasticity in Crystalline Solids, Institute of Metals, 1989.

2. N.E. Paton, C.H. Hamilton, "Microstructural Influences on Superplasticity in Ti–6Al–4V", Met. Trans. A., Vol.10A, 241-250, February 1979.

3. J.Pan, A.C.F. Cocks, "Computer Simulation of Superplastic Deformation", Comp. Mats. Sci., Vol.1, 95-109, 1993.

4. A.K. Ghosh, C.H., Hamilton, "Mechanical Behaviour and Hardening Characteristics of a Superplastic Ti–6Al–4V", Met. Trans. A., Vol.10A, 699-706, June, 1979.

5. M. Zhou, F.P.E. Dunne, "Mechanisms-Based Constitutive Equations for High Temperature Titanium Alloys", Internal Report DMM.93.11, Dept. of Mech. Eng., UMIST, UK, 1993.

6. Hibbitt, Karlsson and Sorenson, ABAQUS User Manuals, v5.2, HKS Inc., Pawtucket, Rhode Island, Providence, USA, 1993.

Figure Captions

Figure 1   Diagram showing a typical uniaxial specimen used for the generation of superplasticity materials data.

Figure 2(a)   Graph showing the computed and experimental true stress-strain curves for Ti–6Al–4V at 927°C with an initial grain size of 6·4μm. The symbols represent experimental data [4], and the lines the predictions obtained from both the uniaxial constitutive equations, and the equations implemented in ABAQUS.

Figure 2(b)   Graph showing the computed and experimental variation of grain size with time for Ti–6Al–4V at 927°C with an initial grain size of 6·4μm. The symbols represent experimental data [4], and the lines the predictions obtained from both the uniaxial constitutive equations, and the equations implemented in ABAQUS.

Figure 3   Diagrams showing the models and boundary conditions employed for the finite element analysis of the uniaxial specimen shown in figure 1, (a) without a loading pin and (b), with a loading pin.

Figure 4   Predicted field variations of (a) loading direction stress, and (b) grain size for the loading conditions shown in figure 3(a), for a bulk gauge length strain of approximately 1·1.

Figure 5   Predicted field variations of (a) loading direction strain, and (b) grain size for the loading conditions shown in figure 3(b) for a bulk gauge length strain of approximately 1·0.

Figure 6   Graph showing the variation of specimen gauge length normalised strain rate with gauge length strain for the loading conditions shown in figures 3(a) and 3(b) for nominal strain rate $2\cdot 0 \times 10^{-4}$ s^{-1}.

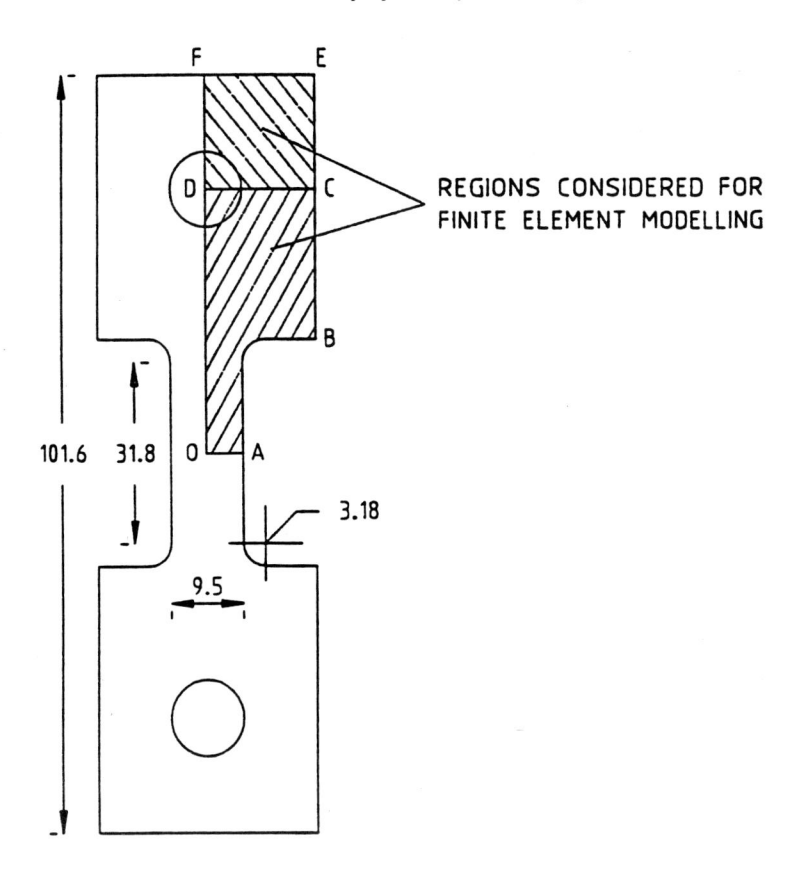

Figure 1   Diagram showing a typical uniaxial specimen used for the generation of superplasticity materials data.

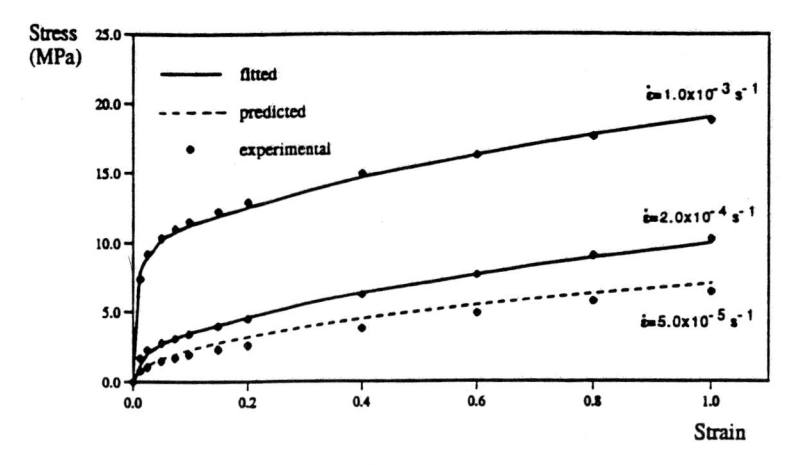

Figure 2(a) Graph showing the computed and experimental true stress-strain curves for Ti–6Al–4V at 927°C with an initial grain size of 6·4μm. The symbols represent experimental data [4], and the lines the predictions obtained from both the uniaxial constitutive equations, and the equations implemented in ABAQUS.

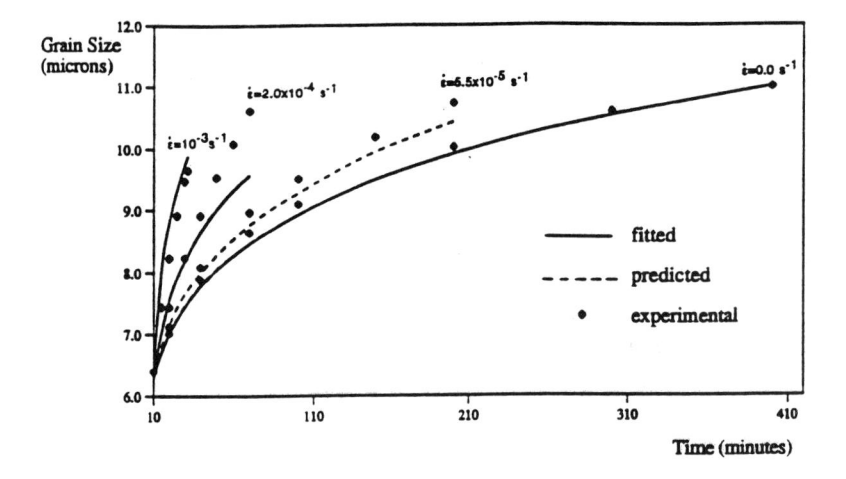

Figure 2(b) Graph showing the computed and experimental variation of grain size with time for Ti–6Al–4V at 927°C with an initial grain size of 6·4μm. The symbols represent experimental data [4], and the lines the predictions obtained from both the uniaxial constitutive equations, and the equations implemented in ABAQUS.

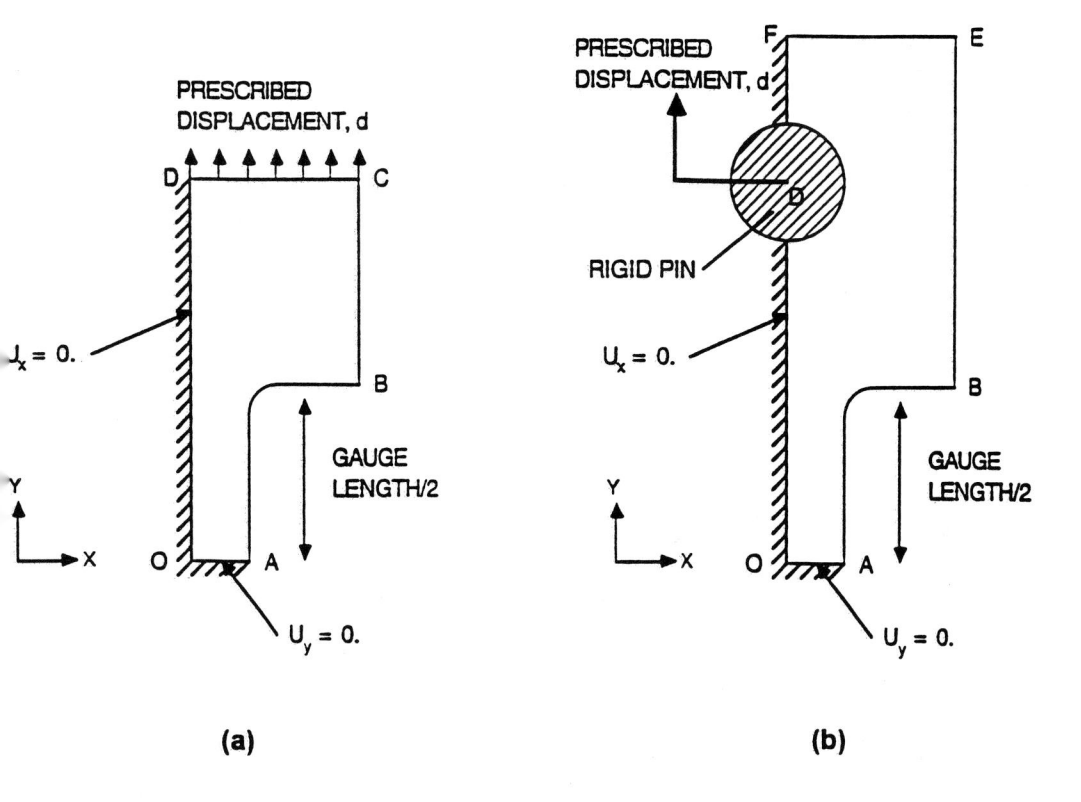

Figure 3 Diagrams showing the models and boundary conditions employed for the finite element analysis of the uniaxial specimen shown in figure 1, (a) without a loading pin and (b), with a loading pin.

248 *Superplasticity: 60 Years after Pearson*

H 12	H .105E-1
G 10.3	G .103E-1
F 8.61	F .101E-1
E 6.92	E .995E-2
D 5.22	D .977E-2
C 3.53	C .959E-2
B 1.83	B .94E-2
A .137	A .922E-2

(a) (b)

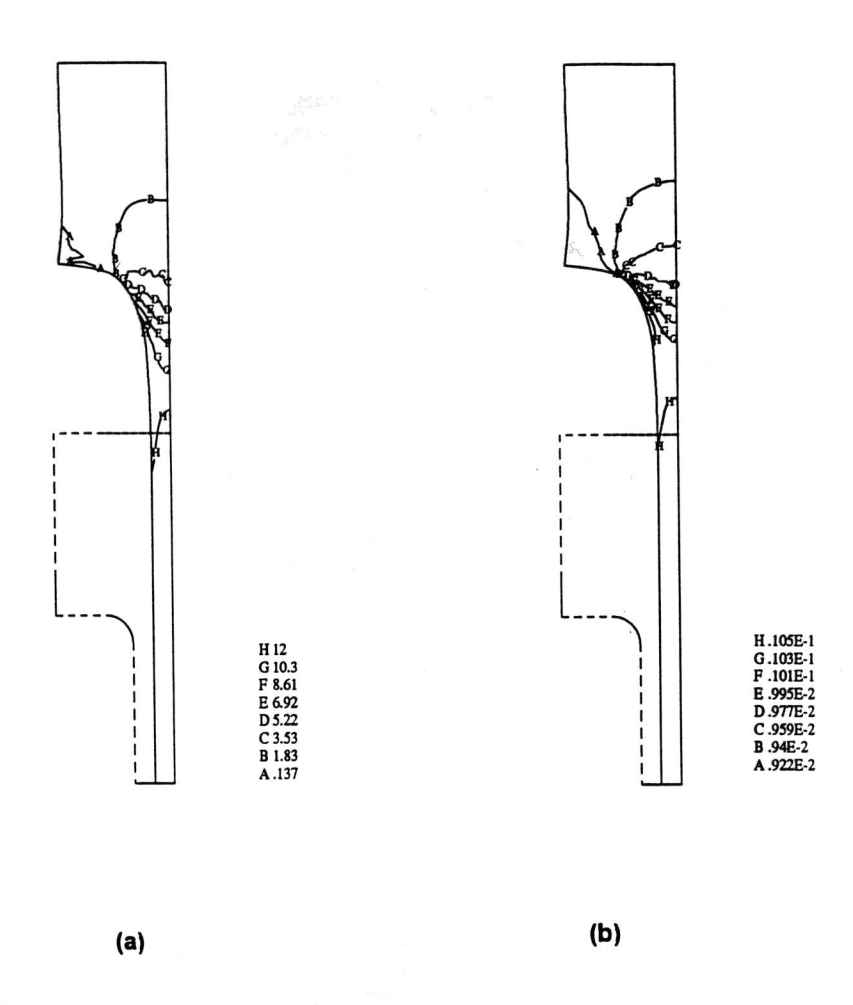

Figure 4   Predicted field variations of (a) loading direction stress, and (b) grain size for the loading conditions shown in figure 3(a), for a bulk gauge length strain of approximately 1·1.

H 1.02	H .105E-1
G .839	G .103E-1
F .659	F .101E-1
E .478	E .995E-2
D .297	D .977E-2
C .116	C .959E-2
B -.643E-1	B .94E-2
A -.245	A .922E-2

(a)   (b)

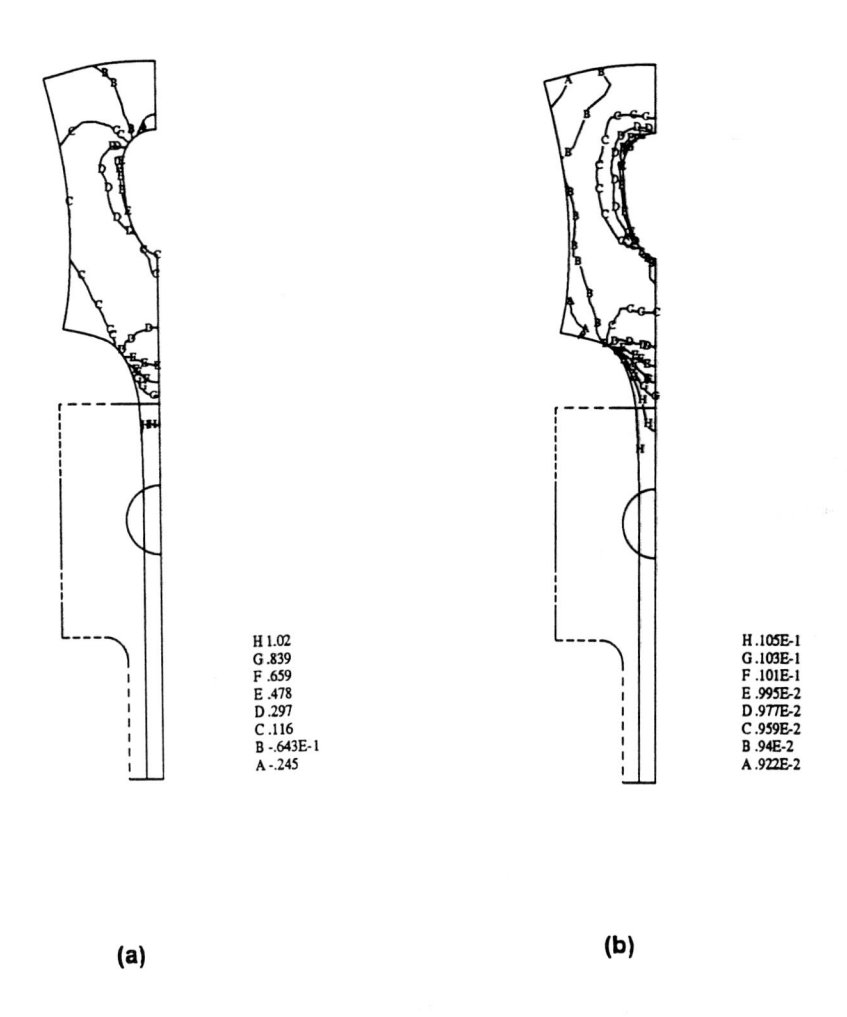

Figure 5   Predicted field variations of (a) loading direction strain, and (b) grain size for the loading conditions shown in figure 3(b) for a bulk gauge length strain of approximately 1·0.

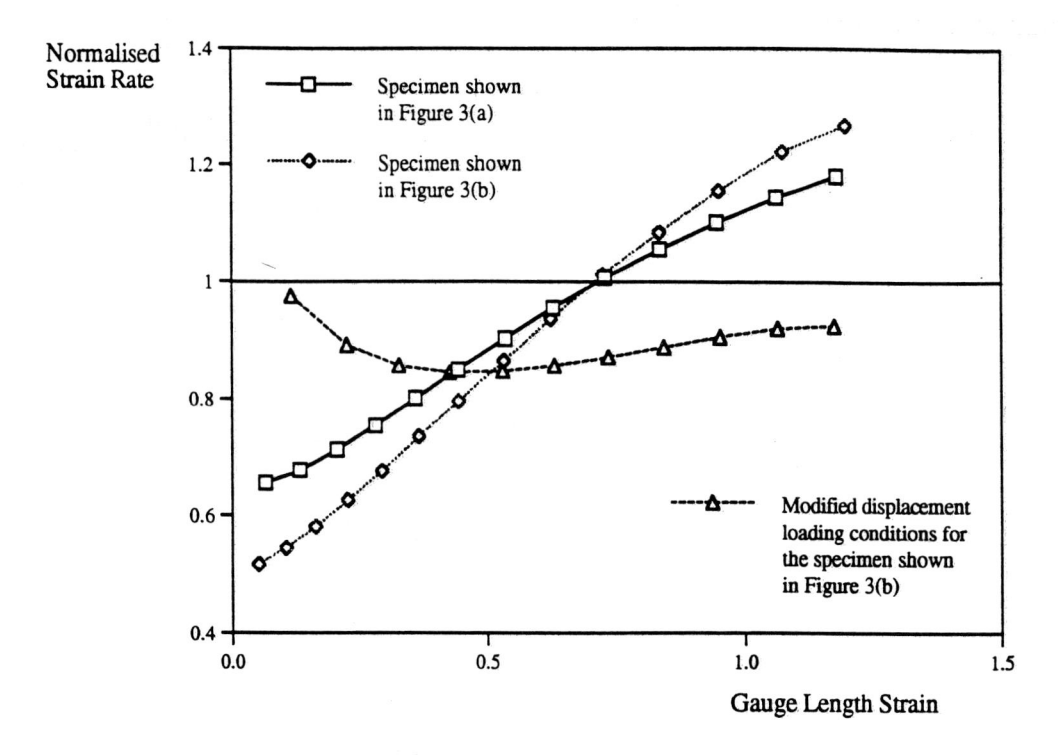

Figure 6   Graph showing the variation of specimen gauge length normalised strain rate with gauge length strain for the loading conditions shown in figures 3(a) and 3(b) for nominal strain rate $2\cdot0 \times 10^{-4}$ s^{-1}.

Superplastic Forming: Equipment, Alloys, Forming and Applications

Design and Manufacture of Hydraulic Presses for Superplastic Forming

Roy Whittingham, C.Eng. F.I.Mech.E.
Managing Director of Chester Hydraulics Ltd.

Abstract

The superplastic forming process (SPF) is a method of forming alloys which have superplastic properties at elevated temperatures such as titanium and aluminium.

SPF offers huge benefits and brings a totally new range of possibilities into the manufacturing arena. Exploitation of the process is radically changing design, production and commercial concepts in meeting the challenges of the future.

- Much more complex shapes can be produced and stress concentrations can be avoided, allowing designers to conceive one piece components instead of multi-part assemblies.
- Manufacture becomes more cost effective and components become lighter. Reductions of 20% on costs and 10% on weight are typical.
- Consistently accurate forms can be produced in the stress free condition.
- A forming shape is required on one die only.

SPF involves the use of heated dies in which injected gas provides pressure to form the work material against the die form. During the process the die and platen temperatures, also the rate of forming, must be maintained within close tolerances to match ideal process values.

Forming is controlled by varying the gas pressure during the production cycle according to the permissible strain rate and thickness of the material, also the flow stress. In addition, the status of the seal between the sheet material and the dies must be accurately maintained.

Diffusion Bonding (DB) occurs when two or more pieces of clean superplastic material are brought together in intimate contact and held under pressure at elevated temperatures for a pre-set period of time. By combining the bonding ability with superplastic forming, it is possible to produce even more complex shaped hollow structure with in-built stiffeners. Dependent upon the design of the component, the SPF/DB process can be carried out concurrently or at different times during the manufacturing cycle.

1. Introduction

Chester Hydraulics Ltd., hydraulic metal forming press builders for more than 150 years have, for the past 20 years, been involved in the design, development and manufacture of superplastic forming presses. Considerable experience has been gained during this period in the design of these machines and in dealing with the very exacting requirements of the SPF/DB process.

"HYDRAULIC CHESTER" SPF machines provide unique computerised control over all these complex inter-related parameters. Designed specifically for the process and possessing unequalled capabilities, they are used throughout the world and are playing a significant part in taking SPF to the very forefront of forming technology.

SPF cycles are managed by a computer which integrates time related control over gas pressure, platen and tool temperatures and tool clamping pressure to requirements imposed by such parameters as the material permissible strain rate, forming temperatures and flow stress.

The computer system provides a keyboard for input and a colour monitor to display the gas management cycle, also other operating and maintenance data, with up to 40 screen formats available. There are also hard copy printout facilities so that key operational information form each individual production cycle can be produced as a quality record.

2. Press Frame Construction

2.1 4-column design (See Fig.1)

This form of structure provides the necessary rigidity and furthermore space for the accommodation and access to the furnace enclosure panels. It also provides for the large platen areas and high tonnage requirements and for the long stroking where the moving slide is required to open to provide adequate clearance for the removal of the work piece. The frame structure comprises of fabricated steel head, base and moving slide sections of robust construction to minimise deflection and to sustain the high loads which are developed for long periods.

2.2. Shuttle Press (See Figure 2)

This concept originates from the simple premise that superplastic forming does not really require a press as such but merely a clamping frame that will react to the forces exerted by the forming gas pressures inside the die.

The machines have a frame structure of more compact design with the shuttle table mounted on an extension of the press frame and driven into and out of the frame by either a long stroking hydraulic cylinder or a screw drive powered by an electric motor. The main drawback with this type of machine is the loss of heat from the tool and platens when the table is out of the frame.

Chester Hydraulics Ltd. prefer to offer the conventional press structure, perhaps fitted with a shuttle table mainly for tool loading and removal.

3. Furnace Enclosure (See Fig. 3)

All machines are fitted with a furnace enclosure comprising of front and rear doors and side walls. The front/rear door may be raised and lowered to give access to the work area and powered in each direction by an electric motor drive. Both the doors and side walls are powered by cylinder type actuators to seal over the hot platen system to prevent the loss of heat during the forming cycle. Apertures in doors allow access for passing gas lines to the tools and the operation of the enclosure forms a part of the operational cycle of the machine.

4. Hot Platens

Hot platens may be manufactured from alloy steel or from ceramic materials, cast to the required form, or utilising standard blocks.

Typically - RA330 by Rolled Alloys.
- P1061 by Richard Carrs.
- HR6 by Cronite Ltd.

Generally speaking the alloy steels are substantially more expensive than ceramics but have the following advantages:

- much more tolerant to mechanical abuse.
- they can be utilised for tool fixing i.e. both lower and upper, for which purpose the platens are made in several sections or segments with tee-grooves.
- metal platens could have a useful life period of 4/6 years.

The main disadvantage is the levels of relative expansion between the tools and the platens. However, sound design and careful operation of the press can minimise these problems i.e controlled heat-up rate and provision to guard against the possibility of over-pressurisation of the tools and imprinting the tools into the platens.

4.2. Ceramic Platens

Their chief advantage is their low cost compared with metal platens. Ceramic platens may be cast or standard blocks can be used. In both cases the heated elements are set within the blocks.

Ceramic blocks have a compressive strength of 5000 p.s.i. up to temperatures of approx. $2500^{\circ}C$ and they are not prone to thermal shock or warping. Blocks may crack but this is not a catastrophic failure. However, the chip and wear on the working surface reaches a point where they become uneven and unusable. Ceramics have poor tensile strength and it is not possible to cast the slots in the blocks for hanging tools. That problem can be overcome by fitting a caul plate supported by bolts passing through the platen and along their edges.

Chester Hydraulics use metal platens almost exclusively for their better wear life and they are more convenient for tool mounting. However, ceramic mediums are used to insulate the metal platen in conjunction with water cooled support platens.

5. Heating Elements

One of the main objectives is to heat the platens as evenly as possible and so the platens are divided into a number of temperature control zones. This is achieved by using multi-zone heaters in each platen segment. Platens measuring 3m x 2.5m may have 21 control zones whereas a platen measuring 1.25m.square may have 9 control zones (See Fig. 3). Temperature tolerances at $950^{\circ}C$ +/- $10^{\circ}C$ and at $600^{\circ}C$ +/- $5^{\circ}C$ can thus be achieved.

There are two heating systems currently in use i.e. resistance wire in ceramic platens and cartridge type elements in metal platens.

Cartridge heaters are inserted into holes gun drilled into the platens (See Figs. 4 & 5). They are metal sheathed with the wire connections at one end and are a loose fit in the platen holes, typically a 28mm dia. element in a 32 mm. dia. hole. This is important for ease of heater removal.

The heater connections to the power feed should be terminated in a suitably insulated enclosure to one end of the platen usually with a draught system to maintain a temperature of approx. 250°C max.

6. Heating Equipment

The power required to heat platens to the required working temperature will vary according to the platen material. Metal platens need a higher watt density than ceramic i.e. 25watt/inch^2 (5 watt/cm.^{2}) to give a temperature rise in the platens from the cold condition of approx. 150°C/hour.

Heater control systems are interfaced with the control computer to enable the heat up period to occur outside normal production periods. Controllers of the thyristor type can now regulate platen temperature in various control zones of a platen to within very accurate temperature levels.

7. Gas Management System

The gas management system is said to be the most important facility on an SPF/DB press. Its function is to regulate the pressure and flow of the gas on the sheet to blow it into the die form. The theoretical pressure to be applied to the sheet varies to impose a constant flow stress on the material. However, the detail of the component form being developed may dictate a change in flow rate and pressure at various points in time during the forming cycle. Thus the pressure versus time relationship for any given part may be calculated by finite element or other computer tools or by experience in producing similar parts.

Important features of Chester Hydraulics' Gas Management System are as follows:-

- May be used with either air (for aluminium forming up to 300°C) and for argon gas.
- May be used with gas pressure up to 50 bar.
- Up to three control lines to work individually or simultaneously.
- Gas control by - On/Off control
    - Volume control
    - Diaphragm operated proportional flow valves.
- Break through or 'cracking' facility.
- Back pressure forming facility.
- Stainless steel valves and pipework used throughout.

8. Hydraulic System

The hydraulic system is suitable for mineral oil or phosphate ester dependent upon layout of machine and where the hydraulic power unit is to be installed and whether or not the machine is up-stroking or down-stroking.

An important feature is the motorised pressure relief valve which controls the clamping force the press applies on the tooling; as the gas pressure increases so the clamping force increases to maintain the seal between the tools and the component being produced. This device also prevents over-pressurising the tool which may cause damage to the platens.

The hydraulic power unit is adaptable to power all the ancillary movements of the press i.e. shuttle table and heat shield enclosures.

9. Computer Controls

These are based on a Motorola 68030 microprocessor arranged to embrace all the control functions of the press, as follows:-

- Platen heat up rate in all the control zones.
- Platen temperature control during a normal production cycle.
- Tool temperature read out.
- All press movements i.e. moving table, furnace enclosures and shuttle table, etc.
- Control of gas management system.
- Input of all gas cycles with colourgraphics to indicate pressure/time cycle for each control line.
- Up to 40 screen displays for operator information, control data, and fault finding.
  Printer for printing cycle data for use as a component quality record.

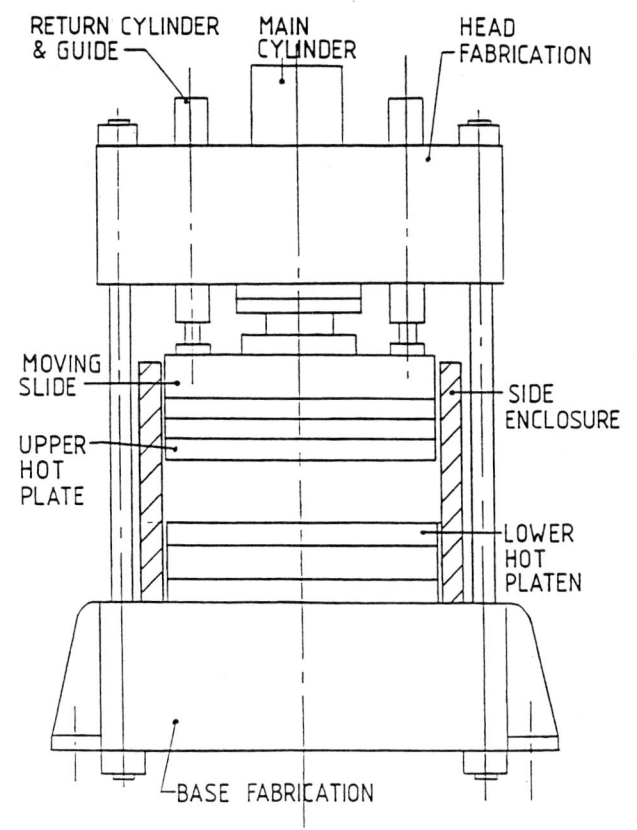

Fig. 1 Shows main features of 4-column press

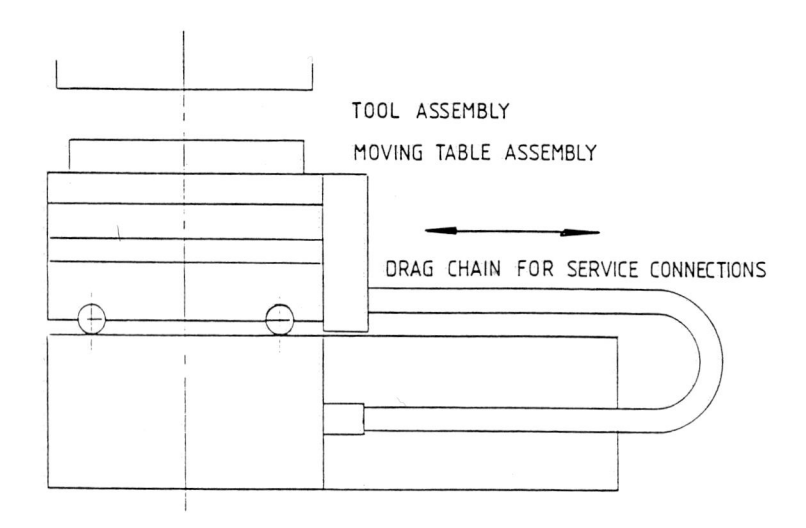

Fig. 2 Illustration of shuttle press

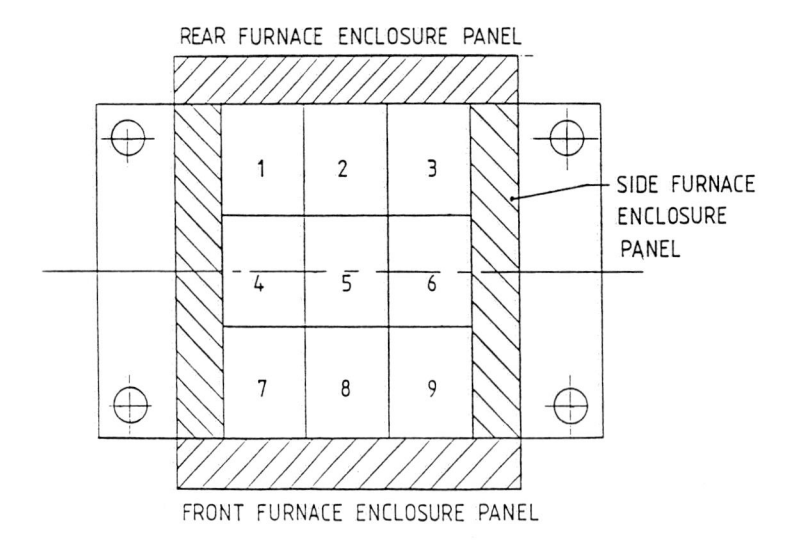

Fig. 3 Furnace enclosure

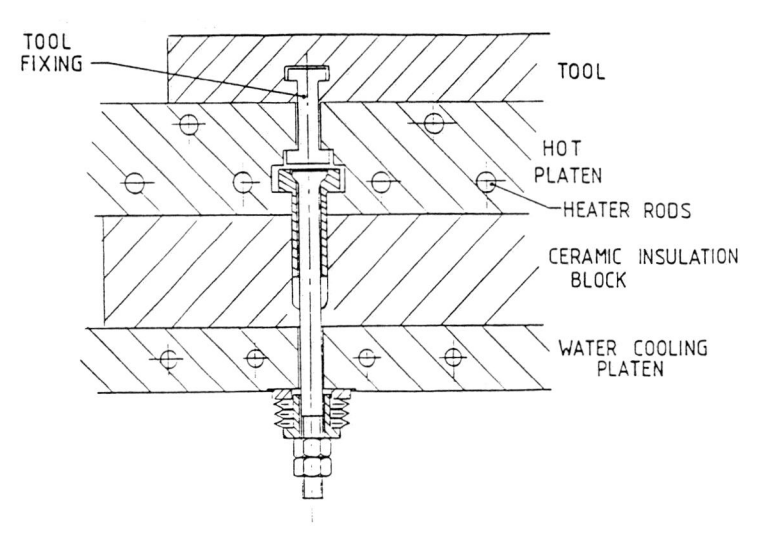

Fig. 4 Shows location of heating elements in hot platen

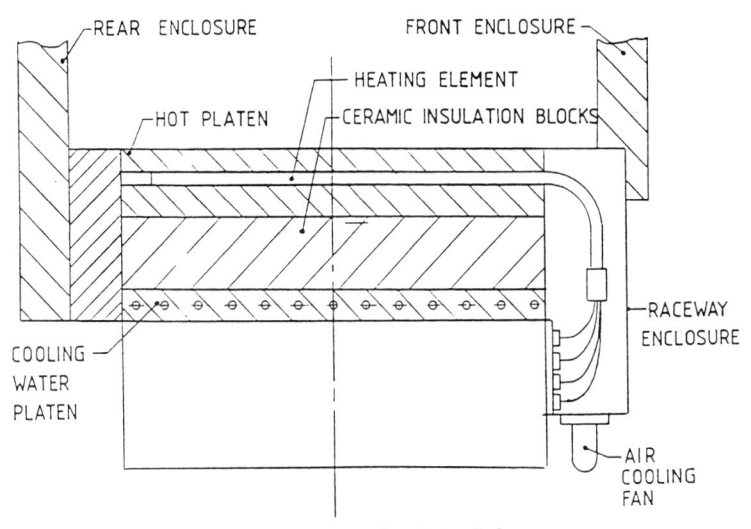

Fig. 5 Location of heating element in platen hole.

Estimation of Strain Rate Sensitivity in Superplastic Compression Test

M. Yoshizawa and H. Ohsawa

Department of Mechanical Engineering, Hosei University,
Kajinocho 3-7-2, Koganei, Tokyo, Japan

Abstract

Plastic deformation processes in metalworking may be classified into only a few categories on the basis of the type of forces applied to the workpiece. These categories are direct-compression, indirect-compression, tension, bending, and shearing processes. In the numerical analysis of superplastic deformation in these metalworking processes, strain rate sensitivity m from tensile test can be involved as unique reliable constant of mechanical properties. Superplastic flow is believed to be simultaneous appearance of grain boundary sliding and some accommodated process. m value is of course a macroscopic material constant, and determined through simple calculation by using tensile test data. It is natural to think that the extent of both grain boundary sliding and accommodation may alter the m value. Each superplastic material has its own m value under specific strain rate range and temperature. The different stress rate from uniaxial stress, however, can naturally change the extent of grain boundary sliding and accommodation therefore the value of m also. This has never been studied. Strain rate sensitivity from compression test appears to be useful for analysis of compressive process such as forging, rolling, extrusion and drawing.

In this study, special attention will be paid to determining strain rate sensitivities in uniaxial compressive deformation. The friction at the tool-workpiece interface leads to a barreled specimen profile, and it is difficult to measure compressive stress and strain. A much more suitable test for determining strain rate sensitivity is Siebel type compression test. Advantage of this test is to eliminate barreling by using the cone shaped punch and corresponding specimen geometry at the same time. The application of this estimation of m value to compressive process was discussed in terms of comparison of the value by this method with other conventional techniques.

1. Introduction

Superplasticity has been found to be associated with a high strain rate sensitivity. The influence of various parameters on strain rate sensitivity is being studied. Backofen, Turner and Avery[1] first recognized that a quantitative index of strain rate sensitivity was obtained by evaluating m value. In their study the flow stress of superplastic material was expressed by the constitutive equation $\sigma = K\dot{\varepsilon}^m$, where σ and $\dot{\varepsilon}$ are the true stress and true strain rate, respectively, and the K is a constant. The m value was determined by the velocity change tests. Subsequently, other workers have confirmed that maximum elongation increases with an increase in m value and significance of m value [2]. From the relationship $\sigma = K\dot{\varepsilon}^m$ Naziri and Pearce[3] suggested that the m value is defined as a form $m = d\ln\sigma/d\ln\dot{\varepsilon}$ and obtained by gradient of log σ vs. log $\dot{\varepsilon}$. Hart [4] proposed the correlation $d\ln\sigma = \gamma d\varepsilon + md\ln\dot{\varepsilon}$, where γ is strain hardening coefficient and indicated that the m value is obtained by stress relaxation test. Furthermore, for the constitutive equation $\sigma = K\dot{\varepsilon}^m \varepsilon^n$, Ohsawa and Nishimura [5] showed the m value is given by $m = \partial \ln\sigma/\ln\dot{\varepsilon}$. These methods was examined by Hedworth and Stowell [6], Arieli and Rosen [7]. In these studies the strain rate sensitivities were only determined in the tension tests, however, these results led a difficult question whether these m values represent appropriate strain rate sensitivity. The object of the present work is to study estimation of strain rate sensitivity in compression test.

2. Description of Methods for Calculating the m Value

The values of the coefficient m were derived by all **the four methods**. All methods for calculating the coefficient m were apply to the data of both compression and tension tests. In each method, the m value of compression was compared with the m value tension. Note that the axial stress and strain are taken positive when its produce tension and negative when its produces compression, and all data was calculated after transforming into absolute value. A brief description of the method for estimating m value is as follows:

2.1 Method 1

The procedure can be involved making sudden changes in crosshead speed. Fig. 1(a) illustrated a schematic diagram of the load-time curves covering change of velocity at time t* from VA to VB. The crosshead speed is suddenly changed to higher beyond the load maximum of lower crosshead speed. The extrapolation of the lower curve is shown by dotted line. Times A and B represent the same strain at the different crosshead speeds. The strain rate sensitivity m value could then be calculated from

$$m = \log(P_B/P_A) / \log(V_B/V_A) \qquad (1)$$

where PA and PB are correspond to the loads at A, B, respectively. the m is identified with lower strain rate of $\dot{\varepsilon} = V_A/(\text{gage length})$.

2.2 Method 2

The relationship $\sigma = K\dot{\varepsilon}^m$ enables the value of m to be obtained by gradient of log σ vs. log $\dot{\varepsilon}$ plots. The m value is determined:

$$m = d\ln\sigma / d\ln\dot{\varepsilon} \qquad (2)$$

Strain rate sensitivity in method 2 are defined as $(d\ln\sigma)/(d\ln\dot{\varepsilon})$, which is approximately equal to $(\log\sigma_B/\sigma_A)/(\log\dot{\varepsilon}_B/\dot{\varepsilon}_A)$, and the m value is effective only within narrow strain range, because of variations in strain rate sensitivity. In this method, the constant crosshead speed tests are done by various crosshead speeds, and the data of stress and strain rate made the regression lines which are composed of same strain. A gradient of the regression line is m value.

2.3 Method 3

The m values obtained by method 1 and method 2 are derived by the constitutive equation $\sigma = K\dot{\varepsilon}^m$. However, it is necessary to consider other method for constitutive equation:

$$\sigma = K\dot{\varepsilon}^m \varepsilon^n \qquad (3)$$

where ε is a true strain, n is the strain hardening exponent, and K is a constant. The m value for constitutive equation (3) is given a form:

$$m = \partial\ln\sigma / \partial\ln\dot{\varepsilon} \qquad (4)$$

Fig. 1(b) is a space coordinate system with $\ln\sigma$ axis, $\ln\dot{\varepsilon}$ axis and $\ln\varepsilon$ axis. In this coordinate, the $\sigma = K\dot{\varepsilon}^m \varepsilon^n$ - plane is constructed of two planes, one is constant strain, the other is constant strain rate, and the m value is given a gradient of this plane. It is need to recognize that the m value is independent of strain and strain rate. A decision of the m, n value is to determine $\sigma = K\dot{\varepsilon}^m \varepsilon^n$ - plane, so constant K is the stress at $\dot{\varepsilon} = \varepsilon = 1$. The plane is determined by multiple regression analysis using the data of stress and strain rate from various constant crosshead speed tests. The logarithm of equation (3) is

$$\ln\sigma = m\ln\dot{\varepsilon} + n\ln\varepsilon + \ln K \qquad (5)$$

The m value is obtained coefficient of $\ln\dot{\varepsilon}$ from the results of multiple regression analysis.

2.4 Method 4

Hart [4] has proposed a method to obtain m value by stress relaxation test. His arguments lead to the relation:

$$d \ln \sigma = \gamma \, d \ln \varepsilon + m \, d \ln \dot{\varepsilon} \tag{6}$$

where γ is a strain hardening parameter. It is readily seen that this criterion is agreement with equation (2) for superplastic materials with negligible strain hardening dependence, and with the usual criterion for viscous materials (γ is nearly equal to zero). Furthermore, the correlation $\dot{\sigma} = -K\dot{\varepsilon}$ between stress rate and strain rate in stress relaxation test is derived from the geometrical constraint, and write the correlation as

$$d \ln \dot{\sigma} = d \ln \dot{\varepsilon} \tag{7}$$

where $\dot{\sigma}$ is the stress rate ($d\sigma/dt$). Rearranging equation (2) by substituting equation (7), the m value of stress relaxation test can be led

$$m = d \ln \sigma / d \ln \dot{\sigma} \tag{8}$$

Therefore, if $\ln \sigma$ is plotted against $\ln \dot{\sigma}$, the m value is determined as gradient of the progressive liner curve at these data points.

3. Experimental Procedures

Superplastic zinc base alloy SPZ was selected for the study. The eutectoid composition 78% Zn - 22% Al of this alloy shows superplastic behavior approximately at 523K. Both compression and tension tests were done by Instron machine (20tonf) at 523K. The m value in compression tests was evaluated by comparing with the results of the compression tests and the results of tension tests. The details of the two tests for obtaining m value are explained in the following paragraphs.

3.1 Compression tests
3.1.1 Specimen geometry

Specimens for compression tests were cut from the plate of SPZ and machined to cylindrical shape. A diagram of the compressive specimen used is shown in Fig. 2 (a) and Fig. 3(a). Initial diameters of these are all $d_0 = 5$ mm and longitudinal direction of these specimens is always aligned with rolling direction. Both specimen edges are machined to fit punch geometry. The compression specimens have various tangents of angle and specimen heights. The tangents of angle of circular cone are ten types from 0 to 0.45 every 0.05, and the specimen heights are 6 types from 4 to 12 mm every 2 mm.

3.1.2 Siebel type Compression test

The Siebel type compression tests were done to permit construction of a series of stress vs. strain rate curves without the effect of friction. It is possible to measure the strain rate sensitivity without the influence of friction of tool-specimen. In this compression test the conical concave surface is compressed by conical punch (Fig. 2(b)). Fig. 2(c) indicates a compression system in equilibrium. Under the action of axial force P, the reactive axial force will be produced. Each axial force divides into two forces the normal force and frictional force. These forces are given respectively N=Pcos θ and F=Psin θ. Let μ be a coefficient of friction and the relationship between normal force and friction force is F = μ N. From this relationship, coefficient of friction μ is equal to tan θ. It may be seen that tan θ chosen to equal to μ that combined stress is corrected to axial direction.

Fig. 3(b)-(d) schematically illustrates some of the type of compressive deformation. The compressive deformation is divided into the following classes. When μ >tan θ or μ <tan θ, the effect of frictional resistance is appeared and this leads to a barreled specimen profile (b) or constrictive specimen profile (d). If μ = tan θ, compressive deformation progress keeping on straight specimen profile (c). The m value in compressive tests is obtained at a state with μ = tan θ. In the experiment, the MoS2 paste was lubricated to the interface at tool-specimen for balancing between μ and the tan θ which is limited from a degree of 0 to 0.45.

The gage length (G.L.) of Siebel type compression test is given the distance between the tops of conical punch. Since the displacement of crosshead becomes elongation in compression tests, it is easy to determine strain and strain rate.

Strain rate change tests (Method 1) were made in a manner that the crosshead speed was changed in increments where the ratio was 2 or 2.5. The velocity of crosshead was changed at the point which shows the steady state of the stress. Constant crosshead speed tests (Method 2,3) were done from a degree of the velocity 0.5 to 50 mm/min. Stress relaxation tests (Method 4) were done by stopping the crosshead motion after the steady state of stress.

3.2 Tensile tests

Tensile test specimens were machined from the same SPZ sheet as the compression specimens with the rolling direction parallel to the tensile axis. Initial sheet thicknesses of these are all t0 = 7 mm and gage lengths are all G.L. = 20 mm. The gage lengths were marked reference lines, and reference lines were observed and recorded by 8 mm video camera until termination. Recorded image of extension process was analyzed by image analyzer, which facilitated the determination of strain and strain rate between gage length. The data of tension tests are obtained in the range of crosshead speed from 2 to 100 mm/min. Experimental procedures in tension tests were the same as compression tests except the point of changing or stopping crosshead speed. The motion of crosshead was changed beyond load maximum point.

4. Results and Discussion

4.1 Set up the compression test

Experimental process is setting to choose optimum of compression tests which making possible to measure the strain rate sensitivity without the influence of friction of tool-specimen at first. Fig. 4 shows the correlation between specimen shape and angle of conical punch. In these tests the angle of conical punch agrees with angle of cone of specimen. Degree of transformation of specimen profile is represented by parameter $(d_2-d_1)/d_0$, where d_0 is initial diameter, d_1 and d_2 are diameter of edge and diameter of center, respectively. If this parameter has a positive or negative value, it is a barreled or constrictive specimen profile. If this parameter is equal to zero, it means independent of friction. Since the parameter is nearly equal to zero at $\tan \theta = 0.1$, the compression tests using conical punch ($\tan \theta = 0.1$) is considered to be optimum. The compression tests were therefore made on at the optimum.

4.2 The m value in method 1

The Alternation of m value with changing strain rate is shown Fig. 5. These m values were given by method 1. The two regression lines correspond to compression tests and tension tests respectively. The m values in both compression tests and tension tests are increase with increasing strain rate. Still it is a little difference in comparison of two lines. The slope of compression line is lower than the slope of tension line, and the m values of compression tests are higher than the m values of tension tests within a range of lower strain rate. Thus, it is supposed that the each m values derived by method 1 are different. However, in the case of comparison is chosen as the subject of whole region of strain rate, it is satisfactory to consider comparison as using averages. Since the averages of m values of compression and tension are 0.40 and 0.39, respectively, it seems that the average of m value in compression test agrees with the average of m value in tension tests.

4.3 The m value in method 2

In method 2, m values were obtained the gradient of liner curve which is constructed same strain on log strain rate vs. log stress plots. Example of results of compression tests is shown in Fig. 6. The range of absolute strain rate is about from 10^{-3} to 10^0 in both compression and tension tests. In this figure each regressive line has different strain that was selected at equal interval. If the these results are rearranged, Fig. 7 was obtained. This figure shows the influence of strain on change of the m value. The m values decrease a little with increasing strain. This tendency is evident at other specimen height. It is considered that this tendency is due to change of strain rate during deformation although constant cross speed. Thus, it is clear from the above that the m value is effective only within a narrow strain range. Fig. 8 indicates the effect of slenderness ratio (diameter/length) on m value and comparison of m value in method 2. All the points in this figure are averages of the region of absolute strain from 0.139 to 0.693. In case of compression tests, because of the m value **does** not change with slenderness ratio, it can be considered that there is no effect of difference of specimen height. In comparison of m value in method 2, the measured m values by two different tests almost agree.

4.4 The m value in method 3

The m values derived as ($\partial \ln \sigma / \partial \ln \dot{\varepsilon}$) were calculated by regression analysis using the same data as method 2. Calculations for compression tests were done every specimen height. Fig. 9 shows the comparison of the m value of tension and compression. The m values of each specimen height are the almost the same value and are nearly equivalent to the m value of tension tests simultaneously. Still more, it is found from comparing with method 2 and method 3 that no matter m value in method 3 is independent of the effect of strain and strain rate, the levels of m values are similar with the m values in method 2.

4.5 The m value method 4

It was noted in section 2 that the strain rate sensitivity may be obtained from the slope of log $\dot{\sigma}$ vs. log σ. An example of the variation of stress rate against stress is shown in Fig.10. In the stress relaxation tests the stress rate generally continued to decrease with time during crosshead in rest. Since the changes of rate of stress rate are constant in both two tests, the m values were given by **the slopes within** this stress rate range. The result of stress relaxation tests is Fig. 11. According to this figure, the levels of all m values are extremely lower than the levels of the other methods. The reason to make a difference of levels is without regard to study in present work. However, the m values both compression and tension are almost the same so far as comparing in method 4.

5. Conclusions

The averages of m values obtained in all methods are rearranged to Table 1. If the magnitude of the m value is discussed, it is obviously that the m value given by method 4 is the smallest, and the m values obtained in the other method 1-3 are almost the same. However, because the purpose of this study is to compare the m values in both compression test and tension test, it is not important that the difference of the magnitude of the m value. The each difference of the m values in compression and tension tests is seem to be so small, and both the m values in compression and tension test are similar in each method. Therefore it possible to consider that this small difference may be the effect of stress state on the m value. This estimation will predict that the accuracy of analysis in various plastic deformations may depend on selecting method for calculating of m value and the stress state of material test.

Acknowledgment

We wish to thank to Takashi Azuchi and Sumito Nishimura for their help in the mechanical testing, and performing the measurements. This work was also supported by AMADA foundation for metal work technology and the light metal educational foundation.

References

1. W. A. Backofen, I. R. Turner and D. H. Avery, Trans. ASTM, 57, 1964, 980-990.
2. W. B. Morrison, Trans. ASM, 61, 1968, 423-434.
3. H. Naziri and R. Pearce, J. Inst. Metals, 97, 1969, 326-331.
4. E. W. Hart, Acta. Met., 15, 1967, 351-355.
5. H. Ohsawa and H. Nishimura, Prepr. of Jpn. Soc. Mec. Eng., 807-2, 1980, 119-126.
6. J. Hedworth and M. J. Stowell, J. Mater. Sci., 6, 1971, 1061-1069.
7. A. Arieli and A. Rosen, Scrip. Met., 10, 1976, 471-475.

Fig. 1 Explanation of Calculating of strain rate sensitivity for method 1 and method 3.

Fig. 2 Illustrations of compression test and specimen geometry.

Fig. 3 The effect of friction on barreling.

Fig. 4 Alteration of specimen profile with changing angle of conical punch.

Fig. 5 The relation between strain rate and m value in strain rate change tests.

Fig. 6 The liner regression between true strain rate and true stress.

Fig. 7 The strain rate sensitivity is a variation of strain.

Fig. 8 The comparison of m value influenced strain and strain rate.

Fig. 9 The m value is independent of strain and strain rate.

Fig. 10 Logarithmic plot of relaxation rate against stress.

Fig. 11 The comparison of m value in stress relaxation tests.

Table 1 Strain rate sensitivity m values.

Superplasticity: 60 Years after Pearson 269

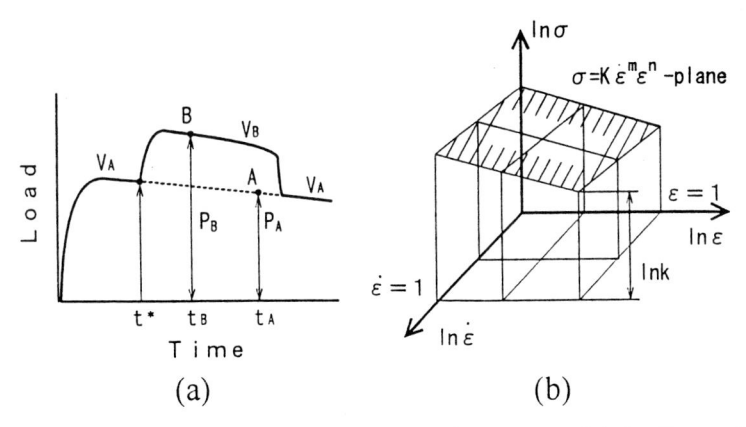

Fig. 1 Explanation of Calculating of strain rate sensitivity for method 1 and method 3.

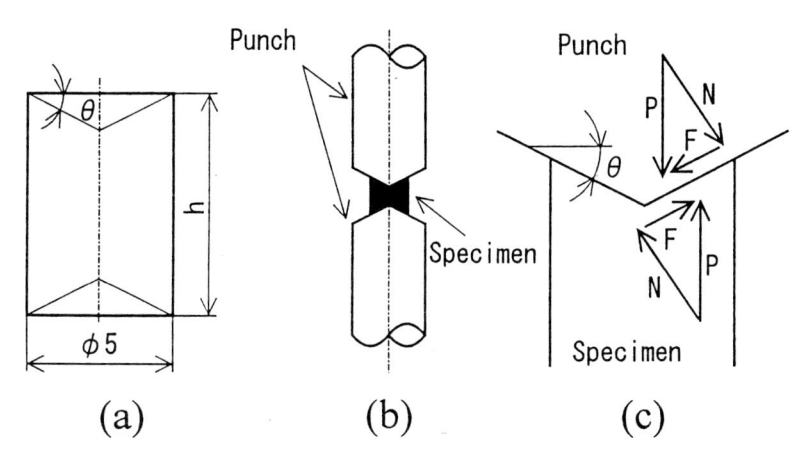

Fig. 2 Illustrations of compression test and specimen geometry.

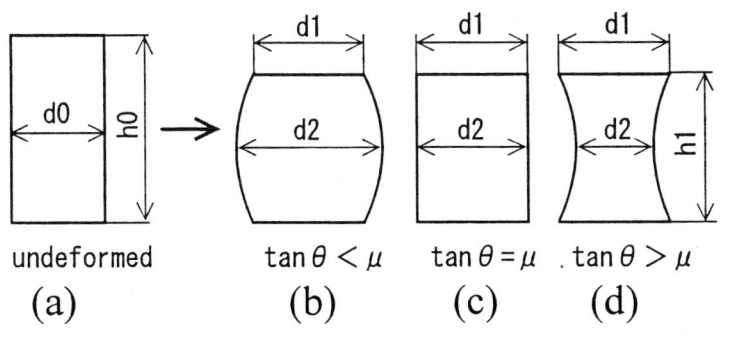

Fig. 3 The effect of friction on barreling.

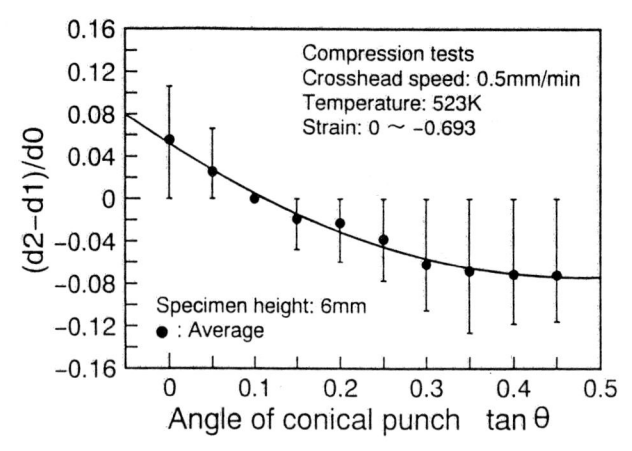

Fig. 4 Alteration of specimen profile with changing angle of conical punch.

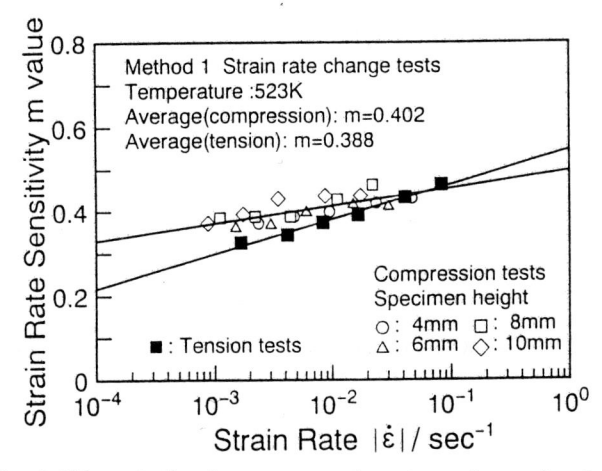

Fig. 5 The relation between strain rate and m value in strain rate change tests.

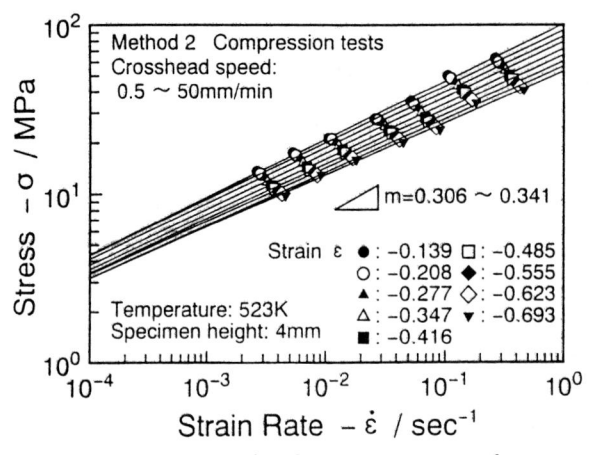

Fig. 6 The liner regression between true strain rate and true stress.

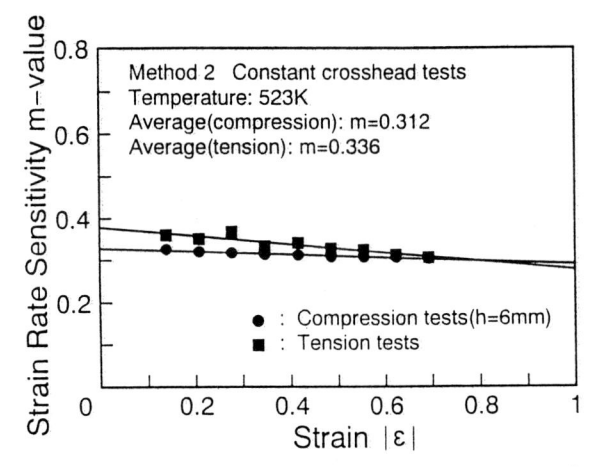

Fig. 7 The strain rate sensitivity is a variation of strain.

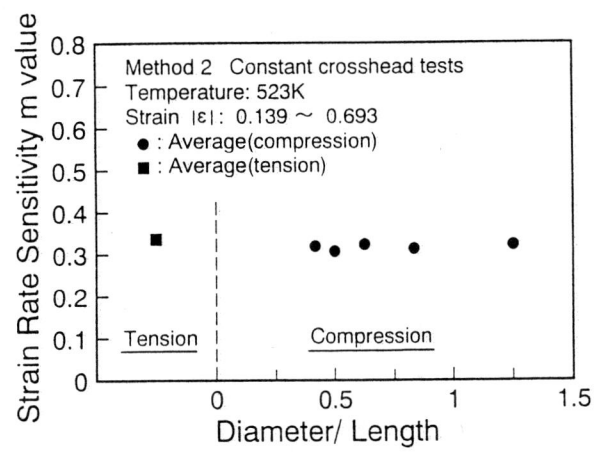

Fig. 8 The comparison of m value influenced strain and strain rate.

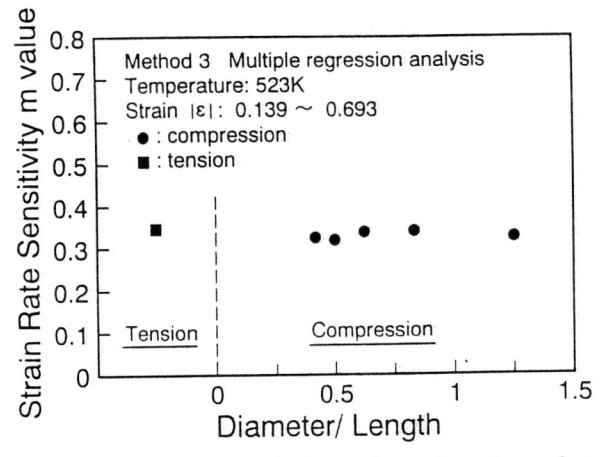

Fig. 9 The m value is independent of strain and strain rate.

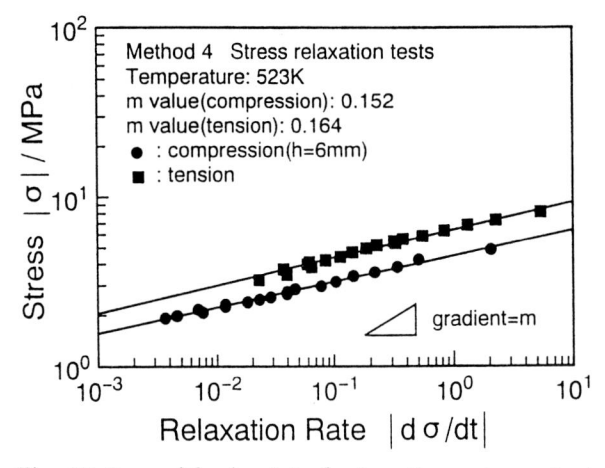

Fig. 10 Logarithmic plot of relaxation rate against stress.

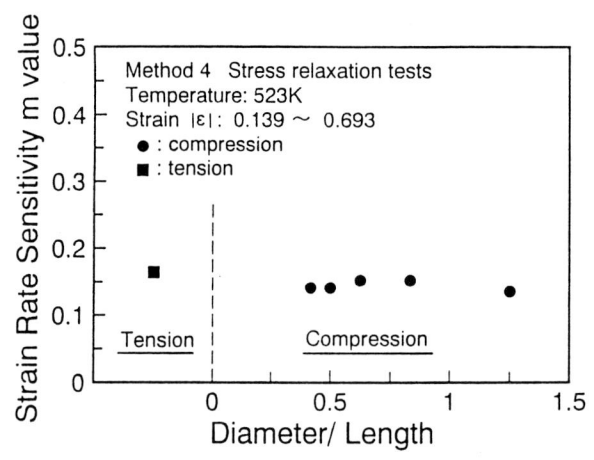

Fig. 11 The comparison of m value in stress relaxation tests.

Table 1 Strain rate sensitivity m value.

Classification of method	Compression	Tension	$\triangle$m
Method 1 Strain rate change tests	0.402	0.388	+0.014
Method 2 Constant crosshead speed tests	0.316	0.336	−0.020
Method 3 Multiple regression analysis	0.329	0.346	−0.017
Method 4 Stress relaxation tests	0.144	0.164	−0.020

$\triangle$m : a difference between compression and tension.

Superplastic Sheet Aluminium Alloys: Their Forming, Application and Future

R.G. Butler and R. J. Stracey

Superform Metals, Worcester, England

Abstract

The superplastic properties exhibited by some sheet aluminium alloys have been exploited to produce components encompassing an impressive range of shapes, for a diversity of industries. The alloys developed to date are discussed, as is the range of forming processes. The product's properties and raw material costs lead directly to its viable application opportunities.

The key to greater future application, for less specialized products, lies in the creation of a significant shift in the cost balance of both raw material and shaping rate. The costs and technology for appreciably increased aerospace applications are already attractive and usage is restrained only by commitment.

1. Introduction

Superform Metals was established to exploit the superplastic aluminium sheet product developed by Tube Investments and British Aluminium in the UK. This paper describes the current alloy range, which has evolved from both what is now British Alcan's developments, and from externally manufactured products. The range of forming processes now available are included. The current market for the products has become fairly specific depending largely upon alloy type.

2. Superplastic Aluminium Alloys

The development and use of a wide range of superplastic aluminium alloys has led to a predominance of process route modified standard compositions, over those developed specifically for their superplastic properties. Thus Al-Ca-Zn, Al-Ca and 5083+Cu are no longer considered, and 5083, 7475 and some Al-Li alloys predominate. The reasons for this are the unwillingness of rolling mills to add to their alloy range, the service property acceptability of 'known' alloys, and the property degradations caused by some superplastic property developments. The exception to this generalisation is Supral, 2004, which has found a niche, due to its considerable superplastic abilities, despite being of novel composition.

The alloy choice is technically dependant upon property compatibility for particular applications, as well as superplastic performance, but price, availability, specification status are equally important in order to secure a commercial contract . Superform Metals have worked closely with the Aluminium Corporation, Dolgarrog, Wales, in order to provide cost effective products with an acceptable overall lead time.

3. Superplastic Forming of Aluminium Alloy

The forming techniques employed use gas pressure to create shapes defined by static or moveable tooling. The standard techniques are female cavity, male tool in female cavity (drape), and male tool moved into sheet. There are a number of variations involving superimposed back pressure , diaphragms, preforms, deliberate tears and chemically milled blanks.

Table 1 Alloy composition and strength

Alloy	Composition	Temper	0.2PS (MNm^{-2})	UTS	El%
2004	Al-6Cu-0.4Zr	T6	300	420	7
2004	Al-6Cu-0.4Zr	0	150	250	12
5083	Al-4.5Mg-0.7 Mn	0	150	280	18
7475	Al-5.7Zn-2.3 Mg-1.5Cu	T76	500	550	10
8090	Al-2.4Li- 1.2Cu-0.7Mg	T6	350	450	4

The forming tools may be ferrous, aluminium or ceramic. They may be in one piece, or fit into standard containment or bolt-up surrounds. The design of tooling greatly influences the final thickness distribution of the component. The 'run-off' areas are very important in determining material flow, and the clamp-line or surround may be profiled to bend the blank prior to clamp up. Methodology has diversified and improved, broadening the scope of application.

The gases employed for superplastic forming may be air, nitrogen or argon. The choice is product dependant with air being cheapest but reactive and likely to contain moisture, where as argon offers the technical preference being inert, but at the cost of price and safety considerations.

The tool/blank interface is lubricated by graphite or boron nitride. Graphite is the easiest to use and the least expensive, but can cause post forming surface-attach if not completely removed. Boron nitride is non-reactive, expensive, and must be applied and monitored in a carefully controlled manner to avoid build-up on the tool.

Forming machines have developed with cycles being automatically followed, with the possibility of gaseous mass flow and sheet movement measurement being possible. The precise process control requirement is determined as much by application requirement eg. aerospace, as by the need to optimise cycle times for economic reasons.

The current status of commercial forming is such that it is still an art, as much as a precise science. Involvement of computers is emerging, but a cost effective track record has yet to be established. 'Forming experts' are few in number and welcome the introduction of software to complement their activities. The total automation of the process must allow for raw material batch variability and perhaps even within batch variance.

Finishing

The trimming of superplastically formed shapes often offers a greater challenge than their forming. The method employed depends upon quantity and quality requirement. Items may, for small quantities, be hand trimmed, and range up to mechanically, laser or water jet cut by numerically controlled machines for larger volumes. The trimming of a three dimensional sheet metal forming product generally involves the use of a formed surface datum which can permit some variability particularly if such a datum is taken from the non-tool side of the component.

Post solution treatment distortion correction is an additional challenge for quench sensitive alloys.

4. Applications

The established primary applications for superplastically formed aluminium sheet components are in the areas of aerospace and rail vehicles. Additional opportunities have been found in architectural panels, electronics boxes, medical equipment and automotive panels.

The rail market involves both the body shell of the current aluminium vehicles, and the interior fitments. Body shell panels offer a good corrosion resistance and may be welded into integral structures. Interior items replace plastics so overcoming fire hazards and offering a neater appearance. The rail market is very dependant upon commitment to maintain and improve rail way networks.

The aerospace applications for sheet metal items cover the range from trim and fairings to class 1 components. The back pressure forming of high strength alloys permits the production of cost competitive items, offering lower weight and reduced part counts.

The market limiting factors are a) inertia: a resistance to change - felt to different degrees in the various companies/sites, and b) the reduced build rate of current aircraft and a reduction on new projects.

The application of superplastic components in general is restrained by cost. This is determined by the price of raw material and the processing rate.

In order to continue to widen the range of applications, the superplastic process must be exploited more fully. This can only occur by its incorporation at the design stage, as the use of superplastic forming for an existing part is generally a compromise.

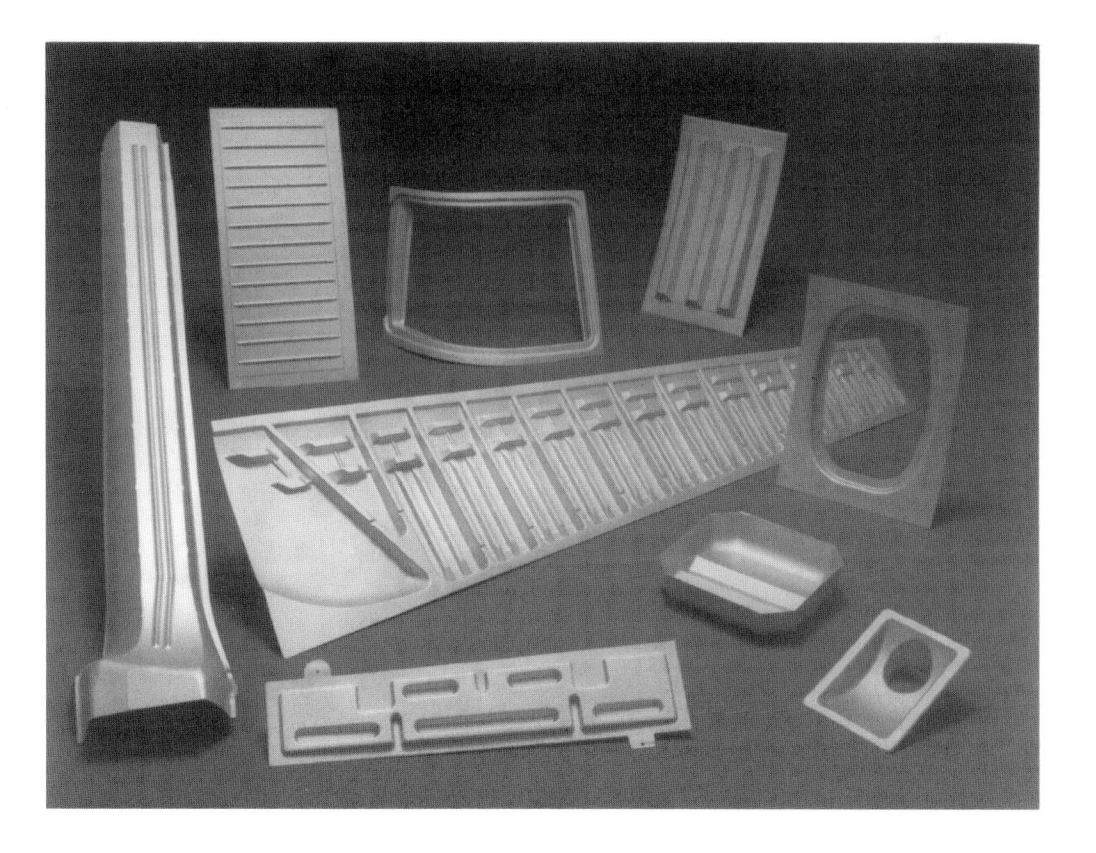

Fig.1 A variety of components produced by Superform Inc. USA.

5. Future

Significant future growth into volume markets will be achieved by reducing the cost balance of the raw material and the forming process. The greatest promise lies in the development of very fast forming alloys, produced at a cost not markedly different to current materials. New developments in the forming process give indications of significantly reduced component specific forming costs. The opening of larger part of the general engineering volume market is seen as having great potential.

The aerospace market will continue to grow as more experience of structural components is gained. The establishment of a significant property data base will enable design and stress engineers to be confident in wider application on an international scale. The growth of current commercial activities will continue by effort, commitment, improved efficiency and by continued innovation.

Properties of superplastic 5083 alloy and its applications

M. Matsuo

Research Laboratory SKY Aluminum Co., Ltd.
1351 Fukaya Saitama, Japan

Abstract

Superplastic 5083 alloy (ALNOVI-1) was developed. It shows a superplastic elongation of 300% or more, at temperatures of 783-823K and strain rates of 2×10^{-4} - 1×10^{-3}. The effects of grain size and of the iron and silicon contents of the alloy on superplastic elongation and cavitation were investigated. It was observed that a finer grain size is preferable for larger elongations and a high purity base alloy is better for larger elongations and lower cavitation levels. It was also revealed that cavitation was small at lower forming temperatures and faster forming rates. Microscopic investigation shows that grains were elongated after superplastic forming which suggests that strain within the grains plays an important role on cavity formation in this alloy. The superior post-forming properties of SPF 5083 alloy such as corrosion resistance and weldability were confirmed.

1. Introduction

It is well known that some aluminum-magnesium alloys show good superplastic elongations, for example Al-6%Mg-0.15%Zr [1] and Al-10%Mg-0.1Zr [2]. Generally, aluminum-magnesium alloys have moderate strength with superior corrosion resistance, weldability, and surface treatability such as anodizing. Therefore, these alloys are important for practical applications. Alloy 5083 (Al-4.5%Mg-0.7%Mn-0.1%Cr) in particular has many practical industrial applications as a medium strength material. It is also noted that non-heat treatable 5083 alloy is not so expensive as the heat treatable 7475 alloy.

For wider application of superplastic forming it is believed that commercial Al-Mg alloys such as 5083 will become important materials. From this point of view, interest in superplastic 5083 seems to have increased recently [3,4]. Superplastic properties such as elongation to failure and cavitation of commercial superplastic 5083 alloy (ALNOVI-1), and some factors affecting superplastic properties, have been investigated. Also, post-forming properties have been examined and several examples of superplastic applications will be introduced.

2. Properties of superplastic 5083

2.1 Experimental procedure

In a series of experiments, tensile tests were conducted at a constant strain rate, and m-values were measured by the method of rapid strain rate change.

Air blow forming was conducted to investigate practical superplastic behaviour. Sphere bulge forming was adopted to evaluate the superplasticity. Bulge height and equivalent elongation were mainly measured. The blank size of the sheet specimen was 160mm x 160mm and the diameter of the sphere was 100mm. Forming of cylinders of diameter 100mm was also carried out to obtain flat specimens, and some large square boxes of base 250mm x 250mm were formed for preparing mechanical test pieces and welding-test specimens.

Here, equivalent elongation was defined as follows:
Equivalent elongation (%)=$[(t_o -t)/t] \times 100\%$ where t_o; thickness before forming, t; thickness after forming.

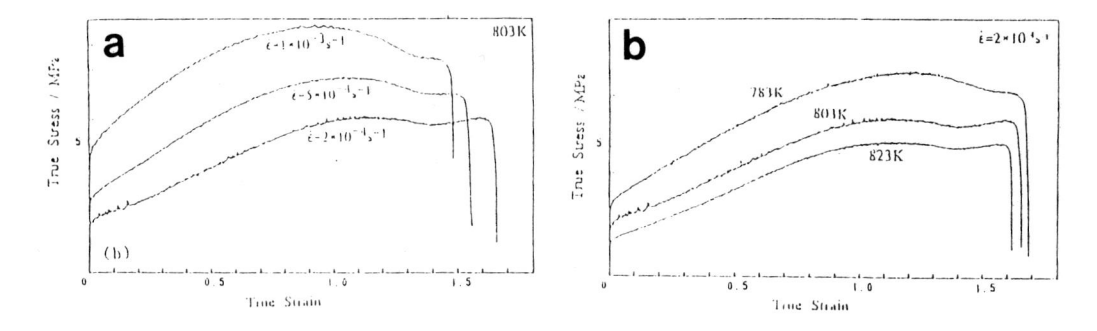

Fig.1 Stress strain curves for superplastic 5083 alloy, (a) Effects of strain rate and (b) Effects of temperature.

Cavitation was measured mainly by an area method. Specimens were polished perpendicular to the rolling direction and area fractions were determined using an area analyzer.

2.2 Superplasticity of 5083

Fig.1 shows typical stress-strain curves for superplastic 5083 alloy. It is recognized that flow stress is increasing with increased strain which means it is a strain hardening material. Generally, the strain hardening of a superplastic alloy is believed to be due to grain coarsening during deformation. However, the strain hardening tendency of this alloy is seen to increase with higher strain rate and lower forming temperature. It is reported that dislocations are observed within the grains in 5083 alloys[5]. Therefore, it can be explained that the stress increase during deformation is due to dislocation interaction within the grains, and both slip within the grains and grain boundary sliding might play an important role in this alloy. The elongated shapes of grains were observed after forming which means that slip within the grains has taken place.

Fig.2 shows superplastic elongation. It was shown that a larger elongation was obtained at a lower temperature and slower strain rate. The same tendency was observed in experiments involving air bulge forming. The m-value of this alloy ranged from 0.45-4.55 at peak condition at about 3×10^{-4} to 1×10^{-3} s^{-1}, and the m-value is higher at higher forming temperatures. This does not always correspond with the maximum in elongation and bulging tendency.

2.3 Effect of grain size on superplasticity

It is generally accepted that a smaller grain diameter is preferable for increasing the superplastic elongation because deformation is accumulated mainly by grain boundary sliding. Since 5083 alloy contains 4.5% magnesium, a lot of nucleation sites were introduced by multiple shear bands formed during cold rolling. Fine intermetallic precipitates containing manganese and chromium will act as obstacles for grain boundary movement. Therefore grain refining of 5083 alloy was achieved by adjusting the size and distribution of fine precipitates, amount of cold strain and heating rate on recrystallization .

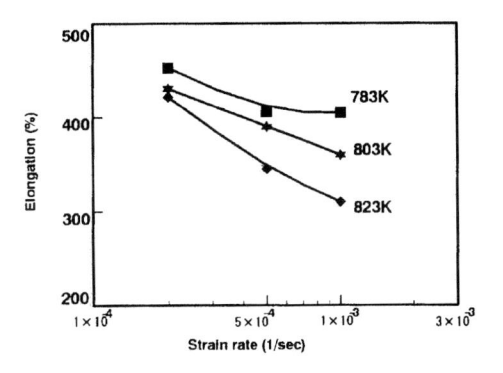

Fig.2 Effects of temperature and strain rate on superplastic elongation.

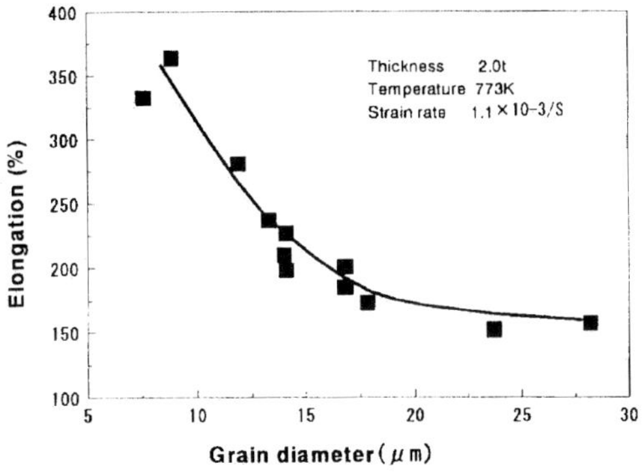

Fig.3 Effect of grain diameter on superplastic elongation.

Fig.3 shows the effect of grain size on superplastic elongation of 5083 alloy. It can be seen that tensile elongation was strongly influenced by grain size. Grain sizes below 15 microns, or more preferably below 10 microns, were desirable to obtain large elongations.

2.4 Effect of chemical composition on superplasticity

Lower cavitation is also required together with larger elongations for superplastic materials. It is believed that fracture will be result from the interconnection of cavities. Therefore, it is important to reduce cavitation.

It is well known that the large intermetallic compounds on grain boundaries will be the sites for cavity nucleation during superplastic deformation and it will be useful for reducing cavitation to decrease the size of intermetallic compounds. Intermetallic compounds in 5083 are known mainly to be Al_6(FeMn), α-AlMnFeSi and Mg_2Si [6]. Manganese is a necessary element for refining the recrystallized grains and stabilizing them during superplastic forming at high temperature. It is concluded that it might be useful to decrease the iron and silicon content so as to reduce the size of intermetallic compounds. As shown in Fig. 4, the size of intermetallic compounds is smaller in the higher purity base alloy.

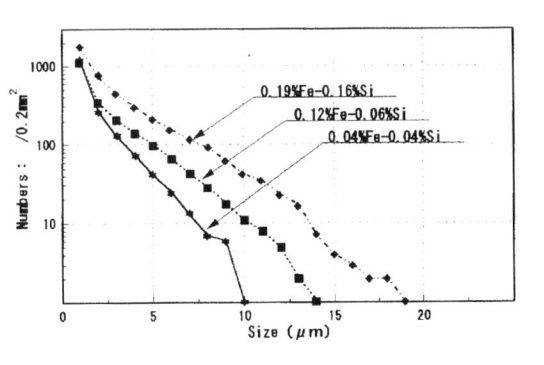

Fig.4 Effects of Fe and Si content on size distribution of intermetallic constituents.

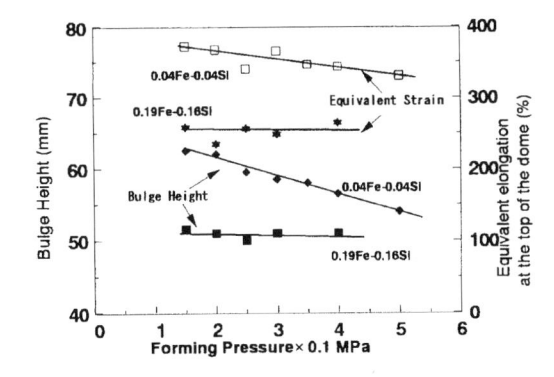

Fig.5 Effects of Fe and Si on superplastic properties.

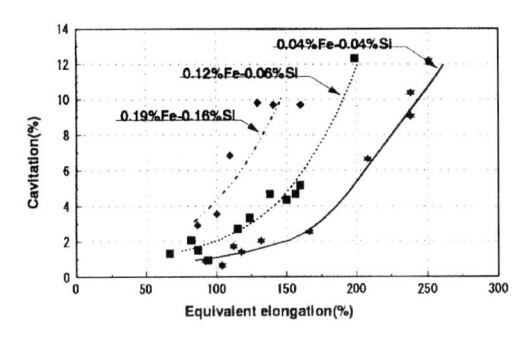

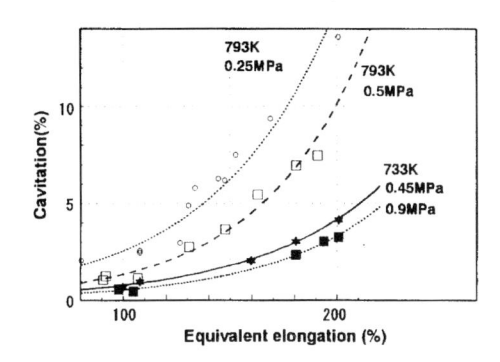

Fig.6 Effects of Fe and Si content on cavitation. Cylinder blow forming at 773K.

Fig.7 Effects of temperature and pressure on cavitation during bulge forming. Bulge diameter 100mm.

Fig.5 shows the bulge height of two 5083 alloys with different iron and silicon contents. It is clearly seen that the higher bulge height was obtained for the high purity base alloy.

Cavitation was also strongly affected by iron and silicon levels and tends to decrease as iron and silicon contents decrease, as is illustrated in Fig.6. The results show that a high purity base 5083 alloy is preferable in order to obtain superior superplastic properties.

2.5 Effects of temperature and forming pressure on cavitation

Cavitation is understood to be caused as a result of incomplete accommodation of grain boundary sliding which is characteristic of superplastic forming. Therefore cavitation was increased as the amount of strain was increased, as was already shown in Fig.6.

Fig.7 illustrates the effects of forming temperature and forming pressure on cavitation during bulge forming. The size of the bulge forming was 100mm in initial diameter. A number bulge tests were carried out with various bulge heights to obtain different amounts of strain. Specimens were collected from the top of the dome for cavitation assessment..

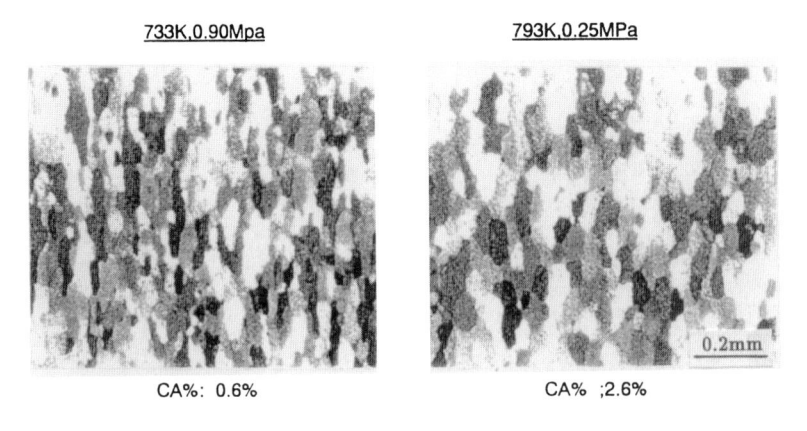

Fig.8 Grain shapes and cavitation after different forming conditions; equivalent elongation 100%.

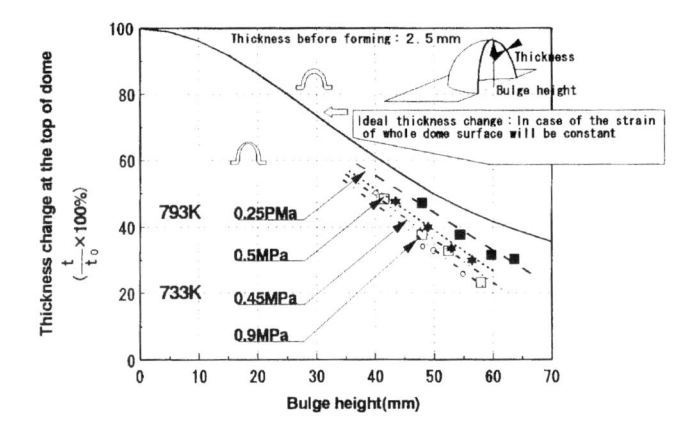

Fig. 9 Effects of forming conditions on thickness change at the top of the dome.

It can be seen that cavitation is decreased at lower temperatures and higher forming pressures. A low forming temperature such as 733K is particularly effective in reducing cavitation.

Experiments using tensile specimens were also conducted and similar results for the effect on temperature were obtained.

It is generally accepted that cavitation is reduced at higher temperatures and slower strain, rates for example in 7475 alloys. However, the tendency for cavitation formation was quite different in this alloy. This might be associated with the mechanism of superplasticity for the material.

It is estimated that the contribution of deformation by slip within the grains to the total deformation might become larger at lower forming temperatures and faster strain rates. This means that less accommodation of grain boundary sliding will be necessary, resulting in reduced cavitation.

Fig.8 shows the grain structures after superplastic bulge forming at several temperatures and pressures, at an equivalent strain of 100%. It can be seen that the recrystallized grains were more elongated at the lower temperature and higher pressure which suggests that more slip was taking place under these conditions.

2.6 Thickness distribution

The thickness changes at the top of the dome for various bulge heights are illustrated in Fig.9. Dome top thickness was least for the lower temperature and higher pressure, compared with the thickness if the whole bulged sphere was deformed evenly. As a result, maximum bulge height was reduced at this temperature and higher forming pressure. This tendency for thickness variation must be related to m-value.

Since the tendency to thickness change and cavitation are quite opposite to each other, the forming condition should be selected according to which property is most important.

3 Post-forming properties

Mechanical properties and fatigue strength after superplastic forming are shown in Fig.10. Static strength and fatigue strength were decreased gradually as the superplastic elongation was increased.

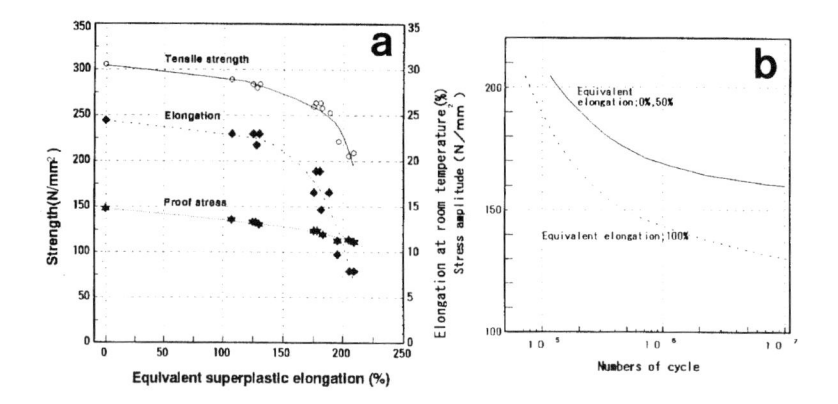

Fig.10 Post-form properties (a) Post-form mechanical properties and (b) post-form fatigue strength of sheets.

TIG arc welds of post-form sheets were investigated and were shown to have no mechanical strength drop after welding.

Corrosion resistance of superplastically formed parts was also examined by the salt spray test. Weight loss of post-form sheets was equal to that of non-formed sheets even after 4000Hr salt spray exposure. The 5083 alloy is likely to be surface-oxidized during forming at high temperatures because of the high magnesium content of 4.5%. Lubricant powder such as boron nitride or graphite is often employed during superplastic forming and these powders are likely to remain on the deformed surface.

These surface oxides or lubricant powders are usually inconvenient for surface treatments and welding, especially spot welding and adhesion. The surface oxide or lubricant powder can easily be removed by acid etching or chelate etching. Preferred properties for practical use were confirmed through examination of post-formed specimens.

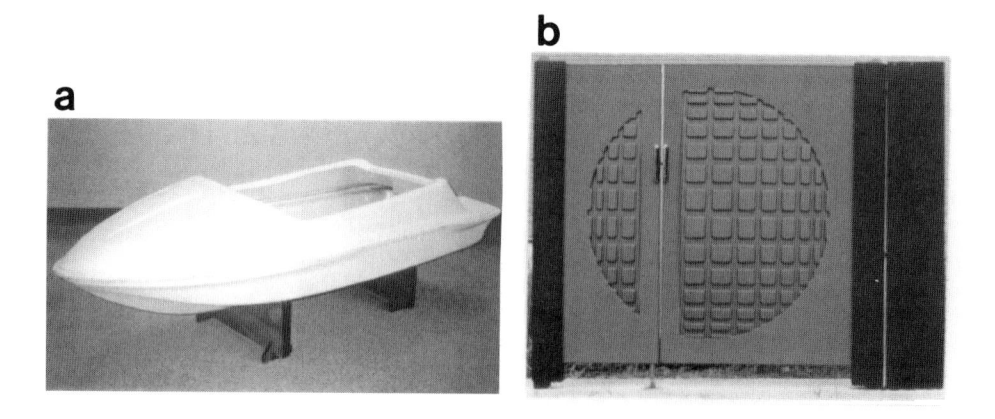

Fig.11 Examples of superplastic forming, (a) Superplastically formed boat and (b) Gate door.

4. Examples of superplastic forming

Examples of superplastic forming are introduced in Fig. 11. Fig.11a shows a hot blow-formed aluminum boat which consists of three parts. The inner and outer panels were bonded by adhesion through a vacuum assisted process. Chelate etching with EDTA and boemate treatment in warm water were adopted for surface treatment. Fig. 11 (b) shows an example of gate doors. Surface treatment in this case was paint coating.

5. Conclusions

Superplasticity of 5083 was investigated and superplastic 5083 (ALNOVI-1) was developed by adjusting chemistry and fabrication practice. The characteristics obtained are as follows.

- Many application experiences exist because the chemical composition is within the 5083 chemistry range.

- Superior superplasticity was developed by adjusting the chemical composition and grain size.

- Lower temperature forming at below 773K and a faster forming rate at about 10^{-3} s^{-1} are possible.

- Moderate strength is obtained in the as-formed condition. It is convenient for shape freezing.

- Corrosion resistance, weldability and surface treatability are excellent.

- Fabrication practice is not so complicated.

References

[1] K. Matsuki, Y. Uetani, H. Yamada, Y. Murakami, Metal Sci. 10, 1976, 235-242.

[2] J.E. Wise, Master's Thesis, Naval Postgraduate School (1987)

[3] H. Iwasaki, K. Higashi, S. Tanimura, T. Komatsubara and S. Hayami, in 'Superplasticity in Advanced Mater., ICSAM'91 (Ed. S. Hori, M. Tokizane and N. Furushiro), 447-452, 1991. JSRS, Osaka.

[4] J.S. Vertano, C.A. Lavender, M.T. Smith and S.M. Bruemmer; Scripta Metall. Mater., 30, 1994, 565.

[5] F. Li, W.T. Roberts anf P.S. Bate; Scripta Metall. Mater. 29, 1993, 875.

[6] ASM Handbook Committee: Metals Handbook 8th Edition, Vol .7 Atlas of Microstructures of Industrial Alloys (1972) ,244

NEW MARKET AREAS
FOR SUPERPLASTIC ALUMINIUM

by R.J. Stracey and R.G. Butler
Superform Metals
Worcester
England

Abstract

The predominant market areas for superplastic aluminium are "traditionally" aerospace, and more recently railway vehicles. This paper describes examples of new components, some in the early stages of development which provide an indication of a widening of application into larger and sometimes simpler components. These new products are involved in the areas of marine vehicles, telecommunications and architecture, radiography, bulk substance transportation, and travel sleeper accommodation. A tendency is demonstrated for a demand for a basic durable "engineering" product, which without incurring the cost of further laboratory based technological innovations, takes full advantage of the superplastic process.

Introduction

It can be seen from the following examples that with projects which carry some commercial or technical risks it may be prudent to produce demonstration or trial components on prototype tooling before embarking on more expensive production tooling. Whilst this proving stage, even with computer aided analysis, is sensible in many selling/production activities, in SPF the approach is enhanced by the ease with which single surface trial tooling can be produced.

SPF Aluminium Boat

Aluminium boat hulls have been in existence for many years. Usually these are fabricated from single curvature sections welded or riveted together. A small number of one piece hulls have been made; some by the technique of explosive forming which is expensive, time consuming and somewhat unpredictable.

At Superform Metals a development program to create what is believed to be the first SPF one piece aluminium boat is currently nearing completion.

The development has revolved around a one third scale model forming, Fig. 1, enabling the necessary level of process confidence to be achieved, whilst facilitating the 'fine tuning' of the boat design before committing to a full size forming tool.

The challenge during the design process was to produce a shape that was sufficiently rounded to be readily formable whilst not having the aqua dynamics or market appeal of a bath tub. See Fig. 2 There is a very high level of confidence that this has been achieved and full size hulls will be produced very early in 1995.

Fig. 1 In order to determine the optimum hull design a one third scale forming tool was manufactured and development formings produced.

Fig. 2 The one third scale hull forming achieved the design and process requirements.

Telephone Kiosks

A new range of British Telecom telephone kiosks is currently being installed in a number of railway stations throughout the country. Superform has played a part in the design of these through the application of SPF to the roof area.

Three variants will be available: wall mounted single and double telephone, and free-standing circular four telephone. Fig. 3 shows a prototype assembly for a single telephone installation.

The roof of each variant consists of a deep upper section containing the illuminated telephone sign and a shallow lower or closing section containing the user lighting. Just two forming tools were required as different CNC trim patterns are used to create the three different sizes of component from basic upper and lower formings. Various components are shown in Figs. 4 & 5.

Fig. 3 The prototype assembly of the single telephone kiosk shows the SPF canopy.

Fig. 4 The canopy for the double telephone kiosk model is two station formed and CNC trimmed.

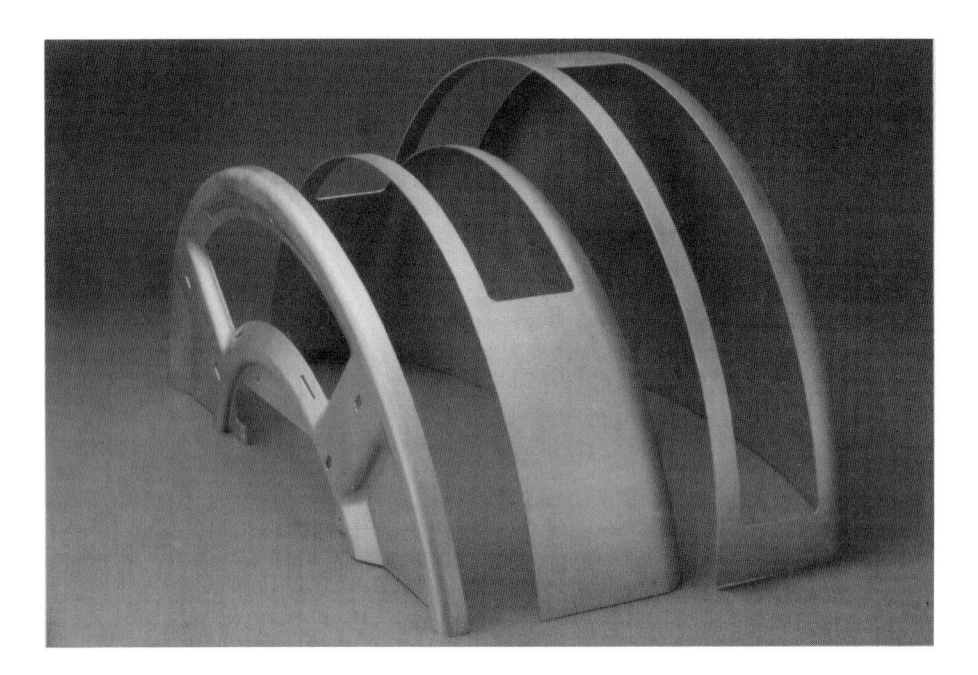

Fig. 5 The single and double telephone canopies with the single lower closing panel which houses the user lighting.

Radiography

Superform have manufactured components for medical diagnostic and test equipment for many years. These components have in the main been covers and finishing items with an aesthetic bias. A development program is currently underway which should exploit the benefits of SPF in a more functional application within the field of radiography.

In this application aesthetics remain important as the component will be on the exterior of the equipment, but in addition there is a requirement to support a static load and also presence of a vacuum requires it to withstand atmospheric pressure.

With these service requirements and the cost of a full size forming tool a cautious approach involving scaled down testing has been adopted, Figs. 6 & 7 show the component and form tool that represent the second phase of this development.

Results so far are very encouraging and following a possible third phase of development it is anticipated that work will begin on the full size tool early in 1995.

Superplasticity: 60 Years after Pearson 289

Fig. 6 A trial forming was produced for evaluation and testing by the customer prior to manufacture of the full size component.

Fig. 7 To achieve the trial forming a fabricated form tool was produced.

Road Bulk Tank Vehicles

Aluminium is used extensively for the manufacture of bulk tankers, Fig. 8, where the benefit of weight (and hence energy) savings and corrosion performance are significant. The main tank structure consists of a large diameter cylinder or quasi cylinder produced by joining single curvature rolled sheets using automatic welding techniques. Other more complex sheet components such as dished ends, tapered hoppers and transition panels are welded onto or into the cylinder to create the whole tank structure. The challenge is to form some of these shapes in a 5000 series alloy, which work hardens, and achieve the required dimensions and radii necessary to avoid residual stresses in the tank due to welding components with poor fit-up. Figs. 9 & 10 indicate some of the these problem areas.

To date these shapes have been produced by hand working using traditional tapered rolling techniques and many hours of skilled metal workers time. Superplastic forming has been identified as offering a possible solution to achieving some of these complex shapes with a higher degree of accuracy and repeatability.

Initial trials have shown that SPF will probably be employed in future designs to improve the fit-up tolerances easing any required cold working to achieve a consistent weld gap.

Fig. 8 Transportation of bulk materials by road is generally by large trailer mounted aluminium containers.

Superplasticity: 60 Years after Pearson 291

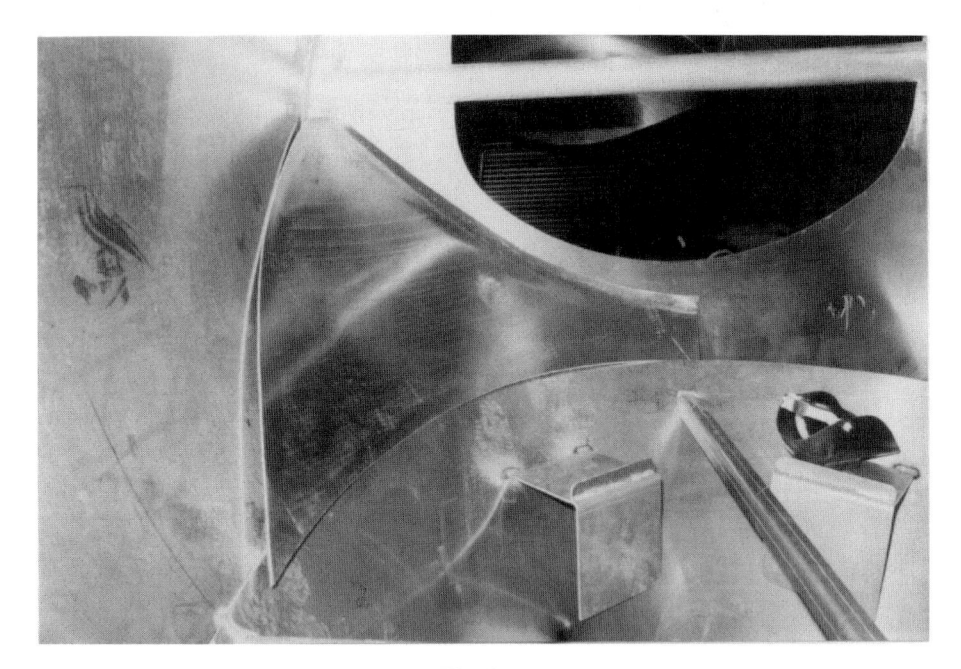

Fig. 9

Fig. 10

Figs. 9 & 10 show the interior of a bulk tanker at the junction of the tank structure and the hopper and weir plates. The problems with fit up and weld tolerances are shown in Fig. 9 and a fully welded assembly is shown in Fig.10.

Rail Passenger Transport Beds

A great number of SPF aluminium components have been employed in the construction of railway vehicles. Whilst SPF rail seats, Fig. 11, have been demonstrated to show a technical viability, the economics of this area are still being examined.

Rail beds are however a less complex forming and the lower associated costs indicate that there is a good potential for SPF.

Figs. 12 & 13 show the proposed installation for a project currently being evaluated. Both beds are required to pivot upwards; the lower reverting to a seat back.

Full size prototypes have been formed and are currently under test, Figs. 14 & 15. The challenge is to produce a shell that has sufficient strength to withstand the rigors of use without the addition of extrusions or secondary SP formings.

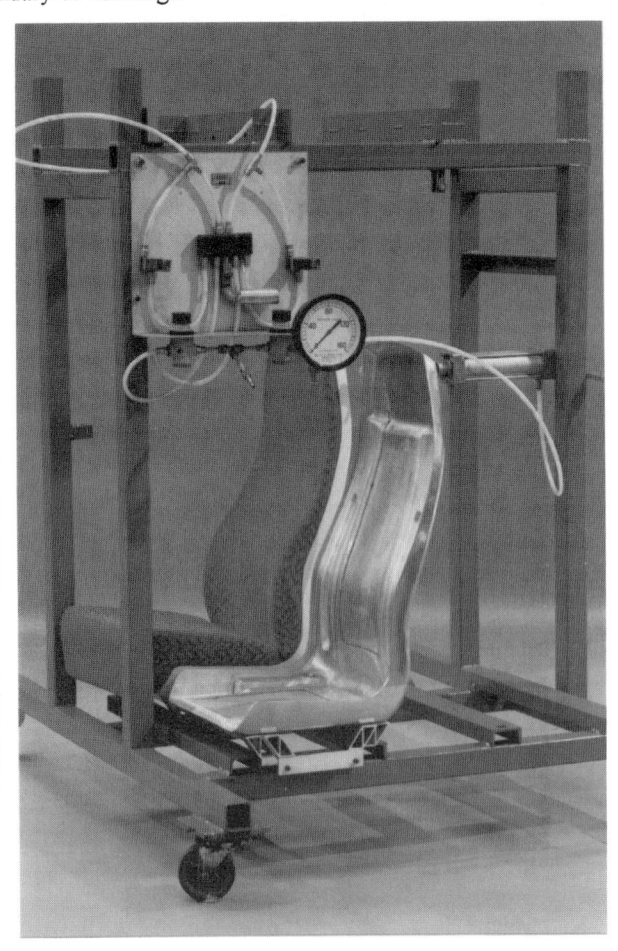

Fig. 11 SPF aluminium rail seats have achieved the static and fatigue loading requirements as well as showing excellent fire performance and weight savings.

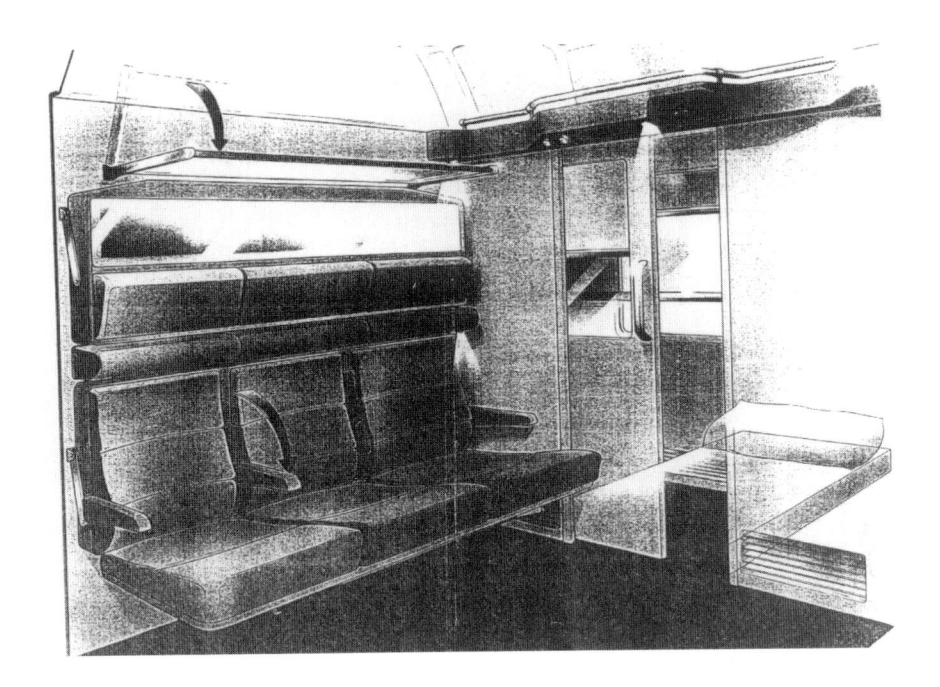

Fig. 12

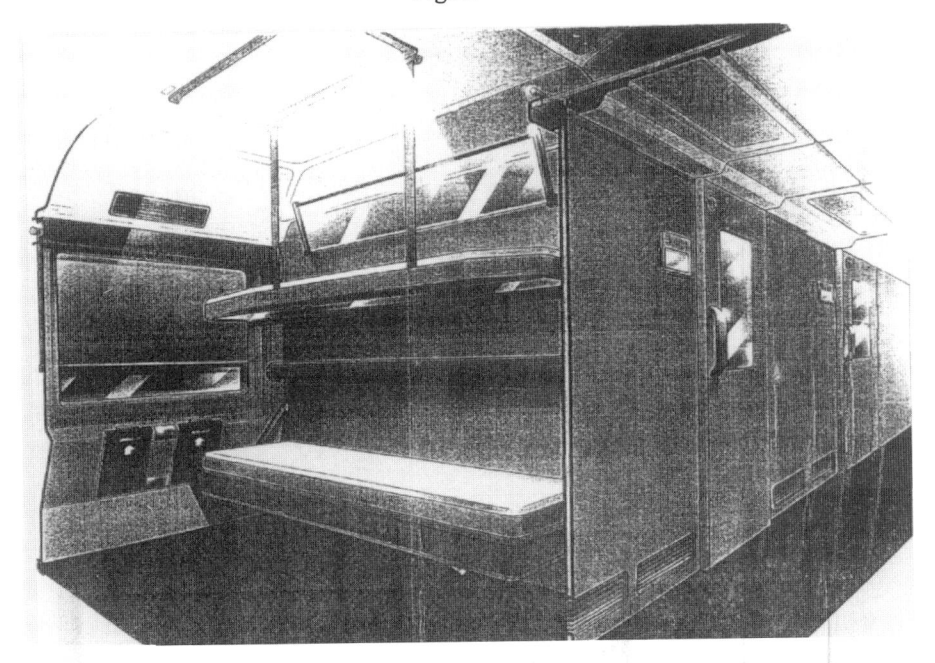

Fig. 13

From the design layouts in Figs. 12 & 13 it is clear that the bed base must meet the strength and aesthetic requirements.

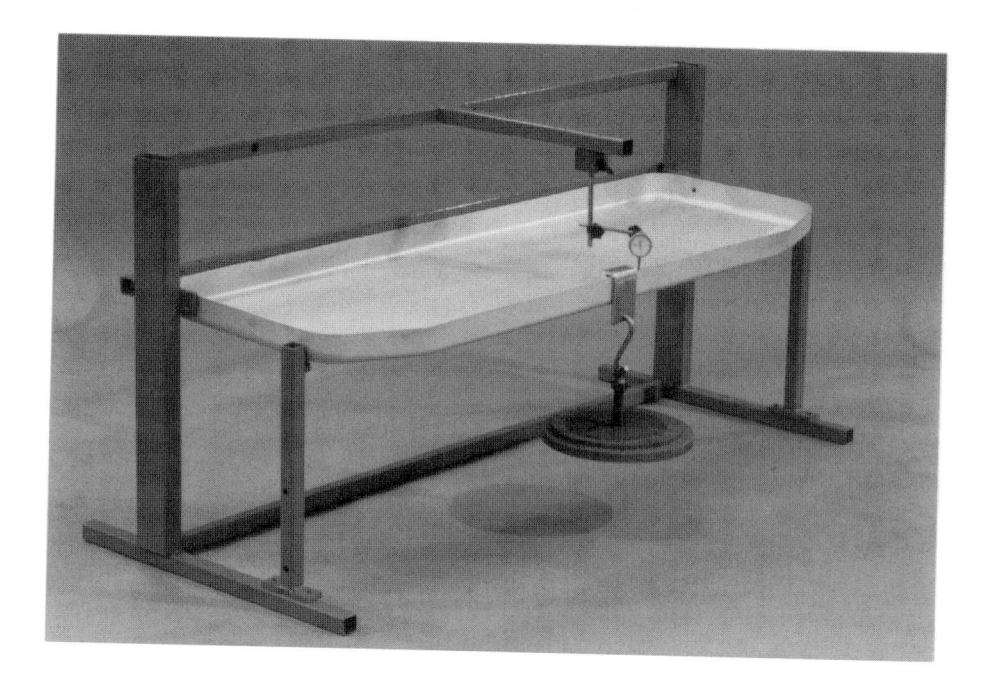

Fig. 14 In order to monitor deflection and stress levels a test rig simulating the specification support and load conditions was produced.

Fig. 15 The prototype bed base panel forming prior to trimming.

Conclusions

The new applications detailed in this paper indicate that SPF aluminium forming continues to offer benefits in a wide range of markets.

To increase usage in the future we must examine the possibilities for further innovation in SPF process technology. The development of new, sometimes unusual, applications often benefit from close liaison with the product engineers and a degree of risk taking tempered with a cautious approach.

Roll Bonding / Superplastic Forming (RB/SPF) of Superplastic Aluminium Alloys

Satoshi Furihata and Hiroaki Ohsawa
Department of Mechanical Engineering, Hosei University,
Kajinocho 3-7-2, Koganei, Tokyo, Japan

Abstract

Superplastic forming (SPF) of titanium and aluminium alloys, and concurrent superplastic forming/diffusion bonding (SPF/DB) of titanium alloys, has developed to a state of relative maturity over the past twenty years [1]. These processes allow the aerospace industry to produce lighter, more cost-effective components. SPF/DB, however, is currently limited to titanium alloys; an effective way to diffusion-bond aluminium alloys has not been developed yet because of the strongly adherent oxide layer on the surface.

On the other hand, it is well known that if the reduction ratio, for instance during the course of rolling clad metal sheets, is high enough so that bonding at the interface can be firmly achieved, roll bonding between similar/dissimilar metals will have been successfully carried out. In this rolling process, if a shaped area in the plane of the rolled sheet is left unbonded and also if some material is located within that area which will play a role of a gas-pressure source during superplastic forming at an elevated temperature, an external gas supply will not be needed. Roll bonding with an unbonded area preferentially selected by stopping-off, followed by heating, can then make the superplastic aluminium alloy sheets undergo superplastic forming using a simple die assembly. The procedure may be termed RB/SPF (Roll Bonding/Superplastic Forming). The selection of a gas-pressure source which acts as a STOP OFF when roll bonding, and the incorporation of this source in the interface layer, may be the key in this new processing.

1. Introduction

The combined application of superplastic forming and diffusion bonding(SPF/DB) to Ti alloy sheet is a well established manufacturing technique in the aerospace industries leading to high cost and weight savings. Superplastic Al alloy sheet can be blown or vacuum formed at elevated temperatures and the formed Al structures or mechanical parts now find world-wide use. Diffusion bonding of superplastic Al alloy, on the other hand, may not be carried out successfully, as for Ti alloys, because there always exists a tenacious Al oxide on the surface. A number of studies, however, are still being under taken, with the object of increasing the strength of joints by optimization of some of the bonding factors.

In the manufacturing of clad metals, roll bond processing plays a significant role. If the thickness reduction during rolling is high, roll bonding can be successfully carried out and yield a high strength in the interface. This is also true for clad Al components. Instead of diffusion bonding, roll bonding may be used as a way of obtaining a firmly bonded interface. In fact, the method of Roll Bonding is actually adopted in the manufacture of evaporators for refrigerators made of Al alloys. In this process, the very complex and curved path in which cooling gas flows is made in the plane of the wall. Before rolling, STOP OFF is printed on the interface surface, the geometrical shape change of the printed pattern due to rolling being known in advance. Air blowing into the un-bonded area after roll bonding results in the forming of the tube surrounding the wall of the evaporator.

In this study, tentative work on RB/SPF of superplastic Al sheets is presented. This processing is composed of several stages and involves cladding of similar/dissimilar metals with an un-bonded area in the interface plane. The geometry and dimension of this area should be calculated before roll bonding according to the structure of the component being studied, then STOP OFF is applied. In the un-bonded interface region, a small amount of the substance which will become a pressurizing agent during hot forming is left. The cladding is assembled in a simple die and can be heated to the superplastic forming temperature for the metals. As soon as this temperature is reached, gas in the interface of the un-bonded area begins to cause blow-forming automatically. If the gas pressure, or the type and the amount of substance is properly selected, a structural component of the desired shape with an internal cavity is superplastically formed.

The aim of this work is described as follows;
1) to show the action of simple and ordinary materials such as water and air as examples of gas blowing sources in closed space.
2) to examine which kind of substance can play a role of STOP OFF in roll bonding and in turn cause gas pressure bulge forming.
3) to demonstrate an easy way of locking the utilized substance in the interface.

2. Materials

High strength Al alloys such as A5083, A7475 which can be superplastic at relatively high temperature were tested tentatively. However, hot roll bonding of these alloys could not be carried out successfully and bonding strength was found to be extremely poor. Other ways of cladding, different from our method had to be investigated. A eutectoid alloy of Al and Zn with addition of Cu (SPZ1) and pure Al A1050 were selected for this purpose because of the relatively low superplastic forming temperature of SPZ1 (250-260°C) and the high strength of the bond for these alloys [2].

3. Free bulging with using pressurizing agents

A very simple die and specimen assembly (Fig.1) was used for free bulging testing of SPZ1. This apparatus was set into the furnace, preheated, and with increasing temperature, gas pressure due to steam or air dilatation, began to deform the sheet(Fig.2). Fig.3 demonstrates the relationships between processing time and non-dimensional bulge height (H: dome height, R: radius of die cavity, 20mm) Fig.4 shows an example of a superplastically deformed dome where the gas blow source is simply water or steam, with an ruptured portion near the apex.

4. Experiments on 5-Phenyltetrazole powder

A substance which acts as STOP OFF during roll bonding but then acts as a gas source, for bulge-forming at the superplastic temperature was sought. In the manufacture of foamed plastics which contain countless holes, use is made of 5-phenyltetrazole powder.[3] In Fig.5 some chemical property changes in this material with increasing temperature are shown. At approximately 260°C, N_2 gas is evolved due to thermal decomposition. This temperature happens to be the same as the superplastic temperature for SPZ1. As a consequence this is why this material was used in the present work.

Fig.6 shows the experimentally measured relations between non-dimensional bulge height vs the amount of the powder. The heating rate in Fig.2 and the die set (Fig.1) were used again but with two different depths of die cavity, h=11.5mm and h=0mm (zero-clearance).The latter case simulates cladding metal. It can be seen that a higher bulged dome can be obtained for zero-clearance. This fact also helps our work.

A three sheet sandwich structure, which is of a common type made by SPF/DB of Ti alloy sheets(Fig.10(c)), was selected as an typical example of this approach. The flow chart diagram for the processing of RB/SPF is shown in Fig.7. The 5-phenyltetrazole powder scattered on the metal interface can prevent bonding, but since it will move freely in the bond plane during rolling, roll bonding could not be effectively done. However, wrapping the powder in very thin Al foil, spread of the powder during rolling was prevented.

Peel strength vs rolling reduction was measured also. This was based on JIS K 6854 (Japanese Industrial Standard) which describes the measurement of joint strength obtained by adhesives. The results observed (Fig.8) show that the peel strength of SPZ1 and A1050 becomes higher almost linearly with rolling reduction. In this cladding, the metal rolling reduction is 67% and this gives a bond strength in the interface which is high enough. Fig.9 illustrates STOPPING OFF schematically and Fig.10(a) shows the appearance of roll bonded and then hot formed sheet. As seen in Fig.10(b), inside the structure, traces or broken pieces of Al foil are left. Cleaning by air blowing of the holes in the expanded structure may be needed in the final stage of processing.

5. Summary

Will there be the possibility of completely new developments in superplastic forming and concurrent diffusion bonding, especially for Al alloys? High strength Al alloys which are superplastic at elevated temperatures might be expected to find widespread use in various fields of modern technology. Superplastic forming and if possible, concurrent diffusion bonding of Al alloys will help the increasing design flexibility. However, the extremely poor bond strength of Al alloys is thought to be the greatest obstacle. Diffusion bonding does not work for Al alloys but use of roll bonding instead will lead to a new technology. This may be possible as the processing described here demonstrates. The fundamental idea on RB/SPF was first published as Japanese patents [4][5] in the early of '60s, but a very limited number of components has been manufactured since then. The area is still remains to be explored.

This work was supported by AMADA foundation for metal work technology and the light metal educational foundation

References

1. E. J. Tuegel, M. O. Pruitt and L.D. Hefti, "SPF/DB TAKES OFF", Advanced Materials & Processes, (7), 1989, pp. 36-41.

2. H.Ohsawa and H.Nishimura, "Uniaxial Tensile Behavior of Combined Superplastic Sheets", TRANS. of JSME, (47),1981,pp.654-664.

3. K. T. Collington, "Fundamental Consideration on Chemical Foamed Agents", Vinyl Chloride and Polymer,(6),1979,pp.20-25.

4. Japanese Patent, 1959-10458.

5. Japanese Patent, 1961-13615.

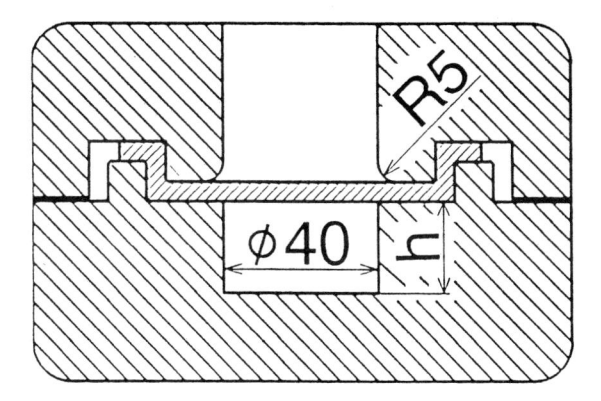

Fig. 1 Assembled dies and specimen for free bulge forming with water, air, silica gel and 5-phenyltetrazole as pressure source.

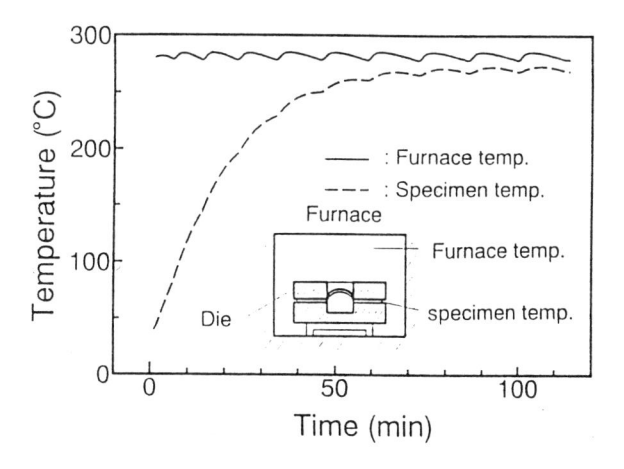

Fig. 2 Temperature changes of furnace and die.

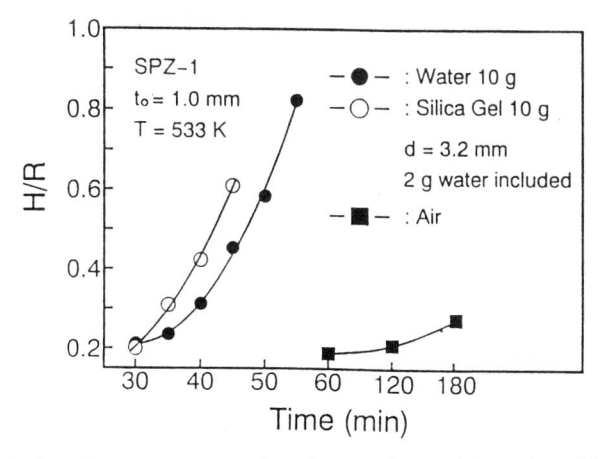

Fig. 3 Relation between processing time and non-dimensional bulge height.

Fig. 4 Free bulge dome of SPZ1 (t_0=1mm), blow formed with 2g water, 260°C x 50min.

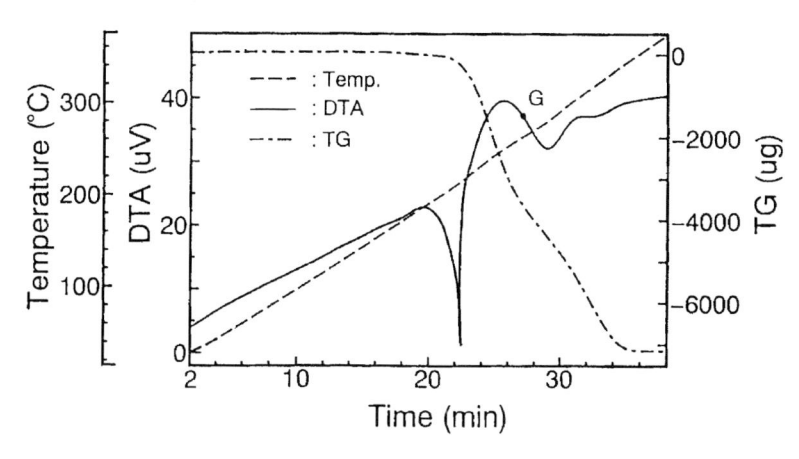

Fig. 5 Mass, Phase and chemical change of 5-phenyltetrazole with increasing temperature.

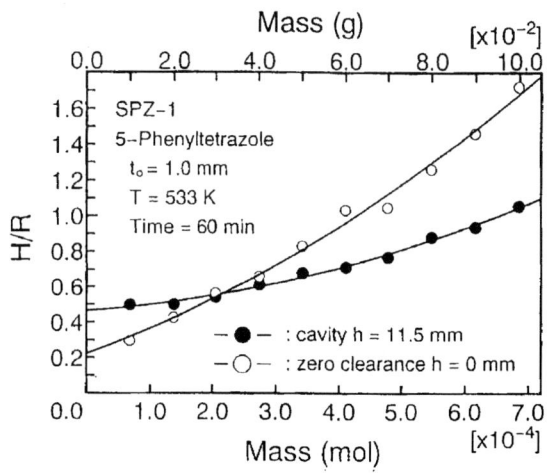

Fig. 6 Higher bulge height can be obtained by zero-clearance.

Superplasticity: 60 Years after Pearson 301

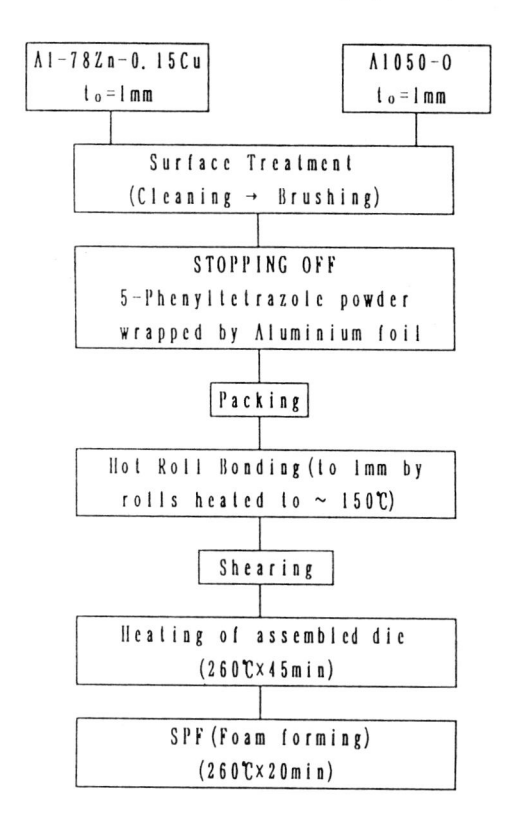

```
┌─────────────────┐      ┌─────────────────┐
│ Al-78Zn-0.15Cu  │      │    A1050-O      │
│    t₀ = 1mm     │      │   t₀ = 1mm      │
└────────┬────────┘      └────────┬────────┘
         │                        │
         └───────────┬────────────┘
                     ▼
         ┌─────────────────────────┐
         │   Surface Treatment     │
         │  (Cleaning → Brushing)  │
         └───────────┬─────────────┘
                     ▼
         ┌─────────────────────────┐
         │      STOPPING OFF       │
         │ 5-Phenyltetrazole powder│
         │  wrapped by Aluminium   │
         │          foil           │
         └───────────┬─────────────┘
                     ▼
              ┌──────────┐
              │ Packing  │
              └─────┬────┘
                    ▼
         ┌─────────────────────────┐
         │ Hot Roll Bonding (to 1mm│
         │ by rolls heated to ~150℃)│
         └───────────┬─────────────┘
                     ▼
              ┌──────────┐
              │ Shearing │
              └─────┬────┘
                    ▼
         ┌─────────────────────────┐
         │ Heating of assembled die│
         │     (260℃ × 45min)      │
         └───────────┬─────────────┘
                     ▼
         ┌─────────────────────────┐
         │    SPF (Foam forming)   │
         │     (260℃ × 20min)      │
         └─────────────────────────┘
```

Fig. 7 Flow chart of Processing of the RB/SPF.

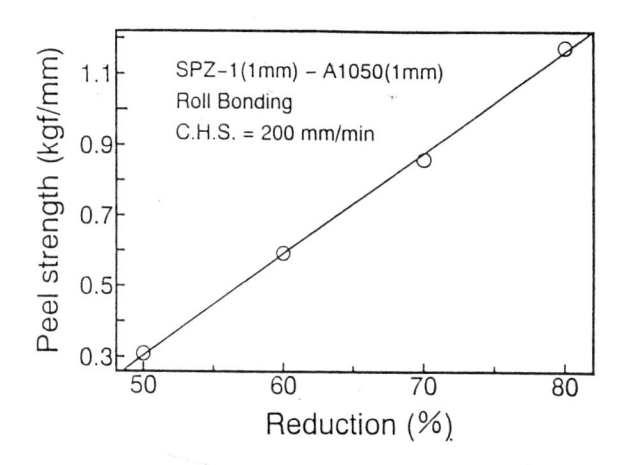

Fig. 8 Peeling strength vs. Reduction in rolling showing a linear relationship.

302 *Superplasticity: 60 Years after Pearson*

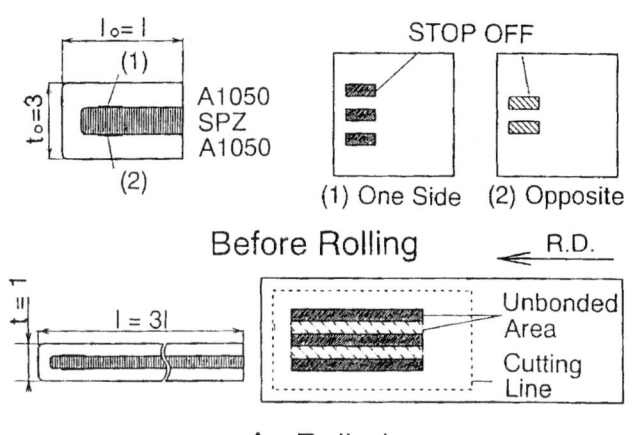

Fig. 9 STOPPING OFF (ASA-67).

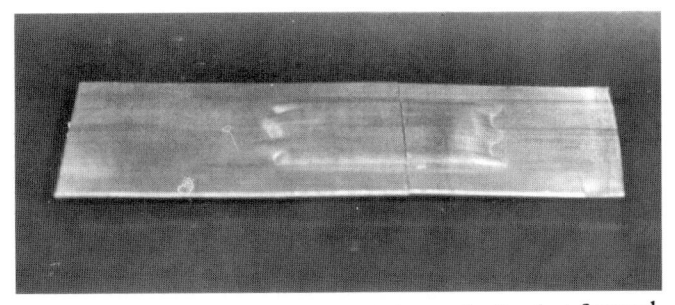

Fig. 10 (a) An appearance of clad metal after hot formed

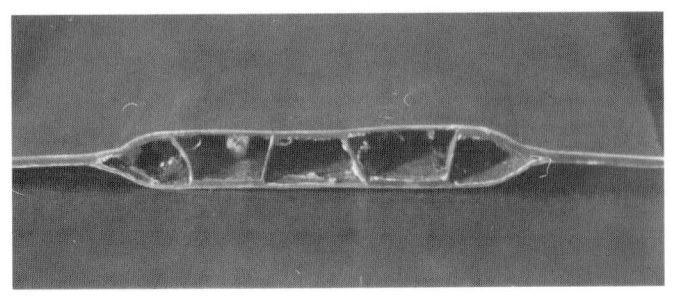

Fig. 10 (b) Al foil should be removed away.

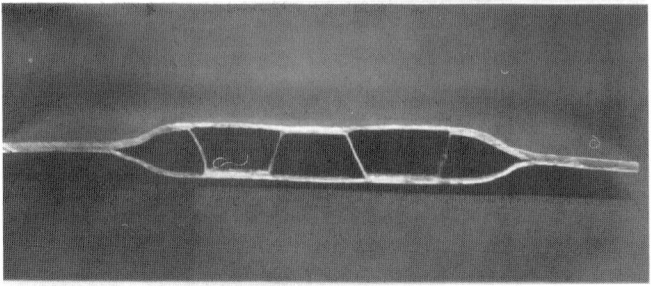

Fig. 10 (c) After removal of Al foil by air blowing.

Fig. 1 Assembled dies and specimen for free bulge forming with water, air, silica gel and 5-phenyltetrazole as pressure source.

Fig. 2 Temperature changes of furnace and die.

Fig. 3 Relation between processing time and non-dimensional bulge height.

Fig. 4 Free bulge dome of SPZ1 (t_0=1.0mm), blow formed with 2g water, 260° c x 50min.

Fig. 5 Mass, Phase and chemical change of 5-phenyltetrazole with increasing temperature.

Fig. 6 Higher bulge height can be obtained by zero-clearance.

Fig. 7 Flow chart of Processing of the RB/SPF.

Fig. 8 Peeling strength vs. Reduction in rolling showing a linear relationship.

Fig. 9 STOPPING OFF (ASA-67).

Fig. 10 (a) An appearance of clad metal after hot formed.

Fig. 10 (b) Al foil should be removed away.

Fig. 10 (c) After removal of Al foil by air blowing.

SUPERPLASTIC FORMING AND DIFFUSION BONDING

AN OVERVIEW

D Stephen
Group Director - Inco Engineered Products

In the 1950's the advent of high strength titanium alloys was greeted with great enthusiasm by Aerospace designers. Here was a material which had high specific properties both static and fatigue when compared with high strength aluminium and, in addition, exhibited exceptional corrosion and high temperature properties. The potential to save considerable weight on subsonic airframes was demonstrated at that time, both by theoretical analysis and structural testing. In contrast to the demonstrated structural advantage, parallel exercises exploring the manufacture of airframe components by the methods available at the time concluded that these advantages could not be achieved at an economic cost. As a consequence, titanium was relegated in the minds of designers to high temperature applications and highly loaded components where space was at a premium, e.g. engine mountings, undercarriage fittings and for which the high costs were acceptable..

With the advent of sustained supersonic cruise aircraft in the 1960's and 70's, such as Rockwell B1 and Concorde, a wider use of titanium was envisaged because of the structural temperatures involved and as a result there was a motivation to address the issues of cost in manufacture of titanium components. A number of methods of manufacture evolved from these investigations.

- Welding EB TIG
- Isothermal Forging
- Superplastic Forming (SPF)
- Diffusion Bonding (DB)
- SPF/DB

All of these manufacturing methods are exploited today in the manufacture of airframe and aero-engine components, but at present they are largely used to reduce costs or weight in the manufacture of components in the niche to which titanium was relegated in the 1950's.

This paper reviews the history of the development of SPF/DB components and highlights the structural and economic significance of this process in particular in competition with conventional aluminium fabrications and highly loaded composite components. The analysis indicates the potential to broaden the application of titanium away from its present niche. To achieve this however it is essential that designers are "brought on board" and it is the author's view that a key to success is the provision of suitable process modelling tools which can be used by the designer.

TITANIUM ALLOYS FOR SUPERPLASTIC FORMING

A.Wisbey, M.W.Kearns*

Structural Materials Centre, DRA, Farnborough, Hants. GU14 6TD U.K.
* IMI Titanium Ltd, Witton, PO Box 704, Birmingham. U.K.

Abstract

The major factors affecting superplastic deformation in titanium alloys will be introduced, including microstructural features and process variables. Most of the commercial SPF of titanium alloys uses sheet, however, the possibilities of other product forms with non-ideal microstructures will be discussed. Ti-6Al-4V alloy is by far the most commonly used titanium alloy for SPF, however, a number of factors are leading to increased interest in other alloys and titanium base materials. These factors include higher service temperature capability, increased modulus, a lower forming temperature and increased forming rate compared to Ti-6Al-4V. Current understanding in these areas will be reviewed.

Introduction

Superplastic forming (SPF) in titanium alloys was demonstrated in the 1960's [1] and subsequently has become a significant commercial and technological success. SPF has been mostly associated with the gas pressure forming of sheet components into shaped dies, either a single sheet or multi-sheet structures where SPF has been combined with diffusion bonding (DB). Superplastic deformation may also occur during isothermal forging operations. Titanium sheet SPF and SPF/DB components are in service in numerous aerospace structures [2]. The application of this technology has permitted dramatic cost and weight savings, and a reduction in the number of parts for assemblies, compared to conventional manufacturing routes. The majority of SPF and SPF/DB titanium components are fabricated from the Ti-6Al-4V alloy. This alloy is limited in its applications to service temperatures of up to ~300°C. SPF in Ti-6Al-4V usually requires temperatures of about 900°C and the SPF cycle times are fairly lengthy at several hours. As a result of Ti-6Al-4V alloy service performance and production limitations there is increased interest in other titanium alloys with superior service properties and also alloys that may give production advantages. Hence higher service temperature capability, increased modulus, a lower forming temperature and increased forming rate are reviewed in this study after summarising the major features of superplastic deformation in titanium alloys.

Superplastic Deformation in Titanium Alloys

A fine, stable and fairly equiaxed grain structure is required for SPF. In titanium alloys the small grain size (~10 μm) and equiaxed morphology is achieved readily with conventional sheet production techniques. The two phase nature of the majority of titanium base materials (α+β, α$_2$+β etc) stabilises the grain size at elevated temperature and enables SPF. For conventional titanium alloys it has been found that an α:β ratio of about 50:50 coincides with conditions for optimum superplastic deformation [3] and this largely determines the SPF temperature. The temperature for the 50:50 phase ratio varies with β transus temperature and is a function of the alloy chemistry. Typically little or no cavitation is found in conventional titanium alloys or Ti_3Al based, α_2 alloys after superplastic strain.

Within the superplastic regime grain size particularly affects the flow stress for deformation. Reducing the grain size lowers the flow stresses and increases the strain rate at which the maximum strain rate sensitivity, m, occurs (Fig. 1) [4]; clearly lower flow stresses are useful industrially in requiring lower gas pressures for forming operations. However, grain growth almost invariably accompanies superplastic deformation. Grain growth occurs both due to the thermal cycle associated with SPF and also as a result of superplastic strain, thus the grain size increases with increasing temperature, superplastic strain and strain rate (Fig. 2)[4,5]. After superplastic deformation a reduction in room and elevated temperature tensile properties is often noted (Fig. 3) and this is associated with the increased grain size, the reduced dislocation density after SPF [6]. Changes in crystallographic texture may have an effect on post SPF tensile properties. It is possible to re-heat treat some titanium alloys, like IMI 550, to regain some of the room temperature tensile properties [7], however, heat treatment after SPF may lead to distortion of sheet components and therefore cooling rates for the die also becomes an important issue.

The initial crystallographic texture of titanium alloys has little effect on their superplastic deformation [8]. However, SPF can change the crystallographic texture, a weakening has often been reported [9,10], but increases in texture intensity have also been noted [11]. These effects are related to the deformation mechanisms operating, grain boundary sliding leading to grain rotation would be expected to result in texture intensity reductions. The relatively recent micro-texture measurement technique, using electron back scattered diffraction patterns, may prove useful in gaining further clarification of the role of different phases in superplastic deformation.

Phase shape and distribution have been shown to be of primary importance for stress and strain anisotropy, whilst also affecting surface finish. Flow stresses parallel to aligned contiguous α phase have been found to be higher than those normal to the aligned α phase [12]. However, increasing superplastic strain in the same direction produces an increasingly equiaxed phase distribution, whilst superplastic deformation normal to aligned α leads to preferential deformation in the more equiaxed areas of microstructure with the aligned α constraining deformation and acting as relatively undeformable fibres; an increased surface roughness results (Fig. 4) [13]. This behaviour is thought to be due to the differences in sliding resistance of α/α, α/β and β/β phase boundaries during grain boundary sliding [14]. These non-ideal SPF microstructures in sheet can be produced by inadequate prior thermo-mechanical treatment, however, in product forms other than sheet this type of microstructure may be unavoidable. Despite a non-ideal microstructure useful superplastic strains may be possible in other product forms and high strain rate sensitivities (Fig. 5), along with reasonably uniform tensile elongations of 300% have been obtained in Ti-6Al-4V bar, U shaped extrusions and extruded tube [12]. It has even been demonstrated that electron beam welds in Ti-6Al-4V may be deformed superplastically if constrained by surrounding superplastic material [15].

These basic factors influencing SPF in titanium alloys, particularly Ti-6Al-4V, provide a good basis from which the possibilities and problems of exploiting superplastic deformation in other titanium base materials can be understood.

Higher Temperature Capability

Changes in base alloy composition have produced alloys (the near α alloys) suitable for use up to 600°C, however, this also leads to increases in the β-transus temperature and usually a resultant increase in the SPF temperature (Table 1). Higher SPF temperatures, 950-990°C for IMI 834 for example, may require specialist tooling for production. Changes in alloying composition can also change the slope of the β phase volume fraction versus temperature curve (Fig. 6) [16]. As the alloys move from the near β alloys, through the $\alpha+\beta$ to the near α alloys the slope of the approach curve becomes steeper and thus the temperature range over which SPF occurs is reduced. However, despite the steep β transus approach curve for the near α alloys, superplastic deformation has been successfully demonstrated in several alloys, including Ti-6242 [17], Ti-811 [18] and IMI 834 [10]. Apart from the high processing temperature the superplastic behaviour of the near α alloys is very similar to that noted for the $\alpha+\beta$ alloys like Ti-6Al-4V, in terms of flow stress and strain rate.

Recent developments to further extend the temperature capability of titanium base materials have focused on the intermetallic titanium aluminide alloys, with maximum service temperatures of 700°C for the Ti_3Al (α_2) alloys and 800-900°C for the TiAl (γ) based alloys. In an effort to improve the poor room temperature ductility of this class of materials, alloying the α_2 based alloys has produced a microduplex, $\alpha_2+\beta$ structure, suitable for SPF. SPF of the α_2 alloys, of which Super α_2 is the most mature, is analogous to conventional titanium alloys, the SPF temperature for Super α_2 is very similar to the near α alloy IMI 834 (Table 1). However, a comparison of the flow stresses at the SPF temperature between IMI 834 and Super α_2 shows higher values for the Super α_2 (Fig. 7), possibly the result of reduced diffusion rates in the α_2 phase compared with disordered α [19]. Similarly two phase structures of either $\gamma+\alpha_2$ [20] or $\gamma+\beta$ [21] have been produced in γ based materials. Less characterisation of SPF in γ alloys has been performed, however, some notable differences from conventional titanium alloys and the α_2 alloys are evident. Firstly the maximum superplastic strain recorded for these alloys is significantly less (at 550% for a Ti-48at.%Al alloy) [20] than that observed for either the conventional titanium alloys or the α_2 alloys and secondly cavitation has been noted after superplastic deformation [22]. Two temperature regimes have been identified for the maximum superplastic strain in γ alloys; testing at ~1000°C has given an elongation of 550% in fine grained (0.85 µm) mechanically alloyed Ti-48at.%Al [20], whilst in Ti-47.3at.%Al-1.9Nb-1.Cr-0.5Si-0.4Mn an elongation of 470% was achieved at 1280°C [23].

Increased Modulus

The intermetallic α_2 and γ based alloys offer increases in modulus, of up to 25% for α_2 alloys with a high volume fraction of ordered α_2 phase and ~50% for γ alloys [24], compared with conventional titanium alloys. Alloy chemistries which develop a high volume fraction of titanium - β eutectoid intermetallic phase, Ti-12Co-5Al and Ti-12Fe-4Al for example, have been shown to exhibit higher modulus and strength than Ti-6Al-4V [25]. Alternatively ceramic reinforcement particles can be used to achieve property improvements, however, the production of particulate titanium matrix composites is difficult. A powder metallurgy production route for titanium MMC's has been demonstrated but other processing, mechanical alloying (MA) for example, will probably be required to give a homogeneous particle distribution. Now that MA with low O_2 and N_2 contamination is possible [26] titanium

MMC's of similar quality to the aluminium based examples are feasible, using both conventional titanium alloys and titanium aluminide matrices. An example of this class of materials is a Ti-6Al-4V matrix composite, with TiC or TiN reinforcement, produced via MA and shown to give maximum elongations to failure of ~400% under SPF conditions typical for Ti-6Al-4V (Fig. 8) [27]. The flow stress of these materials was found to increase with increasing volume fraction of ceramic reinforcement.

Lower Forming Temperature

There are three main routes to achieving a lower forming temperature than currently used with Ti-6Al-4V. The first method is to increase β stabiliser content (and/or reduce α stabiliser content) to produce near β or stable β alloys. An SPF temperature of 750°C for Ti-10V-2Fe-3Al has been noted [28], however, in other alloys, Ti-15V-3Cr-3Al-3Sn for example, low optimum strain rates and higher flow stresses can make SPF unattractive [18]. A further disadvantage of the β alloys is their relatively low strength in the post SPF condition and subsequent heat treatment to high strength may not be possible, due to component distortion.

Hydrogen is a potent β stabiliser and has been employed experimentally as a temporary alloying addition to a variety of titanium alloys, eg. Ti-6Al-4V [29], Ti-6.3Al-2.7Mo-1.7Zr [30] and near α Ti-6.5Al-1.3V-1.3Mo-2Zr [31]. Additions of hydrogen have produced reductions in SPF temperature of about 100°C compared with the alloy without hydrogen (Fig. 9). However, this technique requires the initial charging of sheet with hydrogen and after SPF a subsequent vacuum degassing at elevated temperatures to remove the hydrogen and prevent the formation of embrittling hydrides.

Another route to a reduced SPF temperature is to increase β phase diffusivity, by selective alloying with high diffusivity elements. The β eutectoid elements Cr, Mn, Fe, Co and Ni exhibit β phase diffusion rates up to two orders of magnitude greater than titanium self diffusion [32] and thus enhance creep rate. Only relatively low additions are required for dramatic reductions in SPF temperatures of over 100°C. Both Ti-6Al-4V [33] and IMI 550 have been used as the basis for these alloying trials. It is only recently that commercial alloys using this technology have been released, SP 700 Ti-4.5Al-3V-2Mo-2Fe [34] and β-Cez Ti-5Al-2Sn-4Zr-4Mo-2Cr-1Fe [35]. SP 700 can be superplastically deformed at 700°C but the optimum temperature is between 750°C and 800°C [34,36]. This material seems to show very little reduction in the post SPF tensile properties in a comparison with the as received annealed SP 700. An advantage of both the near β alloys and alloys like SP 700 is the wide processing window available, a result of the shallow β transus approach curve as seen in Fig. 6.

The final means to lower SPF temperature is via the production of an extremely fine grain size. This technique, like the hydrogenation route is only at the experimental stage. Grain sizes of 0.06 μm have been obtained in the Ti-6Al-3.2Mo-0.4Si-0.1C (VT8) alloy by upsetting at strain rates between $5 \times 10^{-5} - 5 \times 10^{-3} s^{-1}$ and temperatures of 600-700°C [37,38]. With a grain size of 0.06 μm a tensile elongation of 600% was obtained at 600°C, some 200°C lower than the usual SPF temperature for the VT8 alloy. It is thought that the sharp rise in grain boundary length is responsible for increased diffusion.

It should be noted that concurrent SPF/DB may not be possible at low forming temperatures of around 600-700°C, due to difficulties with the DB. This may necessitate two stage processing with a higher bonding temperature and lower forming temperature.

Increased Forming Rate/Low Flow Stress

The effect of reducing grain size on the SPF temperature has been discussed above, however, it is expected that this is also the route to increasing forming rates and lower flow stresses (similarly the diffusion bonding times and or pressures can be reduced for SPF/DB). Taking the generalised equation for creep rate-

$$\dot{\varepsilon} = \left(\frac{AGb}{kT}\right) \cdot \left(\frac{b}{d}\right)^p \cdot \left(\frac{\sigma}{G}\right)^m \cdot D_0 \exp\left(-\frac{Q}{RT}\right)$$

(A= constant, G= shear modulus, b= burgers vector, k= Boltzmans constant, T= absolute temperature, d= grain size, σ= flow stress, D_0= pre-exponential factor, Q= activation energy, R= universal gas constant.

Experimental measurements can provide data (eg p=2-3, m=0.5-1.0) to allow some predictions and for Ti-6Al-4V it can be shown that a ten-fold reduction in grain size from the currently typical 10 µm down to 1 µm could result in the strain rate being increased by a factor of 100 or 1000 if lattice diffusion or grain boundary diffusion is rate controlling. There are very few experimental measurements of these effects in titanium alloys. However, an experimental Ti-12Co-5Al alloy has been thermomechanically processed to give a fine matrix grain size with intermetallic particles of 0.3-0.6 µm Ti_2Co [39]. This alloy, deformed in tension at 750°C, gave 790% elongation at a strain rate of $5 \times 10^{-2} s^{-1}$ (Fig. 10). This strain rate is significantly higher and the temperature much lower than currently used for sheet structures, however, work with aluminium alloys and MMC's has shown elongations to failure in excess of 1000% at strain rates greater than $10^1 s^{-1}$ [40]. Unfortunately titanium alloys containing high concentrations of cobalt and iron are difficult to produce on a production scale by conventional ingot metallurgy due to a tendency for these elements to segregate markedly. The very fine microstructures required for higher strain rates in titanium are clearly possible by a variety of methods, including high strain isothermal upsetting [38], mechanical alloying [41] and rapid solidification techniques like vapour quenching [42]. Most of these techniques for producing fine microstructures will incur a substantial cost penalty in comparison with current ingot metallurgy wrought products, however, unique materials may be made (eg titanium MMC's) and significant manufacturing cost savings should be possible with reduced SPF cycle times thanks to lower forming temperatures and higher deformation rates in some combination.

Conclusions

Superplastic deformation of titanium alloys, particularly Ti-6Al-4V is being successfully exploited. Performance and process limitations are prompting interest in SPF of alloys with enhanced mechanical properties and materials with improved superplastic properties. Hence, commercially available near α titanium alloys allow the service temperature to be increased from 300°C for Ti-6Al-4V up to 600°C. SPF in these alloys has been demonstrated and may be an area for increasing commercial SPF and SPF/DB in the near future. Further extending

Table 1. Titanium base materials with their respective maximum operating temperatures, SPF temperatures and maximum recorded superplastic tensile elongation.

Alloy (wt.%)	Service Temperature (°C)	SPF Temperature (°C)	Max. SP elong'n (%)	Ref.
Ti-6Al-4V	300	850-925	>1000	17
IMI 550 (Ti-4Al-4Mo-2Sn-0.5Si)	400	~900	>1500	43
Ti-6242 (Ti-6Al-2Sn-4Zr-2Mo)	500	850-940	800	17
Ti-811 (Ti-8Al-1Mo-1V)	370-540	940-1010	>200	18
IMI 834 (Ti-5.8Al-4Sn-3.5Zr-0.7Nb-0.5Mo-0.3Si-0.05C)	600	950-990	>400	10
Super α_2 (Ti-14Al-19Nb-3V-2Mo)	650-700	960-990	1350	44
γ base alloys, eg. Ti-48at%AL, Ti-Al-Cr & Ti-Al-Nb-Cr	800-900	1000-1280	550	20

Figure Captions

Fig. 1 a) Stress versus strain rate for four different initial grain sizes in Ti-6Al-4V obtained by step strain rate test. b) Strain rate sensitivity (m) versus strain rate determined from the slopes of the curves in a) [4].

Fig. 2 Grain growth kinetics in Ti-6Al-4V at four different tensile strain rates compared with static kinetics for an initial grain size of 6.4 μm [4].

Fig. 3 The effect of a superplastic strain of 140% tensile elongation at 925°C on the elevated temperature tensile and 0.2% proof strength of Ti-6Al-4V sheet [6].

Fig. 4 Ti-6Al-4V sheet test piece with contiguous α phase at A machined from highly textured rectangular section bar after superplastic deformation to 273% elongation. Microstructurally the contiguous α phase is at A in b), with more equiaxed grains below.

Fig. 5 Strain rate sensitivity (m) versus strain rate at 925°C for sheet, bar, and extruded Ti-6Al-4V alloy products.

Fig. 6 β approach curves for β (Ti-15-3-3-3), near β (Ti-10-2-3), α+β (Ti-6Al-4V and IMI 550), and near-α (IMI 829 and 834) alloys: slope of curves generally decreases with increasing β stabiliser content, but for IMI 834 slope is reduced by alloying with carbon [16].

Fig. 7 Initial flow stress versus strain rate for IMI 834 and Super α_2 sheet at 990°C and γ+α_2 alloy at 1270°C [23].

Fig. 8 Nominal stress versus nominal strain curves for deformation at 925°C of consolidated Ti-6Al-4V alloy containing a) 10 and b) 20vol.% TiN particles produced via mechanical alloying of powders [27].

Fig. 9 Curves of elongation to failure versus deformation temperature for standard Ti-6Al-4V and Ti-6Al-4V temporarily alloyed with 0.12wt.% H_2. Deformation rate for all tests was $1.67 \times 10^{-3} s^{-1}$ [29].

Fig. 10 Test pieces of a Ti-12wt%Co-5Al alloy as machined and after deformation at 750°C with an initial strain rate of $5 \times 10^{-2} s^{-1}$ [39].

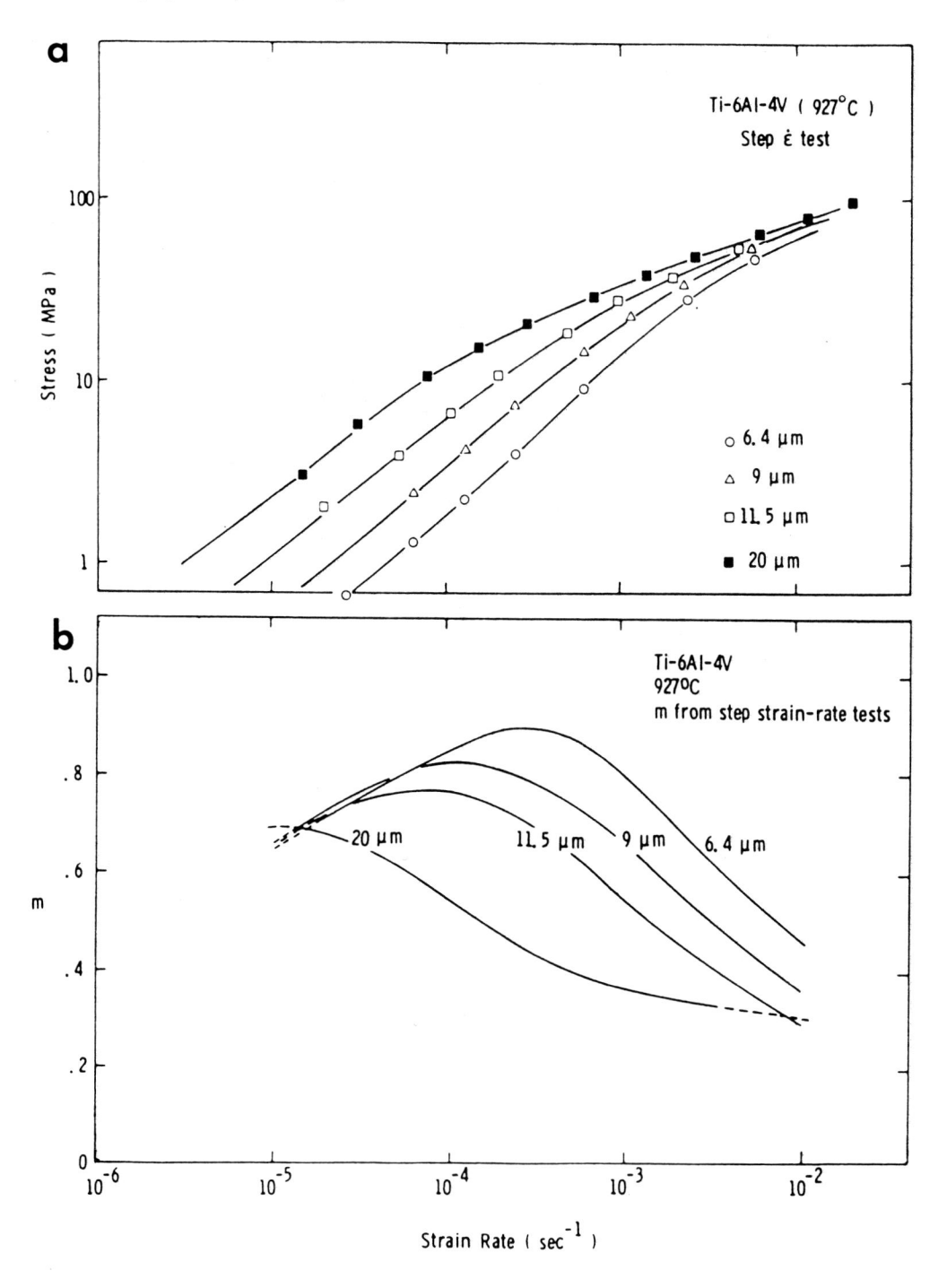

Fig. 1 a) Stress versus strain rate for four different initial grain sizes in Ti-6Al-4V obtained by step strain rate test. b) Strain rate sensitivity (m) versus strain rate determined from the slopes of the curves in a) [4].

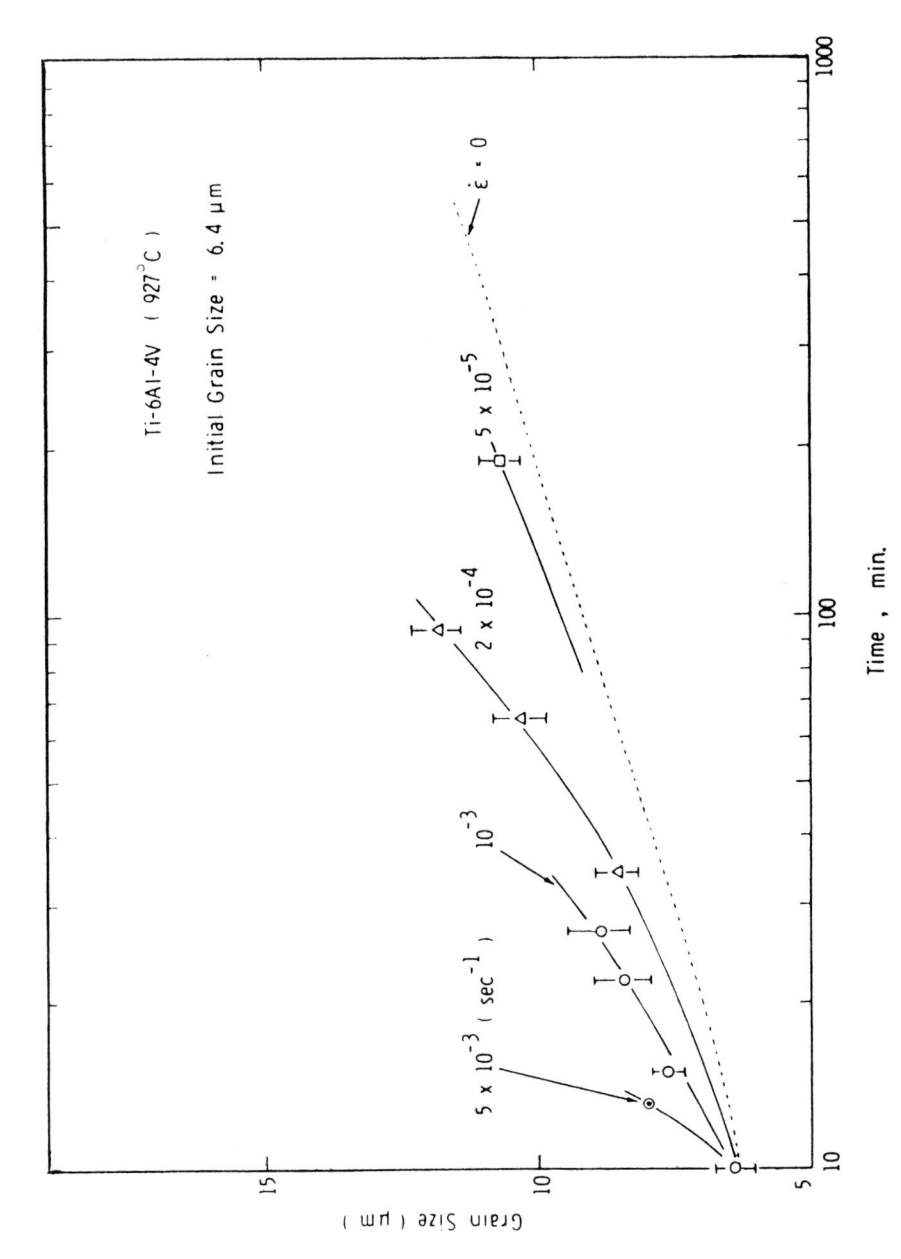

Fig.2 Grain growth kinetics in Ti-6Al-4V at four different tensile strain rates compared with static kinetics for an initial grain size of 6.4 μm [4].

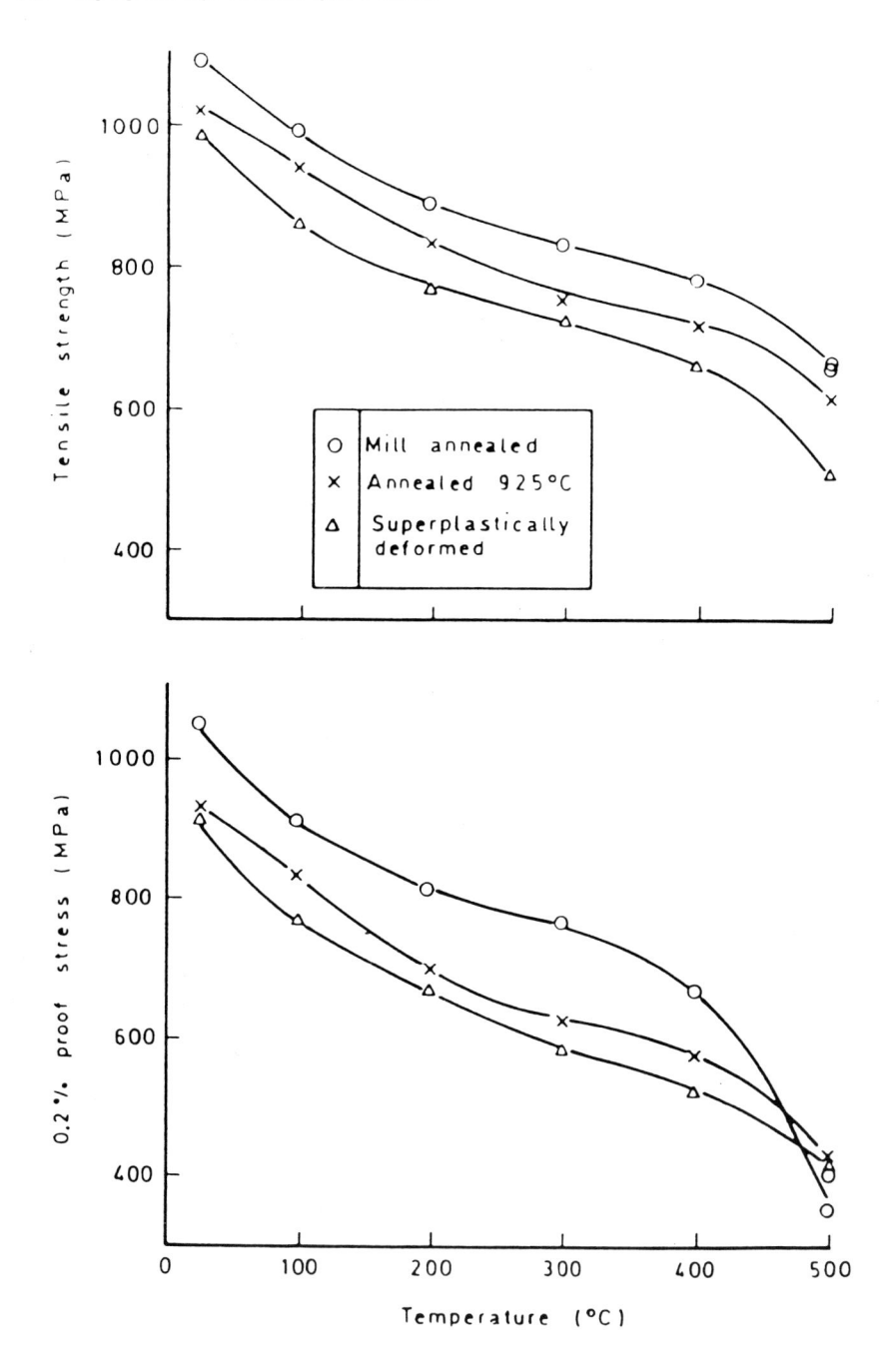

Fig. 3 The effect of a superplastic strain of 140% tensile elongation at 925°C on the elevated temperature tensile and 0.2% proof strength of Ti-6Al-4V sheet [6].

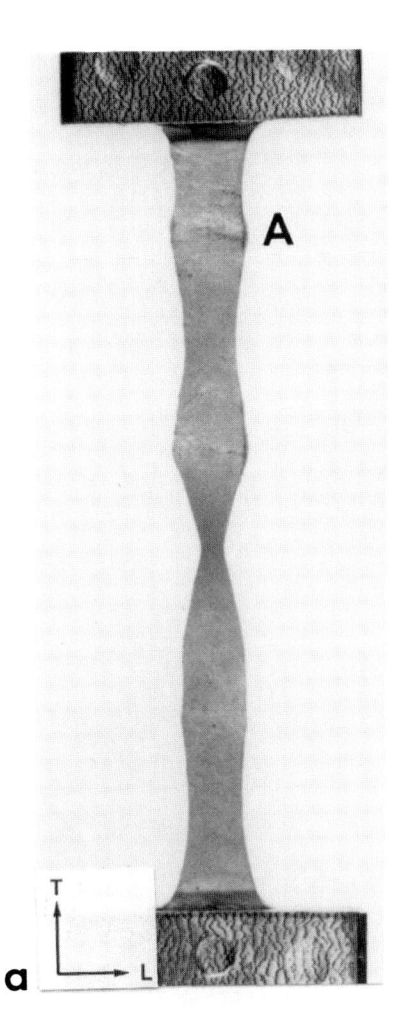

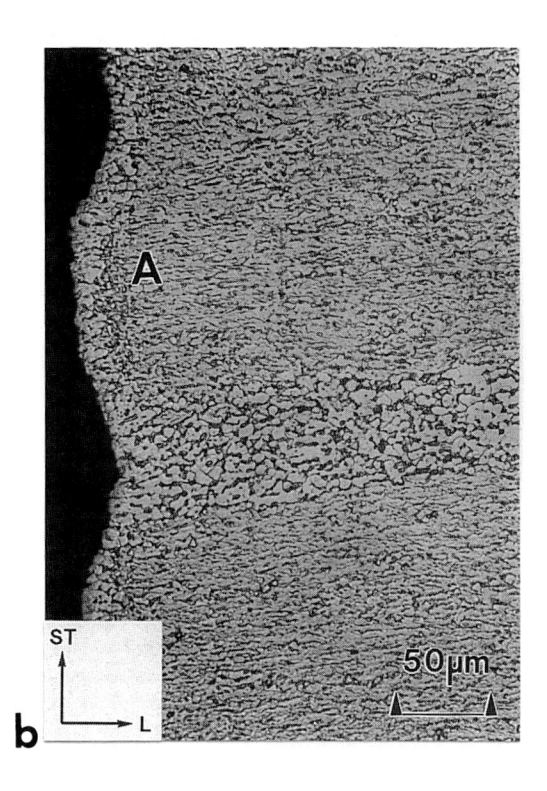

Fig. 4 Ti-6Al-4V sheet test piece with contiguous α phase at A machined from highly textured rectangular section bar after superplastic deformation to 273% elongation. Microstructurally the contiguous α phase is at A in b), with more equiaxed grains below.

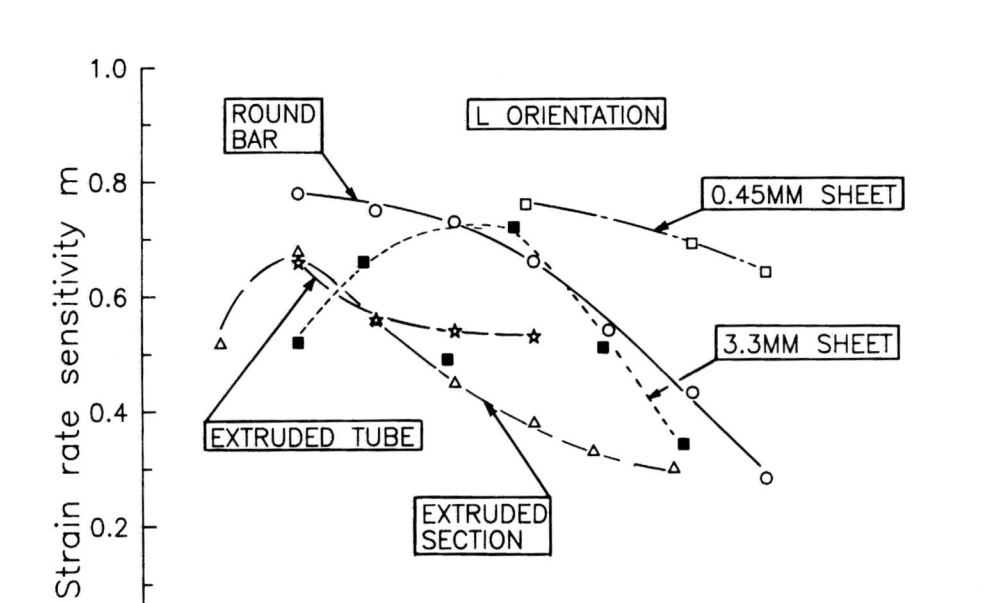

Fig. 5 Strain rate sensitivity (m) versus strain rate at 925°C for sheet, bar, and extruded Ti-6Al-4V alloy products.

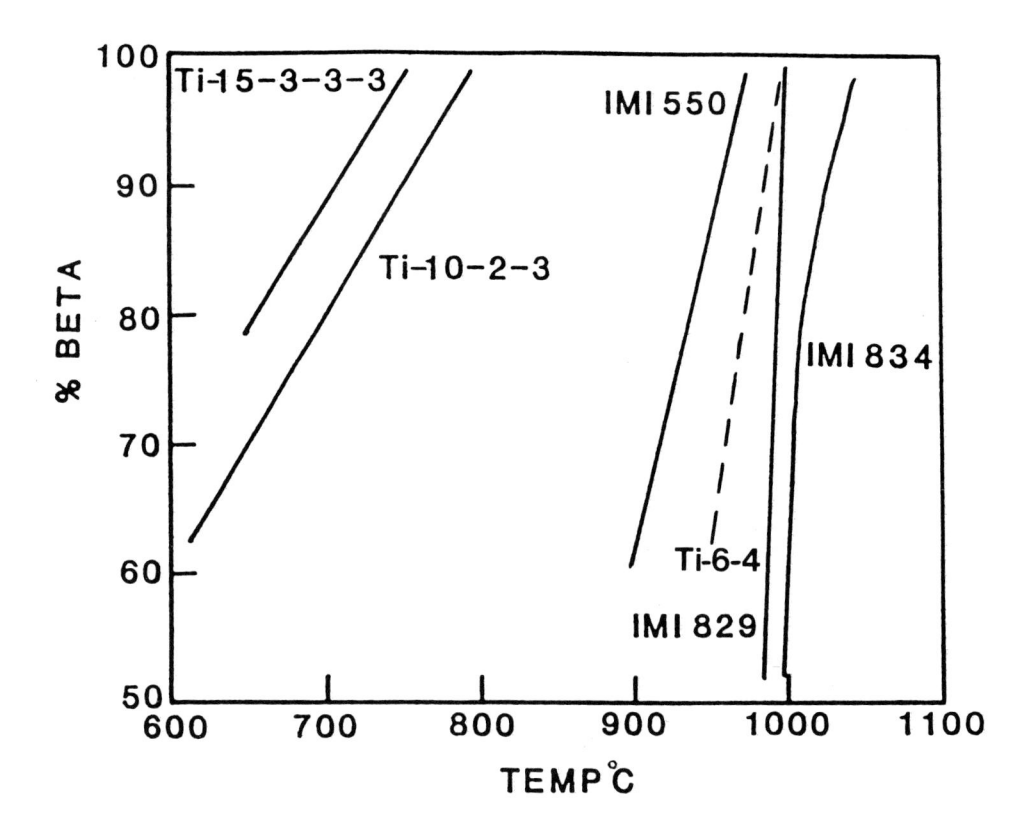

Fig. 6 β approach curves for β (Ti-15-3-3-3), near β (Ti-10-2-3), α+β (Ti-6Al-4V and IMI 550), and near-α (IMI 829 and 834) alloys: slope of curves generally decreases with increasing β stabiliser content, but for IMI 834 slope is reduced by alloying with carbon [16].

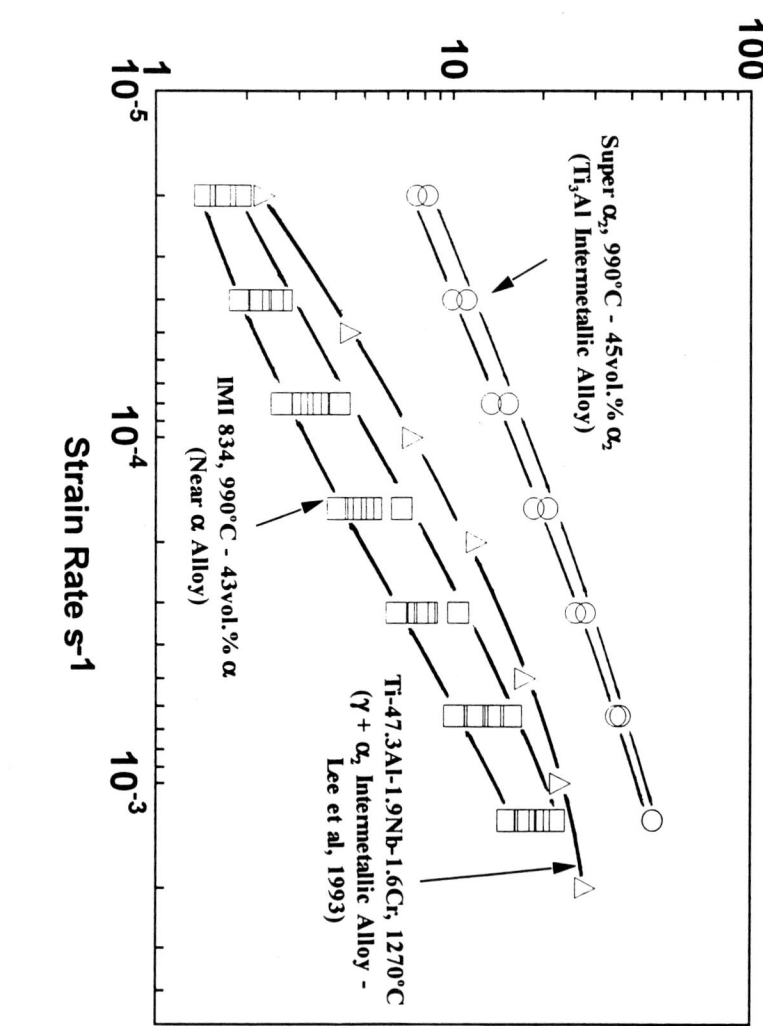

Fig. 7 Initial flow stress versus strain rate for IMI 834 and Super α_2 sheet at 990°C and $\gamma+\alpha_2$ alloy at 1270°C [23].

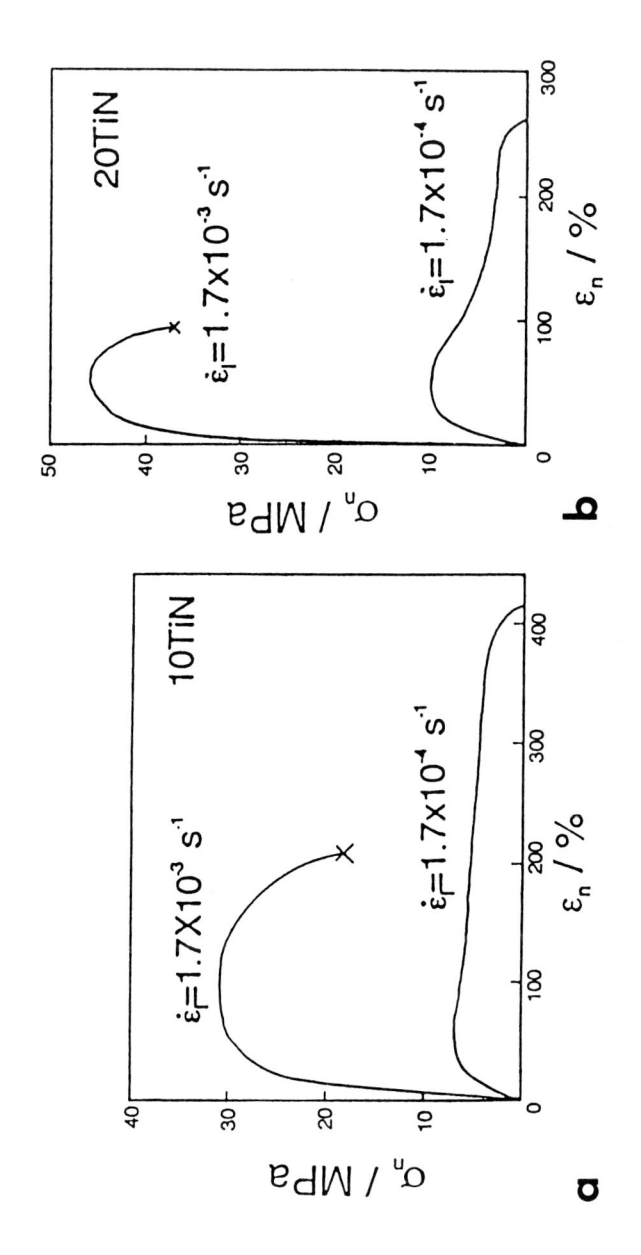

Fig. 8 Nominal stress versus nominal strain curves for deformation at 925°C of consolidated Ti-6Al-4V alloy containing a) 10 and b) 20vol.% TiN particles produced via mechanical alloying of powders [27].

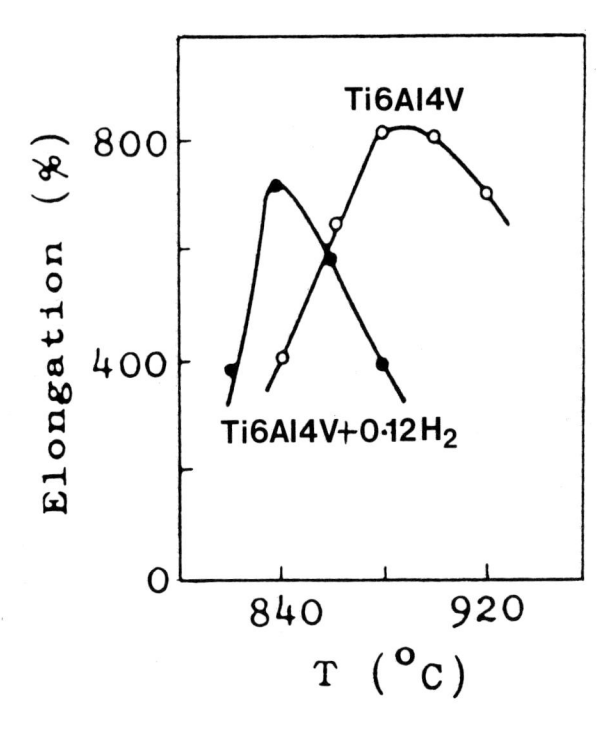

Fig. 9 Curves of elongation to failure versus deformation temperature for standard Ti-6Al-4V and Ti-6Al-4V temporarily alloyed with 0.12wt.% H_2. Deformation rate for all tests was 1.67×10^{-3}s-1 [29].

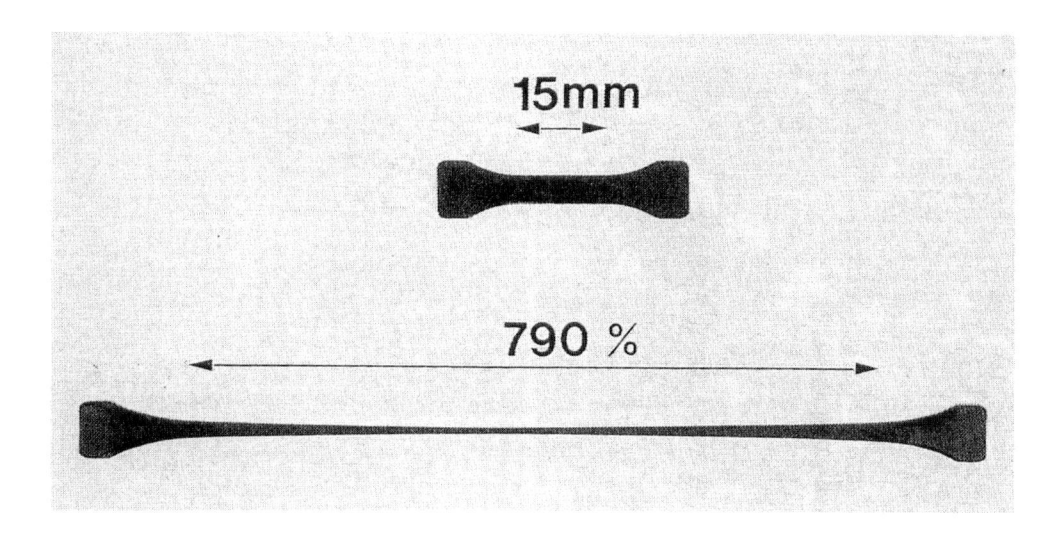

Fig.10 Test pieces of a Ti-12wt%Co-5Al alloy as machined and after deformation at 750°C with an initial strain rate of $5 \times 10^{-2} s^{-1}$ [39].

SUPERPLASTICITY IN THE TITANIUM ALLOY SP 700 WITH LOW SPF TEMPERATURE

A.Wisbey, B.C.Williams, H.S.Ubhi, B.Geary*, D.P.Davies*, C.M.Ward-Close, A.W.Bowen

Structural Materials Centre, DRA Farnborough, Hants. GU14 6TD U.K.

* Westland Helicopters Ltd, Yeovil, Somerset, BA20 2YB U.K.

Abstract

The new high strength titanium alloy SP 700 offers the potential for a reduced superplastic forming (SPF) temperature compared with current titanium alloys and this is investigated here. Excellent formability was demonstrated over a wide range of temperatures at a strain rate of $\dot{\varepsilon}=3\times10^{-4} s^{-1}$. The optimum temperature for SPF was between 750°C and 800°C, about a 100°C lower than for Ti-6Al-4V. The material had a very fine grain size of ~4 µm and the possibility of deformation at higher strain rates than those often used commercially has been studied here. No reduction in the post SPF room temperature tensile properties was noted, compared to as received annealed sheet. The effect of superplastic deformation on the SP 700 microstructure and texture has been investigated using both standard x-ray macro-texture measurements and electron back scattered diffraction (EBSD) patterns to determine the micro-texture.

Introduction

The high structural efficiency and low cost of titanium sheet components manufactured by superplastic forming (SPF) and diffusion bonding (DB) are well known [1], however, to date the majority of components have been fabricated from the Ti-6Al-4V alloy. There is now considerable interest in the use of alloys which may permit lower cost forming operations, combined with higher structural performance. This is particularly so in modern helicopter designs owing to the increasing use of titanium sheet fabrications [2]. SP 700 has been developed to give a wide processing window, a reduced SPF temperature and some hardenability [3]. In this work the superplastic deformation behaviour of SP 700 has been characterised, the effect of superplastic deformation on the post-formed room temperature tensile properties have been measured and the superplastic deformation mechanisms have been investigated using both macro- and micro- crystallographic texture determination techniques. The micro-texture was determined using electron back-scattered diffraction (EBSD) patterns [4] and some preliminary results are presented.

Experimental

The SP 700 alloy (composition wt% Ti-4.5Al-3.0V-2.0Fe-2.0Mo) was supplied by NKK Corporation in the form of annealed 1.0 mm thick sheet. Sheet SPF test pieces (25 mm gauge length, 16 mm wide) were cut with the tensile axis in the longitudinal (L) and transverse (T) directions of the sheet. After SPF the room temperature tensile properties were measured on the deformed test pieces, with a 50 mm parallel gauge length, 6 mm wide machined centrally; the thickness was about 0.42 mm and the variation along the gauge length after superplastic deformation was small (±0.03 mm).

Previous work has suggested that high elongations may be obtained at temperatures from ~700°C [2]. Here the SPF behaviour was assessed at 50°C intervals between 700°C and 850°C, using uniaxial tensile testing in a gettered argon environment. The temperature of the SPF test piece was monitored using three Chromel-Alumel (Type k) thermocouples. Once at the test temperature the SPF test piece was soaked for ~10 minutes and then given a 5% pre-strain at a strain rate of $\dot{\varepsilon}=3.2\times10^{-4} s^{-1}$. The initial flow stresses and strain rate sensitivity indices (m=ln[σ_1/σ_2]/ln[V_1/V_2], where σ_1 and σ_2 were the true flow stresses at the cross-head velocities V_1 and V_2 respectively) were determined within the first 25% elongation over a range of strain rates and then deformation continued at nominally constant strain rate ($\dot{\varepsilon}=3\times10^{-4} s^{-1}$, $1\times10^{-3} s^{-1}$ and $1\times10^{-2} s^{-1}$) up to ε~1.4 (300% elongation). After deformation the plastic strain ratio (R=ln[w/w_o]/ln[t/t_o], where w_o, t_o and w, t are the initial and final width and thicknesses) was measured. The effect of superplastic deformation was assessed metallographically.

Crystallographic texture measurements were made using two techniques. Firstly, standard x-ray macro-texture measurements were made on as received material, and sheet after superplastic deformation to 300% elongation at 750°C and 850°C, with a strain rate of $3 \times 10^{-4} s^{-1}$ in both the transverse and longitudinal sheet orientations. Micro-texture measurements were made on as received sheet and after superplastic deformation, at 750°C and a strain rate of $3 \times 10^{-4} s^{-1}$ to 300% elongation of longitudinally oriented sheet via the observation of electron back-scattered diffraction (EBSD) patterns in a SEM. Each diffraction pattern was indexed after integration and background subtraction (to improve pattern quality). The positions of individual grain orientations were obtained in the form of (0002) and (002) pole figures for the α and β phases respectively.

Results

At all of the SPF test temperatures 300% elongation was easily achieved. Reasonable width and thickness uniformity along the gauge length of test pieces was found at 700°C and 850°C but was excellent at 750°C and 800°C (true area strain $\varepsilon_{true}=1.35\pm0.15$). After testing at 750°C and various strain rates it may be seen that good dimensional uniformity was maintained at strain rates of $3 \times 10^{-4} s^{-1}$ and $1 \times 10^{-3} s^{-1}$ but at $1 \times 10^{-2} s^{-1}$ necking of the test pieces occurred (Fig.1). There was a significant decrease in the flow stresses from 700°C to 800°C but at 850°C the slope of the flow stress versus strain rate curve changes (Fig.2), however, they all follow an approximately logarithmic relationship. During deformation at the constant strain rate, $\dot{\varepsilon}=3 \times 10^{-4} s^{-1}$ the flow stresses decreased until ~150% elongation and then increased (Fig. 3). At 700°C this resulted in an increase from ~33 MPa to ~43 MPa after 300% elongation and at 800°C the equivalent values were ~15 MPa and ~23 MPa. The strain rate sensitivity (m) indices were all very similar at 700-800°C with values of 0.4-0.5 from $\dot{\varepsilon}=2 \times 10^{-5} s^{-1}$-$1 \times 10^{-3} s^{-1}$ (Fig. 4). However, at 850°C the m values were lower at 0.3-0.4 for the same range of strain rates. There was no significant effect of test piece orientation on the width and thickness uniformity or the m values. However, some stress anisotropy was noted with longitudinal test pieces giving generally higher values of flow stress.

SP 700 showed significant strain anisotropy with plastic strain ratios (R) of 0.4-0.6 after 300% elongation at 700-800°C and a strain rate of $3 \times 10^{-4} s^{-1}$ (Fig. 5), indicating a greater thickness strain compared to width strain during SPF. Under these deformation conditions there was a tendency for the T test pieces to exhibit lower R values than the L test pieces. Further reductions in the R values (~0.3) were obtained after deformation at 850°C and there was no significant difference in R value between the two test piece orientations. Similarly after deformation at 750°C with a strain rate of $1 \times 10^{-2} s^{-1}$ there was no effect of test piece orientation on the R values, which were found to be higher at 0.6-0.8.

The room temperature tensile properties are shown in Table 1. In the annealed as received condition 0.2% proof and tensile strengths of 1025 MPa and 1090 MPa respectively were found for both L and T sheet directions, however, there was a decrease of about 100 MPa in the strengths at 45° to the rolling direction. Room temperature tensile properties were obtained for material superplastically deformed to $\varepsilon_{true}=1.35$ at 700°C, 750°C and 800°C. For this post SPF material the 0.2% proof and tensile strengths are similar or higher than in the as received condition, with tensile strengths of up to 1170 MPa. The tensile strengths for L test pieces, deformed at 700-800°C, were lower (50-70 MPa) than for the T direction. There was some variability in the post SPF tensile ductilities, with values of 7-13% elongation, this was a reduction compared to the 15.8% for as received sheet.

As received SP 700 sheet had a very fine grain size, with some evidence of contiguous α phase aligned along the rolling (L) direction; the grain size is ~4 µm and the α phase has a volume fraction of 38% at 800°C (Fig. 6a). After SPF at 800°C to $\varepsilon_{true}=1.35$ the grain size had become ~5.4 µm but there was no change in the α phase volume fraction (Fig. 6b). The grain shape became more equiaxed.

Standard x-ray macro-texture (0002) α phase pole figures are shown in Fig. 7 for as received sheet and after deformation at 750°C or 850°C with a strain rate of $3 \times 10^{-4} s^{-1}$ to $\varepsilon_{true}=1.35$ (300% elongation). The h.c.p. basal poles are mainly found around the sheet normal and there is some weakening of this texture after deformation at 750°C. There is also an effect of test piece orientation after SPF with the poles rotated through 90° for the transverse test piece, compared to the longitudinal. Deformation at 850°C significantly increased the texture intensity, both at the poles around the sheet normal and those oriented perpendicular to the tensile axis. A similar effect of test piece orientation was observed at 850°C as found after deformation at 750°C.

Preliminary EBSD micro-texture measurements for as received sheet and after deformation of a longitudinal test piece at 750°C and $3 \times 10^{-4} s^{-1}$ to $\varepsilon_{true}=1.35$ of both the α and β phases are shown in $(0002)_\alpha$ and $(002)_\beta$ pole figures respectively in Fig. 8. The EBSD patterns in the as received material were diffuse, indicative of deformation, evidence of sub-structure was also found. After SPF the EBSD patterns were sharper, suggesting a reduced lattice strain; sub-structure was also much reduced. While the α phase texture, in agreement with x-ray textures (Fig. 7b & c) showed randomisation with superplastic deformation, there was little randomisation of the β phase with SPF. Micro-texture measurements of the post SPF material at the sheet surface and in the sheet centre (Fig. 8) showed no variation in the texture of either phase.

Discussion

SP 700 sheet showed excellent formability over a wide range of temperatures, the best width and thickness uniformity in the gauge length were obtained at 750°C and 800°C [5]. The wide processing window for this alloy has been noted previously [6]. At 800°C the sheet had the lowest flow stresses, high m values and good dimensional uniformity, this may be the optimum temperature for SPF in this alloy. The flow stresses measured for SP 700 are higher than previously measured for either Ti-6Al-4V [7] or IMI 834 [8] at their respective optimum deformation temperatures. However, the economic benefits of SPF at lower temperatures may mitigate against the higher flow stresses and reduced dimensional uniformity observed with SP 700 at these temperatures. Good width and thickness uniformity was found after deformation at both $3 \times 10^{-4} s^{-1}$ and the relatively high strain rate of $1 \times 10^{-3} s^{-1}$, presumably the result of the fine grain size. However, at $1 \times 10^{-2} s^{-1}$ necking was found and the deformation was becoming less superplastic.

No reduction in post SPF room temperature tensile strength was found. This is potentially very significant, since it may permit the production of structures for applications with loadings higher than those currently contemplated. An approximately 10% reduction in post SPF tensile strengths is commonly observed in Ti alloys [7,8] and is associated with increases in the grain size. For Ti-6Al-4V sheet this results in tensile strengths of ~950 MPa (Table 1) after $\varepsilon_{true}=0.85$ and further decreases in strength may be expected for strains similar to those obtained for SP 700. Thus the post SPF SP 700 tensile strengths represent an increase of ~100 MPa over Ti-6Al-4V. The grain size of SP 700 was found to increase and some strain hardening, usually associated with grain growth was noted during SPF to $\varepsilon_{true}=1.35$. However, some other microstructural features are clearly dominating the room temperature tensile properties. A cooling rate effect has been noted after solution treatment and ageing [6] and the SPF temperatures are similar to that used during solution treatment of SP 700. The test pieces were cooled at ~1.5°C/min. from the SPF temperature to 600°C, however, the cooling rate thereafter becomes much lower. The ageing temperature for SP 700 has been reported as 510°C and it may be possible that the cooling from 600°C is slow enough to allow some ageing and hence strengthening to take place. A post SPF ageing heat treatment may further enhance the room temperature tensile strength. The reported retention of ~30% β phase at room temperature should maintain a good balance of mechanical properties.

Standard x-ray macro-texture measurements have showed texture weakening for the (0002) α phase textures after superplastic deformation, as noted previously [7]. However, after deformation at 850°C a significant increase in the same texture was found and this was presumably a result of a change in the deformation mechanism at this temperature, also evidenced by a change in the plastic strain ratio R.

Some preliminary results of micro-texture measurements have been presented (Fig. 8). These measurements have shown that under superplastic deformation conditions (ie 750°C and $\dot{\varepsilon}=3 \times 10^{-4} s^{-1}$) there is little change in both the α phase texture, as also found via x-ray measurements and in the β phase texture. This is perhaps surprising since it might be expected that much of the deformation is occurring in this "softer" phase, clearly very little grain rotation is occurring even after $\varepsilon_{true}=1.35$ (300% elongation). Also it was found that there was no difference in the textures of α and β phase at either the sheet surface or middle after SPF, these results suggest that identical deformation mechanisms are operating at both surface and centre of the sheet. Diffuse EBSD patterns in as received material were due to lattice strain, presumably the result of cold work during rolling. Sub-structure was also found by EBSD in nominally single grains. After SPF EBSD patterns became sharper suggesting a reduction in lattice strain as may be expected from observations of reductions in the dislocation density after SPF, there was also little sub-structure observed. A sub-grain structure may explain the flow softening observed in the first 100% elongation since a conversion of sub-grains to higher angle boundaries would lead to an apparent reduction in grain size; once this process was complete then strain enhanced grain growth occurs, leading to strain hardening as witnessed in most titanium alloys [7,9]. Even with this limited data available from this technique it would seem that a useful insight into the superplastic deformation mechanisms can be obtained.

High strain anisotropy was observed, with R values of 0.4-0.6 after ε_{true}=1.35 in both L and T test pieces; the anisotropy was slightly more pronounced for the T test pieces after deformation under conditions for superplasticity. This behaviour has been reported previously after SPF of a Ti-6Al-4V extruded section with aligned contiguous α phase in both L and T directions [10]. The contiguous α phase is thought to constrain the width deformation [11]. Crystallographic texture is not thought to have any effect on the strain anisotropy during SPF, however, at 850°C the deformation is less superplastic and increased strain anisotropy was found, this may be texture related. However, the situation is complex since significant increases in the α phase texture intensity have been measured (Fig. 7) after deformation at this temperature and it is difficult to separate cause and effect.

Diffusion bonding has been demonstrated in SP 700 [5] and the combination of this with superplastic deformation should enable high performance structures to be fabricated.

Conclusions

SP 700 has exhibited excellent superplastic formability over a wide range of temperatures. The optimum temperature for SPF in this alloy is probably between 750-800°C, where the lowest flow stresses, moderate m values and good dimensional uniformity are obtained. The room temperature tensile properties are retained and enhanced after SPF but control over the cooling rate from the SPF may need to be exercised to obtain the full benefit. Since lower forming temperatures are required for SP 700 there should be economic benefits in terms of reduced heating and die wear. X-ray texture measurements have shown a weakening of the α phase texture after superplastic deformation at 750°C, however, significant texture sharpening was found following deformation at 850°C. Preliminary micro-texture measurements, via EBSD patterns have shown little or no randomisation in the β phase texture. Diffusion bonding has been demonstrated in this alloy. With these properties SPF/DB structures in SP 700 should be possible, giving higher performance than currently envisaged with other Ti alloys and at reduced cost.

Acknowledgements
The authors would like to thank Mrs M.Wood for the metallurgical preparation and Mrs.S.M.Goodwin for the x-ray texture measurements.

References

[1] B.Rolland, in Sixth World Conference on Titanium Vol. 1, ed.s P.Lacombe, R.Tricot, G.Beranger (Les Editions de Physique, Les Ulis Cedex 1988), pp.399-404.
[2] D.P.Davies, AeroMat' 93, 7-10th June 1993, Anaheim, USA.
[3] C.Ouchi, K.Minakawa, K.Takahashi, A.Ogawa, M.Ishikawa, NKK Tech. Rev. No. 65, 61-67 (1992).
[4] V.Randle, "Microtexture Determination and its Applications", 1992, Institute of Materials, London.
[5] A.Wisbey, B.Geary, D.P.Davies, C.M.Ward-Close, in Proc. Conf. "International Conference on Superplasticity in Advanced Materials, ICSAM'94", 1994 to be published.
[6] M.Ishikawa, O.Kuboyama, M.Niikura, C.Ouchi, in Titanium '92 Science and Technology Vol.1, ed.s F.H.Froes, I.L.Caplan (TMS, Warrendale 1993), pp.141-148.
[7] D.V.Dunford, A.Wisbey, P.G.Partridge, Mats. Sci. & Tech., 7, 62-70 (1991).
[8] A.Wisbey, P.G.Partridge, Mats. Sci. & Tech., 9, 987-993 (1993).
[9] A.Arieli, B.J.Maclean, A.K.Mukherjee, Res Mech.,6, 131-159 (1983).
[10] D.V.Dunford, P.G.Partridge, in Sixth World Conference on Titanium Vol. 1, ed.s P.Lacombe, R.Tricot, G.Beranger (Les Editions de Physique, Les Ulis Cedex 1988), pp. 1197-1202.
[11] P.G.Partridge, D.S.McDarmaid, A.W.Bowen, Acta Met., 33, 571-577 (1985).

Table 1. The room temperature tensile properties of SP 700 and Ti-6Al-4V sheet.

Condition	Dim.	0.2% Proof Stress MPa	Tensile Strength MPa	Mod (E) GPa	Elong. %
SP 700 as received (annealed)	L	1010	1082	116	16.5
	T	1039	1101	119	16.5
SP 700 after SPF at 700°C & $\dot{\varepsilon}=3\times10^{-4}\text{s}^{-1}$ to $\varepsilon_{true}=1.35$	L	1060	1097	124	10.9
	T	1126	1162	130	11.6
SP 700 after SPF at 750°C & $\dot{\varepsilon}=3\times10^{-4}\text{s}^{-1}$ to $\varepsilon_{true}=1.35$	L	1018	1028	114	8.7
	T	1078	1098	120	13.0
SP 700 after SPF at 800°C & $\dot{\varepsilon}=3\times10^{-4}\text{s}^{-1}$ to $\varepsilon_{true}=1.35$	L	1066	1120	123	6.6
	T	1103	1170	116	11.9
Ti-6Al-4V 2 mm sheet as received.	L	973	1048	122	19.6
	T	921	994	114	14.6
Ti-6Al-4V 2 mm sheet after SPF at 925°C & $\dot{\varepsilon}=3\times10^{-4}\text{s}^{-1}$ to $\varepsilon_{true}=0.85$.	L	875	956	113	16.7
	T	854	944	111	17.3

Figure Captions

Fig. 1 SP 700 sheet test pieces, prior to deformation and after deformation at 750°C and various strain rates.

Fig.2 The initial true flow stress versus true strain rate for SP 700 sheet at temperatures between 700°C and 850°C.

Fig.3 The true flow stress versus percentage elongation of the SP 700 test pieces, deformed at temperatures between 700°C and 850°C at a strain rate of $3 \times 10^{-4} s^{-1}$.

Fig.4 The initial strain rate sensitivity index (m) value versus true strain rate for SP 700 deformed at temperatures between 700°C and 850°C.

Fig.5 The plastic strain ratio (R) versus true area strain after deformation at temperatures between 700°C and 850°C.

Fig.6 SP 700 sheet microstructure a) as received and 800°C 15 minutes, water quench and b) after SPF at 800°C and $\dot{\varepsilon}=3 \times 10^{-4} s^{-1}$ to $\varepsilon_{true}=1.35$ and 800°C 15 minutes, water quench. L-T plane.

Fig.7 X-ray (0002) pole figures from SP 700 sheet centre of as received material and after deformation at 750°C and 850°C to $\varepsilon_{true}=1.35$ (300% elongation) with a strain rate of $3 \times 10^{-4} s^{-1}$.

Fig.8 Micro-texture pole figures of both the α and β phases in SP 700 as received sheet and after superplastic deformation to $\varepsilon_{true}=1.35$ (300% elongation) at 750°C and $\dot{\varepsilon}=3 \times 10^{-4} s^{-1}$.

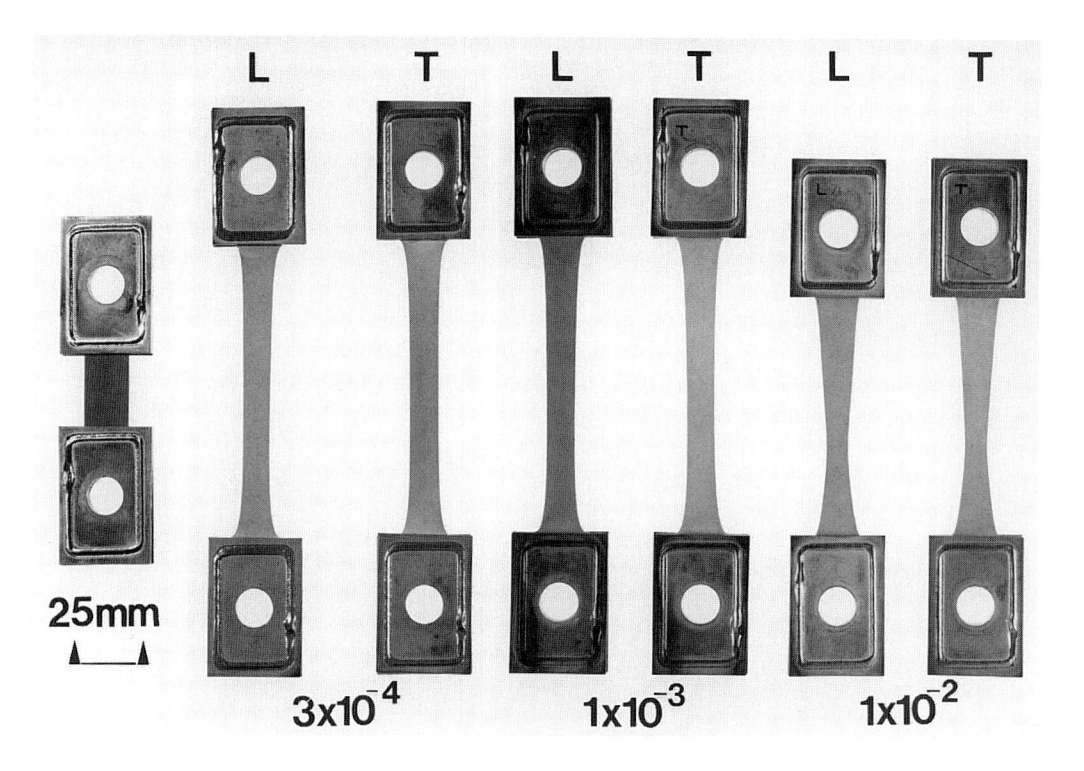

Fig. 1 SP 700 sheet test pieces, prior to deformation and after deformation at 750°C and various strain rates.

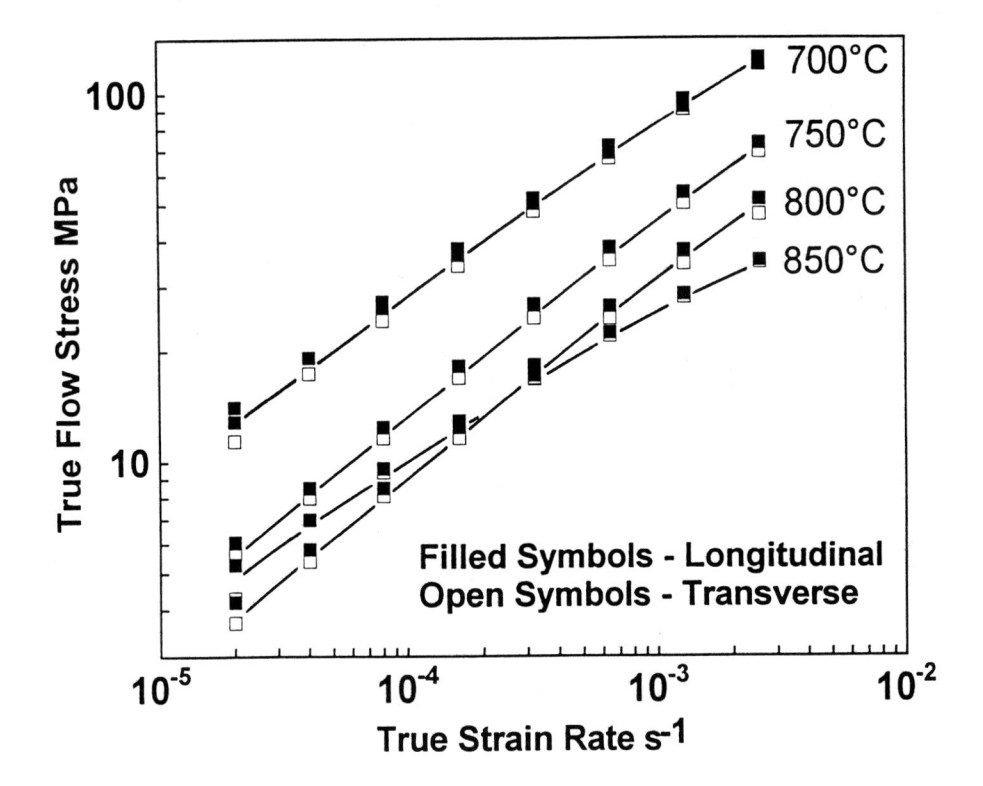

Fig.2 The initial true flow stress versus true strain rate for SP 700 sheet at temperatures between 700°C and 850°C.

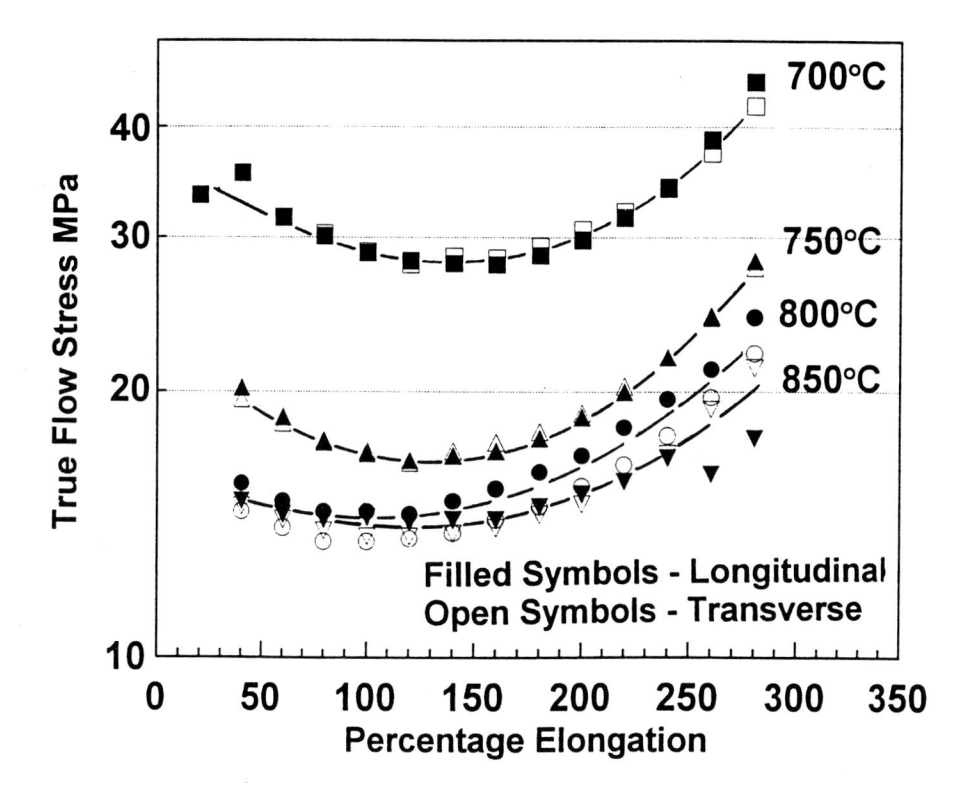

Fig.3 The true flow stress versus percentage elongation of the SP 700 test pieces, deformed at temperatures between 700°C and 850°C at a strain rate of $3 \times 10^{-4} s^{-1}$.

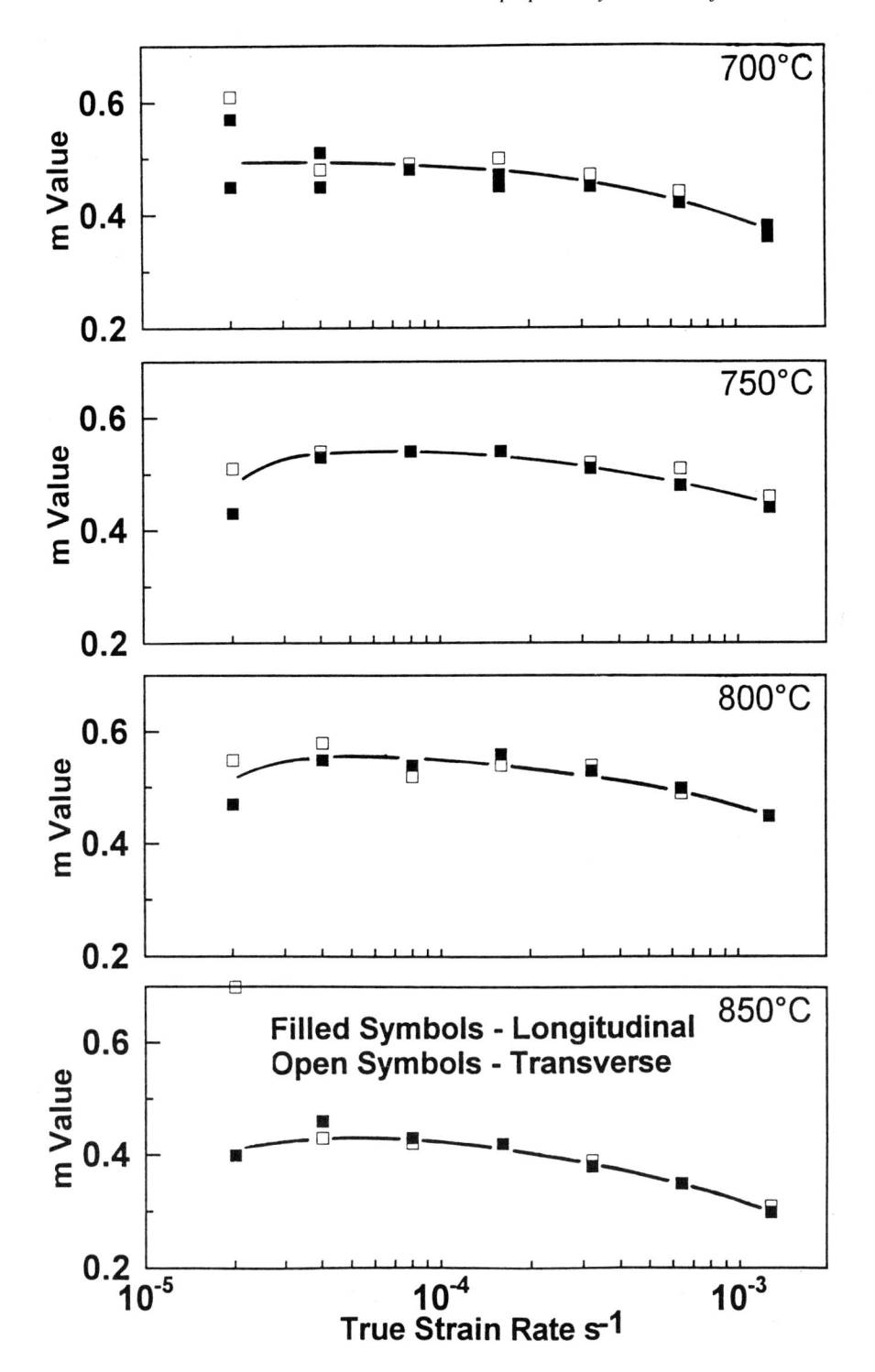

Fig.4 The initial strain rate sensitivity index (m) value versus true strain rate for SP 700 deformed at temperatures between 700°C and 850°C.

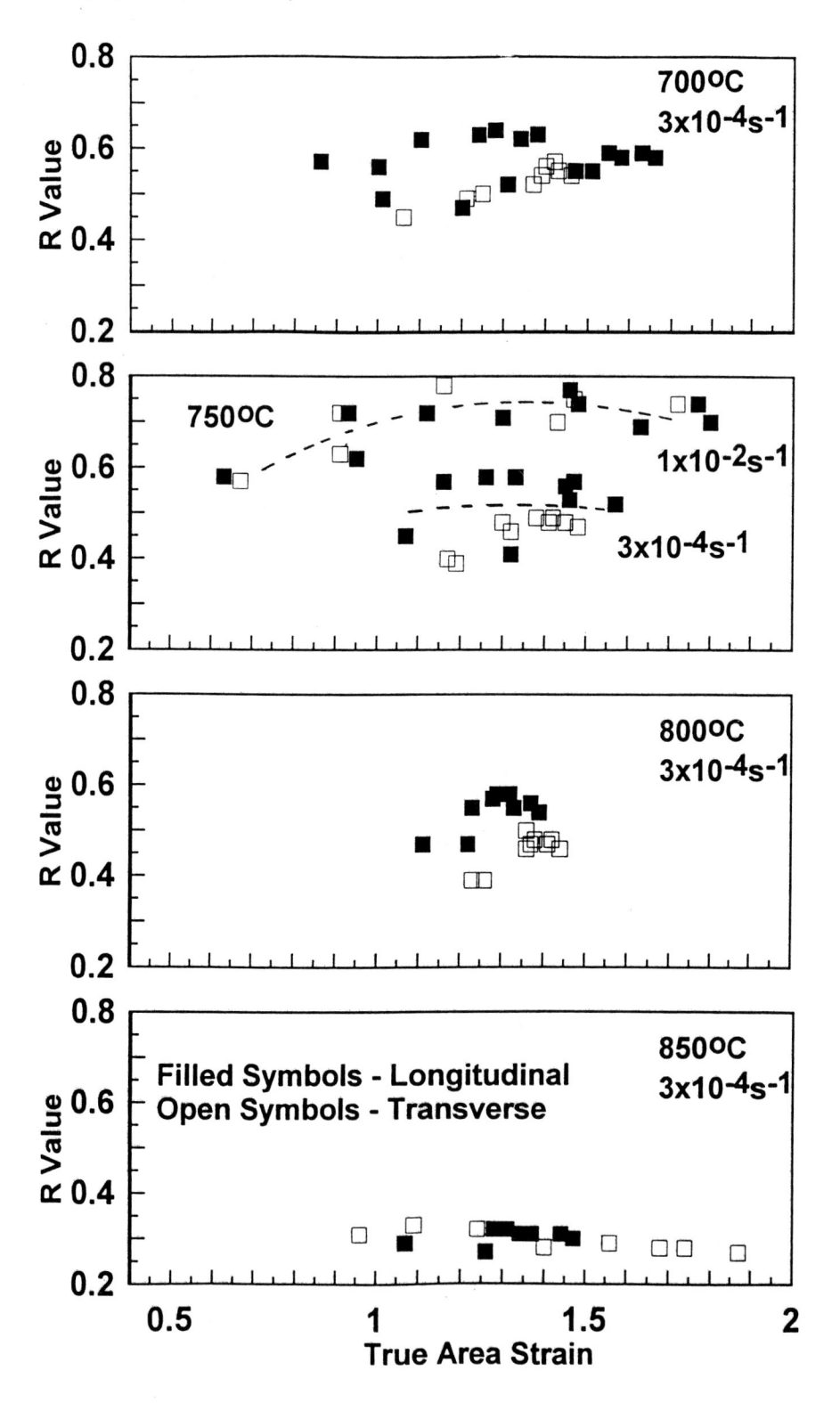

Fig.5 The plastic strain ratio (R) versus true area strain after deformation at temperatures between 700°C and 850°C.

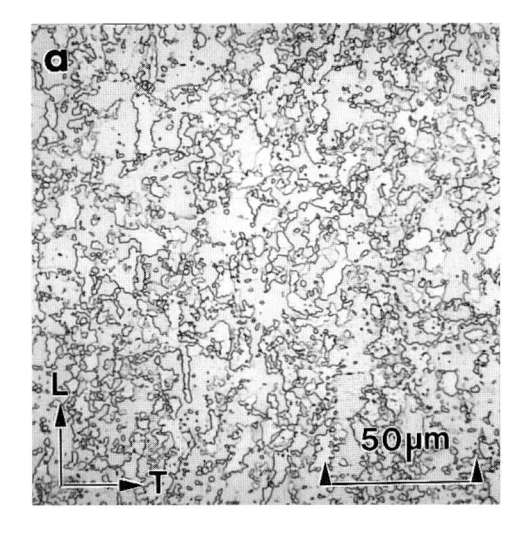

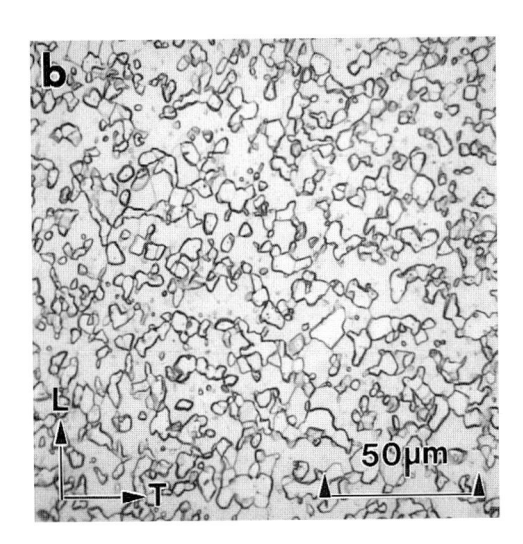

Fig.6 SP 700 sheet microstructure a) as received and 800°C 15 minutes, water quench and b) after SPF at 800°C and $\dot{\varepsilon}=3\times10^{-4}s^{-1}$ to $\varepsilon_{true}=1.35$ and 800°C 15 minutes, water quench. L-T plane.

(0002) Pole Figures

As Received

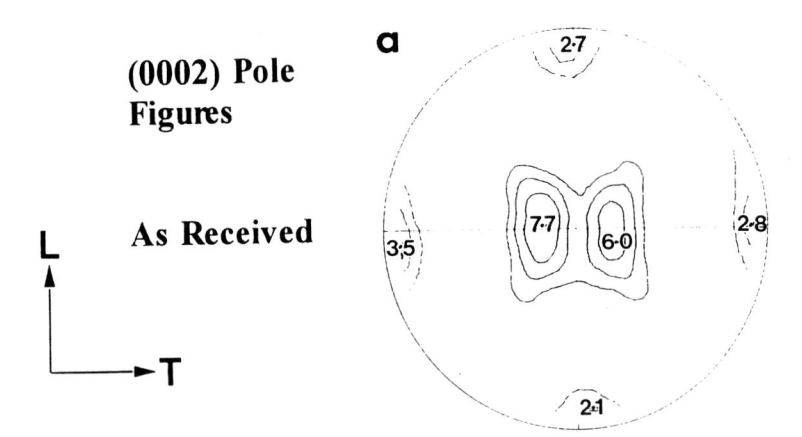

After SPF, $\dot{\varepsilon}=3\times10^{-4}s^{-1}$ to 300% elongation. Tensile axis orientation-
Longitudinal   Transverse

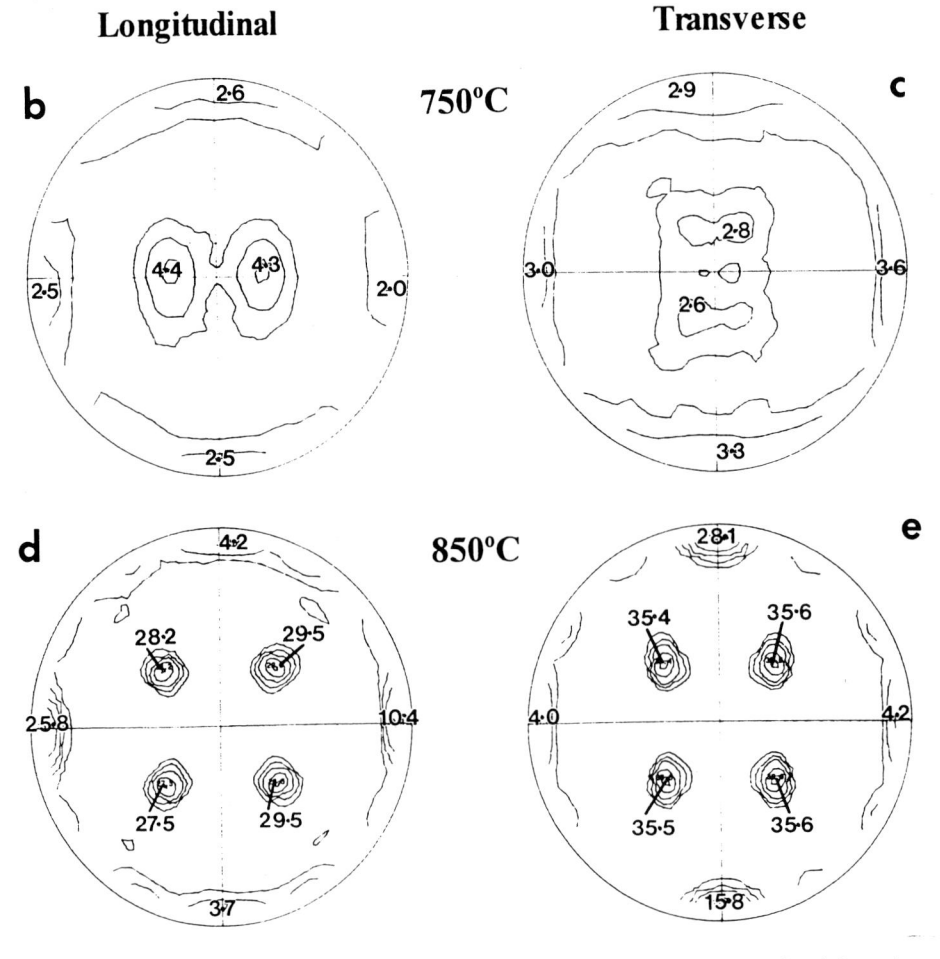

750°C

850°C

Fig.7   X-ray (0002) pole figures from SP 700 sheet centre of as received material and after deformation at 750°C and 850°C to $\varepsilon_{true}=1.35$ (300% elongation) with a strain rate of $3\times10^{-4}s^{-1}$.

| a | As Received | b |

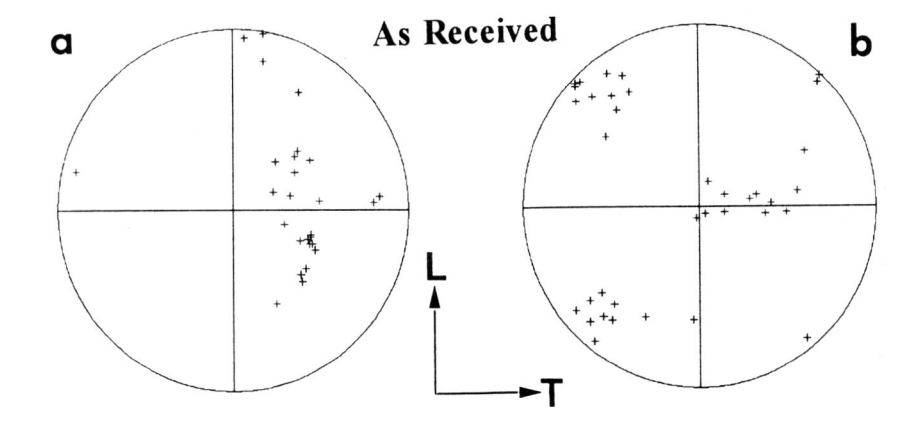

L → T

After SPF at 750°C, $\dot{\varepsilon}=3\times10^{-4}s^{-1}$ to 300% elongation. Tensile axis orientation - Longitudinal.

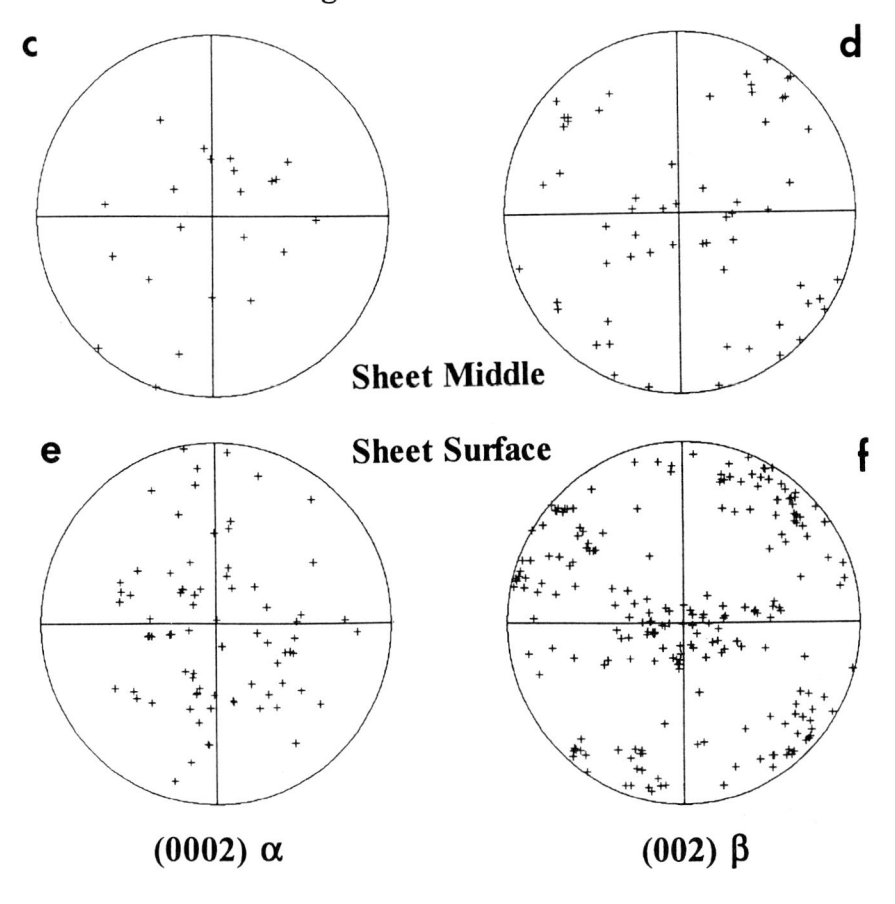

c   d

Sheet Middle

e   Sheet Surface   f

(0002) α   (002) β

Fig.8 Micro-texture pole figures of both the α and β phases in SP 700 as received sheet and after superplastic deformation to $\varepsilon_{true}=1.35$ (300% elongation) at 750°C and $\dot{\varepsilon}=3\times10^{-4}s^{-1}$.

Superplastic Forming (SPF) at AWE - Forming Processes and post-SPF Characterisation of Ti-6Al-4V plate

F J Moran

AWE Aldermaston, Reading, Berks, RG7 4PR

Abstract

Shapes have been manufactured from the Ti-6A1-4V alloy in plate, up to 12.7 mm thickness. The material was supplied in the mill annealed condition. Circular blanks of this material were then manufactured and formed, by superplastic forming (SPF), into various components. Consideration was given to process control techniques such as double forming and blank profiling to reduce the effects of thinning. This evaluation is continuing.

Tensile and metallographic samples were taken, from these plate components, of areas subjected to SPF strain and of areas only subjected to the SPF thermal cycle. Results were compared with the as-received material.

After SPF, the material was stronger than the as-received material, however, the ductility had fallen. Subsequent heat treatment (annealing) of material after SPF gives some recovery of ductility back towards the values for the as-received material. The metallography for the material supports the mechanical property data. The general indications are that post-SPF heat treatment improves mechanical properties. All the post-SPF values for strength and elongation satisfy the relevant standards for this material.

This work was presented to the 60th anniversary conference on SPF at UMIST Manchester on 7/8-12-94.

1. Introduction

Superplastic forming (SPF) is being considered as a possible alternative fabrication technique to forging, for Ti-6A1-4V alloy components. Initial hemispherical type parts have been made, from starting plate thicknesses of up to 12.7mm. A typical hemispherical type component, made from 12.7mm thick Ti-6Al-4V alloy plate, is shown in Figure 1.

This work was to determine the effect of the SPF cycle on the mechanical properties of the Ti-6Al-4V alloy and to investigate the related metallographic structure. The work also considered the effects of further heat treatment on post-SPF material.

The Ti-6A1-4V alloy used was sheet/plate of 10 and 12.7mm thickness. The material was supplied, in the mill annealed condition to the relevant standard[1]. Circular blanks of this material were formed, by SPF, into various components. Samples for metallography and mechanical property evaluation were taken from components after SPF processing, in strained and unstrained regions.

2. Processes

In summary, the process for SPF was a hot open die technique using a 75 ton hot platen press and Ti-6Al-4V plate up to 12.7mm thick. Blank profiling and reverse forming techniques were considered with regard to controlling part thinning. The process is shown schematically in Figure 2. This figure shows the process for manufacture of hemispherical parts at 4 inch diameter, however the procedure is the same for other sizes.

This SPF process can be either by forming the blank into the top die chamber (direct or single forming) or by first forming the blank down onto a radiused insert prior to forming into the top die chamber (reverse or double forming). The reasoning behind the latter process is to attempt to produce a more uniform part thickness profile, due to all parts of the blank contacting the die wall at the same time.

3. Results and Discussions

The mean data for metrology of parts made by SPF from nominal 12.7mm plate are shown in Figure 3. This figure shows polar thinning of 60-70% with formed parts and no apparent advantages of double forming over single forming, with thick plate. Note that the initial thickness reduction (at angular position = 0) is due to part forming over a relatively tight die entry radius.

The metallography of the material during SPF processing is shown in Figure 4. This shows that the initial grain alignment tends to be retained after SPF, with some grain coarsening. Post-SPF heat treatment yields a material structure similar to that for turbine disc quality material with this alloy[2].

The mechanical property data are given in Table 1, with the relevant specification values[3] and typical data for forged material. Table 1 indicates that the SPF material is stronger than the forged material, primarily due to a finer starting grain size and the strengthening effects of the SPF thermal cycle and subsequent quenching from the SPF temperature. As implied above, the post-SPF heat treatment produces strength properties similar to those quoted for turbine disc quality material.
The fall in ductility after SPF has been observed by other workers[4,5] and is primarily due to the SPF thermal cycle.

In summary the tensile tests showed that the SPF material was stronger than the forged stock, with slightly less ductility. This work will continue to determine the effects of SPF strain and post-SPF heat treatment on mechanical properties.

All the values for strength and elongation satisfy the relevant standards for this material[1,3].

4. Conclusions

4.1 Parts can be made from thick plate up to 12.7mm thick, with 60-70% pole reduction on hemispherical type parts.

4.2 Initial work indicates no apparent advantages in double forming techniques with thick plate.

4.2 The SPF cycle causes a reduction in ductility which is recoverable by Post-SPF heat treatment

5. FUTURE WORK

The study of post-SPF mechanical properties will continue. Further parts will be formed by double forming and with profiled blanks once the equipment has been moved to a new location on site. Mathematical modelling work is also being undertaken in support of this SPF. Initial work has been previously reported[6] and work has also been carried out in support of optimising a profiled blank form for SPF[7].

British Crown Copyright 1994 /MOD

Published with the permission of the Controller of Her Britannic Majesty's Stationery Office

References

1. ASTM-B265-89, ASTM, USA, 1989

2. Anon, "Medium Temperature Alloys", Technical Brochure, IMI Titanium, Birmingham

3. BS 2TA12, BSI, London, 1974

4. M T Cope, D R Evetts and N Ridley. Mater Sci and Technol. $\underline{3}$, 455-461, (1987).

5. D V Dunford, A Wisbey and P G Partridge. Mater Sci and Technol. $\underline{7}$, 62-70, (1991).

6. C Hammond and W T Roberts, Materials World, 1, Feb. 1993, 123-124

7. M Al-Khalil and J M Snee, Bae Technical Report, Filton, 1993

Table 1 Pre/post-SPF mechanical properties

PROPERTY	Samples made from 10mm plate					From Forgings	
	As-received	Post-SPF samples taken from				Typical	Standard to BS2TA12
		flange	annular dip 1	annular dip 2	flange (heat treated)		
0.2%PS / MPa mean range	958 -	1073 1034-1131	1021 981-1046	1079 1044-1103	973 953-994	863 847-887	830 minimum
UTS /MPa mean range	1044 -	1165 1140-1198	1143 1102-1166	1177 1143-1199	1026 1007-1047	933 911-950	900 minimum
%El. mean range	16 15-17	10.4 10.2-10.6	10.3 9.9-10.5	9.4 9.2-9.5	13.3 12.9-13.7	19 16-21	8 minimum

Notes

1. The post-SPF heat treatment was 2hrs at 720°C under vacuum.

2. Dip one was at -0.006 thickness strain and dip two was at 0.051 thickness strain where thickness strain = ln (t/t_o) where:-
    t - thickness in base of dip
    t_o - initial mean thickness of plate

3. &El. is on a 20mm gauge length

Superplasticity: 60 Years after Pearson 343

Fig. 1 Fully formed SPF hemispherical domed part (from nominal 12.7mm plate)

100mm

Fig.2 Schematic of SPF process for sheet/plate forming at AWE

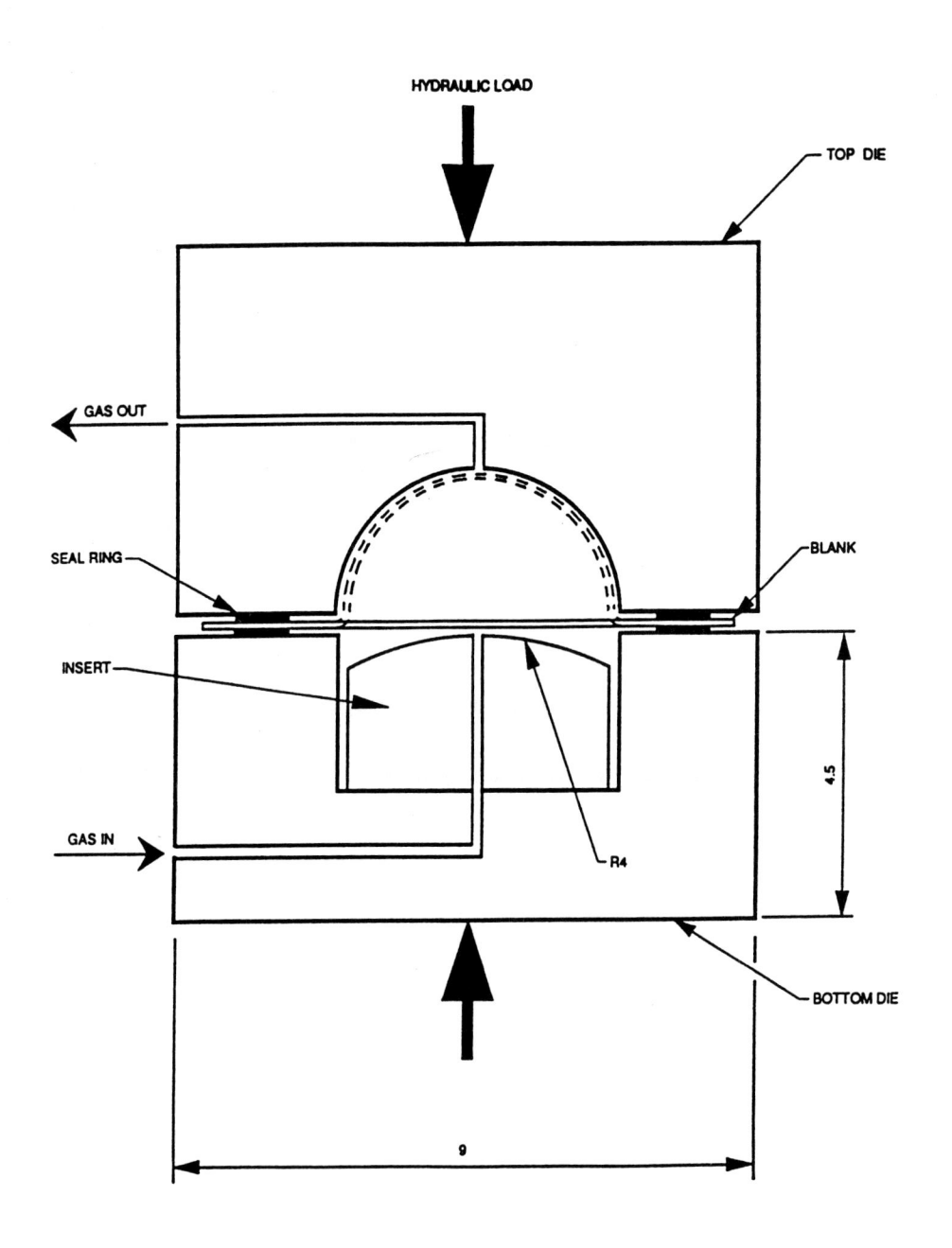

DIMENSIONS IN INCHES

Fig.3 Mean thickness profile of hemispherical type parts made by SPF from nominal 12.7mm plate

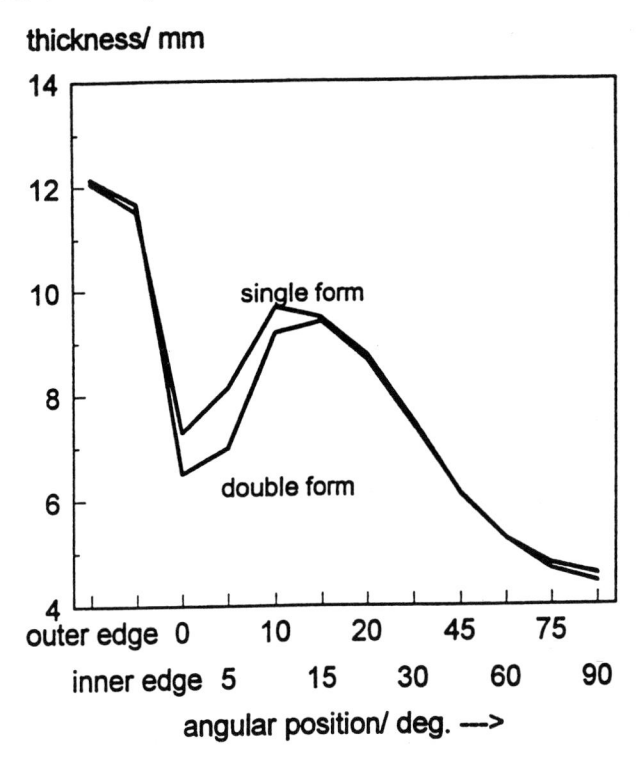

Note

Angular position around a hemispherical type form is defined as shown in the diagram below

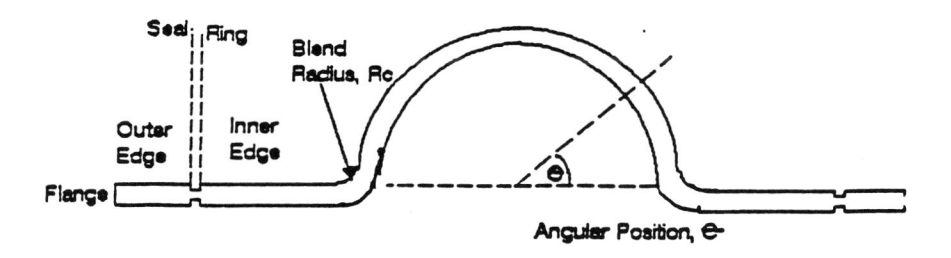

Fig.4 Metallography of Ti-6Al-4V plate used in SPF studies

Scale |———————| 100μm

(a) As-received plate

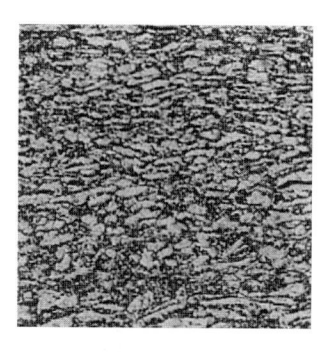

(b) Plate after SPF

(c) Plate after post-SPF heat treatment (2 hours/720 °C

THE EXPLOITATION OF SUPERPLASTICITY FOR ROLLS-ROYCE'S WIDE CHORD FAN BLADE

G A FITZPATRICK
ROLLS-ROYCE plc
P.O. Box 3
Barnoldswick
Colne, Lancs
BB8 5RU
UK

Abstract

ROLLS-ROYCE plc has designed and developed a highly efficient lightweight fan for civil engine applications in the thrust range 22000lbs to over 90000lbs. These wide chord fan designs are hollow and snubberless, and utilise innovative metal forming, metal joining and inspection techniques.

First generation wide chord fan blades are a fabrication of two external titanium alloy panels and a titanium alloy honeycomb core joined by a liquid-phase diffusion bonding process. They initially entered service in the RB211-535E4 powered Boeing 757 in 1984 and have since been incorporated in the IAE V2500 engine for the Airbus Industries A320/A321 and the McDonnell Douglas MD90 aircraft as well as the RB211-524 G/H powerplants for Boeing 747 and 767 aeroplanes.

Second generation fan designs and corresponding manufacturing technologies for Rolls-Royce's larger Trent turbofan engines have evolved to exploit solid-state diffusion bonding for joining the blade fabrication and superplastic forming for the development of the supporting internal core. These wide chord fan blades have both weight and cost advantages over the original concept, and have satisfactorily endured intensive certification integrity and durability testing to permit the application of superplastic forming on a critical rotating component.

1. Introduction

The fan in modern "high by-pass ratio" civil turbofan engines must deliver a high level of performance over a wide range of operating conditions. Thrust, fuel consumption, cost, noise, foreign object impact resistance and mechanical integrity under fatigue conditions are the essential design considerations.

Conventional fan blades are manufactured from solid titanium alloy forgings with mid-span snubbers to counteract aerodynamic instability (Figure 1a). However, the snubbers impede the supersonic airflow causing a loss in aerodynamic performance with corresponding penalties in fuel consumption. Rolls-Royce has, therefore, removed the snubber from the fan to develop an aerodynamically optimised aerofoil, increased its chord for natural stability, and reduced the number of blades per assembly by approximately a third (Figure 1b).

Its position at the front of the engine demands that the fan is capable of developing sufficient power for aircraft safety after suffering impacts, predominantly by birds, during take-off. Fan blades are also subjected to low cycle fatigue stresses during every flight and to high cycle fatigue stresses from air intake disturbances at specific flight conditions. Stresses within the component have, therefore, to be maintained within established limits in order to guarantee adequate fatigue life.

To satisfy these design criteria, Rolls-Royce has developed hollow wide chord fan blades with low density internal supporting cores which can satisfy severe operational requirements (references 1, 2). The constructions of the designs are shown schematically in Figure 2. For both fabrications, the external titanium alloy skins are separated and supported by an internal titanium alloy core. In the latest design, the established internal honeycomb core is replaced by a superplastically formed corrugation which allows the production of a lighter construction as well as reduced manufacturing costs.

2. Wide Chord Fan Manufacturing Technologies

Innovative metal forming, metal joining and inspection techniques have been developed by Rolls-Royce for the manufacture of its wide chord fan blades. First generation designs have been produced as a fabrication of external titanium alloy panels and an internal titanium alloy honeycomb core joined by a liquid-phase diffusion bonding process, Activated Diffusion Bonding (reference 3). The latest designs and corresponding manufacturing methods for the larger Trent fan blades have now evolved to exploit solid-state diffusion bonding for joining the fabrication and superplastic forming for the development of the internal corrugated core (reference 4).

The Trent wide chord fan blade is manufactured from three sheets of the titanium alloy, Ti-6Al-4V. The fabrication is selectively joined in a custom-built high temperature pressure vessel under computer control to generate diffusion bonds with parent material mechanical properties (reference 5). The manufacturing sequence then exploits the inherent superplasticity of these fine-grained titanium alloys. The diffusion bonded construction is inflated at elevated temperature between appropriately contoured metal dies using an inert gas which expands the core by superplastic forming whilst simultaneously developing accurately the blade's external aerodynamic profile in terms of radial bow, axial camber and aerofoil twist. This process is carried out in customised presses in a computer-controlled operation to guarantee the strict tolerances necessary for component temperature distribution as well as the strain-rate of the internal core. Considerable experimentation has been carried out to establish superplastic forming as a viable manufacturing technology for critical hollow aeroengine component designs. In particular, the component has been specifically designed for optimised manufacture via this process whereas the manufacturing process has been designed to recognise the component's design specification as well as NDE capabilities (reference 6).

To satisfy the stringent product assurance standards required for the service environment, all hollow wide chord fan blades are critically inspected for component integrity using a variety of sophisticated non-destructive techniques supplemented by conventional methods.

3. Wide Chord Fan Integrity

First generation wide chord fans have been in service since 1984 and have accumulated over thirteen million hours of operational experience. Second generation wide chord fan blades for the Trent series of aeroengines have now been subjected to intensive engine and rig component tests which have verified the design and associated manufacturing technologies. These test programmes, including bird ingestion and low cycle and high cycle fatigue testing, have been successfully completed, and have confirmed metallurgical integrity as well as the maturity of the superplastic forming process. The first engine specified by Rolls-Royce with second generation fan blades is the Trent 700 which is now certificated to power the Airbus Industries' A330.

4. Summary

First generation wide chord fan designs have demonstrated reduced fuel consumption and increased resistance to component and engine damage over five million hours of Rolls-Royce exclusive operational experience. Second generation wide chord fans have now been specified for the higher thrust-rated Trent aeroengines with inherent weight and cost advantages. This latest design is the first to exploit superplastic forming for a critical rotating component. The Trent 700 version (70000lbs thrust) powers the Airbus Industries' A330 whereas the Trent 800 version (90000lbs thrust) is being developed to power the Boeing 777.

References

1. R E Dawson, D G Pashley, J E Melville. "Technology of the Wide Chord Fan Blade", American Institute of Aeronautics and Astronautics/ASME/SAE/ASEE, 22nd Joint Propulsion Conference, Huntsville, Alabama, USA, June 1986.

2. D J Nicholas. "The Wide Chord Fan Blade - a Rolls-Royce first", Eighth International Symposium on Air Breathing Engines, Cincinnatti, Ohio, USA, June 1987.

3. G A Fitzpatrick, T Broughton. "The Rolls-Royce Wide Chord Fan Blade, International Conference on Titanium Products and Applications, San Francisco, California, USA, October 1986.

4. G A Fitzpatrick, J M Cundy. "Rolls-Royce's Wide Chord Fan Blade - The next generation", Seventh World Conference on Titanium, San Diego, California, USA, June 1992.

5. G A Fitzpatrick, T Broughton. "The Diffusion Bonding of Aeroengine Components", Sixth World Conference on Titanium, Cannes, France, June 1988.

6. M W Turner. "The application of diffusion bonding and superplastic forming technologies to wide chord fan blade manufacture", Thirtieth METRAG Meeting, London, UK, March 1994.

Acknowledgements

The author wishes to thank Rolls-Royce plc for permission to publish this paper. The contributions made by many colleagues at Barnoldswick and Derby to the development of wide chord fan blade technologies are also acknowledged.

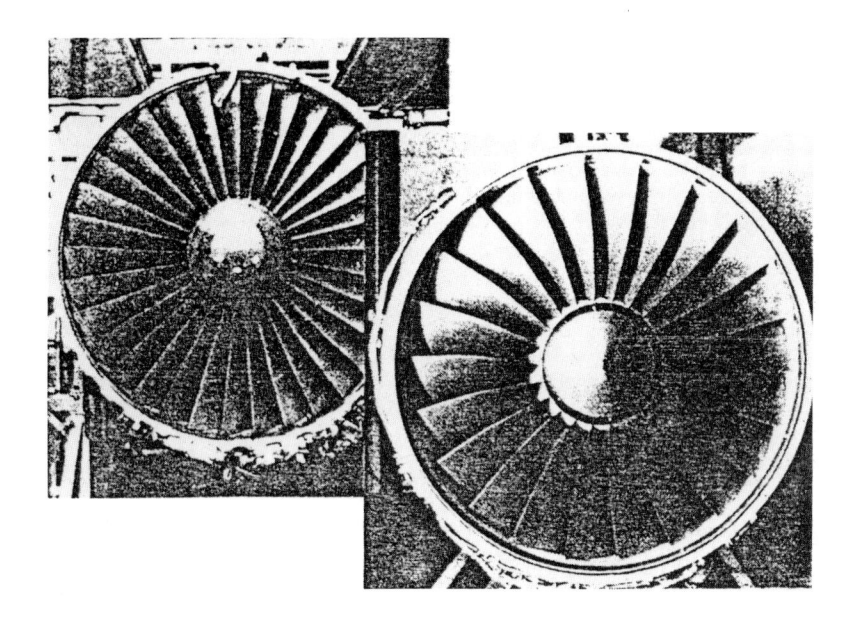

(a) RB211-535C
Snubbered Fan

(b) RB211-535E4
Wide Chord Fan

Figure 1 - Fan Blade Designs

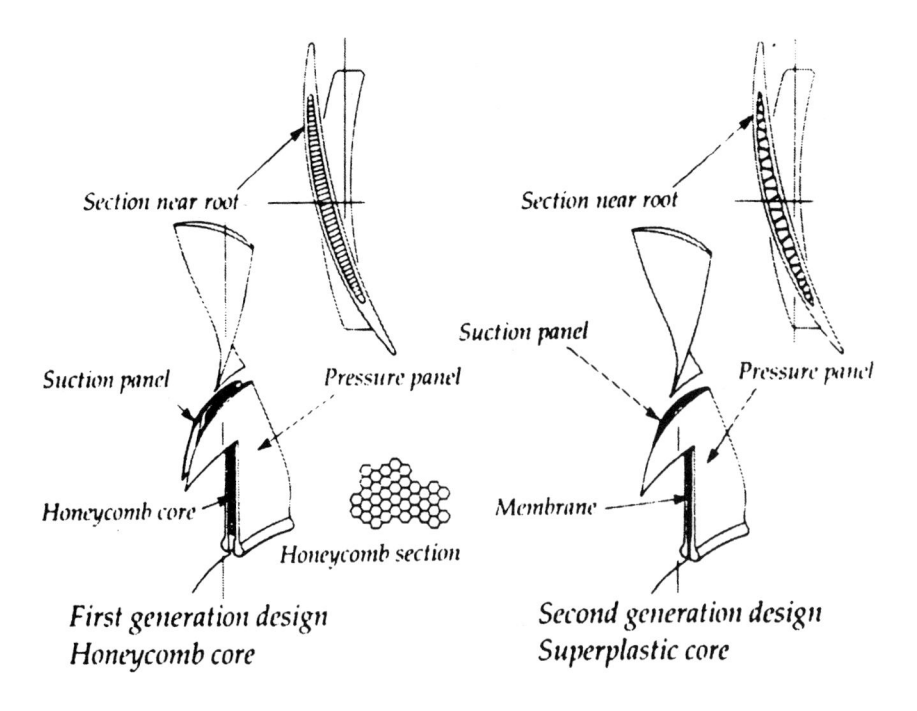

First generation design
Honeycomb core

Second generation design
Superplastic core

Figure 2 - Construction of Rolls-Royce Wide Chord Fan Blades.

SuperPlastic Forming of Large Titanium 6Al 4V and Turbo Prop Nacelle Products

W Swale

TKR International Limited, Watchmead,
Welwyn Garden City, Hertfordshire, AL7 1LT

Abstract

Since the adoption of the titanium SuperPlastic Forming (SPF) process some 20 years ago the Aerospace sector has been constantly challenged to produce cost effective products of ever growing size. Many of the titanium products are to be found in areas where fire containment and minimal weight are essential, such as Doors and Firewalls on Turbo Prop Nacelles.

With the requirement for large diameter by-pass ratio jet engines comes the need to produce very large complex pressings in titanium to be used as structural members in the engine cowl intake area.

SuperPlastic forming is used to produce the complex shapes that these products demand. Amalgamation of parts can also reduce costs and this paper shows how the product size has grown, the types of shapes produced, the tooling and equipment required and the qualification testing demanded by the industry.

1. Introduction

From 1978 TKR International has been involved in the art of applying the SuperPlastic Forming process to aircraft products made from Ti 6Al 4V titanium alloy. During this period many changes have been made to the SPF process in order to obtain the best advantages. Many of these advantages have been driven by design requirements and innovative thinking. The early applications of SuperPlastic forming of titanium for aircraft products utilised and demonstrated the ability to combine several shapes into one forming, thereby making stiffer, lighter, reduced cost products with elimination of fasteners. This design demand has not ceased. The goal for bigger and better and more complex products is almost daily being presented to manufacturing.

2. The Manufacturing Challenge

The early products were relatively simplistic and based upon 0.5M square sheet metal sizes at starting thicknesses of 1.6 to 2.0MM, Fig 1. Manufacturing was expected to produce production quantities at competitive prices. This drove manufacturing to amalgamate various parts into one pressing whilst at the same time maintaining the theoretical parameters which academia had demonstrated to provide optimum SPF results.

Fig 1 shows a pressing with 3 different part numbers included in a typical production batch of 20 pieces.

The next phase almost followed naturally, if various products could be produced in one pressing then a panel with inbuilt stiffness could be conceived. This concept is shown in Fig 2 and Fig 3. Fig 2 shows a fabricated assembly consisting of approximately 70 pieces replaced by a 2 piece fabrication, Fig 3. This shows the same product made from 2 SPF pressings in Ti 6Al 4V titanium.

With the acceptance of this type of design and the knowledge of how to calculate the design load transfer many other products have evolved. The SPF industry is now at a point where the size of sheet metal is the limiting factor and where typically sheet sizes 1M x 3.8M and 1.5M x 2.8M are being used daily.

3. Growth Results

This growth in size has brought many challenges to a multitude of interconnected industries and processes of which the following are examples.

(a) Material Manufacturing

The demand is for titanium alloy Ti 6Al 4V and Ti 6Al-2Sn-4Zr-2Mo in sheet form in sizes up to 1.5M wide x 4.0M long at thicknesses ranging from 0.5MM to 4.0MM.

(b) Heated Platten Presses

These have been developed by specialist companies or in the case of TKR International by its own personnel. In all cases presses from 150 tons to 1000 ton clamping load, with 920°C ± 10°C temperature controlled plattens, ranging in size from 4M x 1M to 2.8M x 1.8M down to 0.5M x 0.5M have been produced to suit a vast range of product sizes.

(c) Tool Designs

Tool designs and materials have been developed to provide uniform heat absorption, high temperature creep resistance, good thermal fatigue properties and high temperature corrosion resistance from materials which can be cast to relatively close tolerances and machined. In addition customers demand that tooling be guaranteed for the life of the project or 300 to 500 products. Fig 4.

(d) Manufacturing Techniques

In order to produce products at optimised temperature and strain rates process control equipment was developed. These ensure repeatable products. In addition, techniques have been developed that enable the removal of very large SPF formed products from heated platten presses whilst still at 900°C without distorting the shape. Fig 5 (EH101 leading edge).

(e) Chemical Milling Processes and Hydrogen Removal

The application of controlled chemical milling techniques for the conditioning and removal of the brittle oxide layer (alpha case) and, where necessary, the removal of absorbed hydrogen by vacuum degassing equipment without distorting the product have also been developed.

In addition to these challenges the component designs demand pressings which conform to profile shape tolerances of 0.5MM with thicknesses at 0.7MM. Also the metallurgical condition must satisfy the customer's material specification requirements.

Product Examples

The type of products that demand these requirements are produced in SPF facilities on a daily basis and in production quantities at TKR International. These products are mainly for Turbo Prop Nacelle structures (stiffened ribbed pressings "waffle", spot welded to plain outer air washed surface skins) and Titanium SPF formed pressings for the new generation of large Turbo Fan Jet engines.

The first large product Fig 6 was made from 2.0M long x 1.5M wide Ti 6Al 4V in tooling weighing 2 tons. The resulting thickness requirement was 1.0MM. There are a very limited number of companies that can roll this width of sheet and at the moment this is the limiting factor with respect to sheet sizes. This principle of "waffle and skin" structure has been applied to various types of pressings for several turbo prop aircraft, all demanding precision forming to 0.5MM of profile tolerance. On recent products the sectional thickness is down to 0.5MM thick to maximise weight reduction.

SPF technology has also been applied to helicopter leading edges 3.5M long produced from a starting thickness of 0.8MM. Fig 5. These are complex critical components with profile tolerance limits of ± 0.25MM along the full length.

As aero engines have increased in size so has the requirement for the product sitting behind the engine intake lipskin, namely the anti-icing bulkhead. This is a structural member, normally made from Ti 6Al 4V, required to withstand the anti-icing temperature. With engines having intake diameters in excess of 10FT these bulkhead pressings require sheet sizes of 2.5M long by 1.5M wide. Fig 7. These products are required to provide a thermal barrier, structural integrity and contribute to control the basic engine intake shape.

In all cases where the SPF process has been used, tensile testing and hydrogen analysis are carried out to qualify the product to the customer's specification. Tensile testing results taken from production products are shown in Fig 8 and hydrogen analysis results are shown in Fig 9. Bend testing is also performed to detect lack of ductility, in particular the presence of a residual brittle oxide layer.

In conclusion not only has the SPF process evolved to enable the manufacture of large complex structures, but it has created its own resource industry.

When Pearson started his investigations into SPF I cannot imagine that he conceived the technology being applied to this extent but all of us I am sure are grateful for the breakthrough in the technology and designers today now have the ability and are using the best of this process to make lighter, complex and more cost effective products.

Fig 1   Production Batch

Fig 2   Conventional Fabrication

Fig 3   Alternative Superplastically Formed Pressing

Fig 4   2.5 Ton SPF Tool Coated with Boron Nitrade

Fig 5   Critical SPF Product

Superplasticity: 60 Years after Pearson 357

Fig 6   Large Superplastically Formed Pressings from 2.0M x 1.5M Sheet Stock

Fig 7   Aero Engine Bulkhead for 10FT I

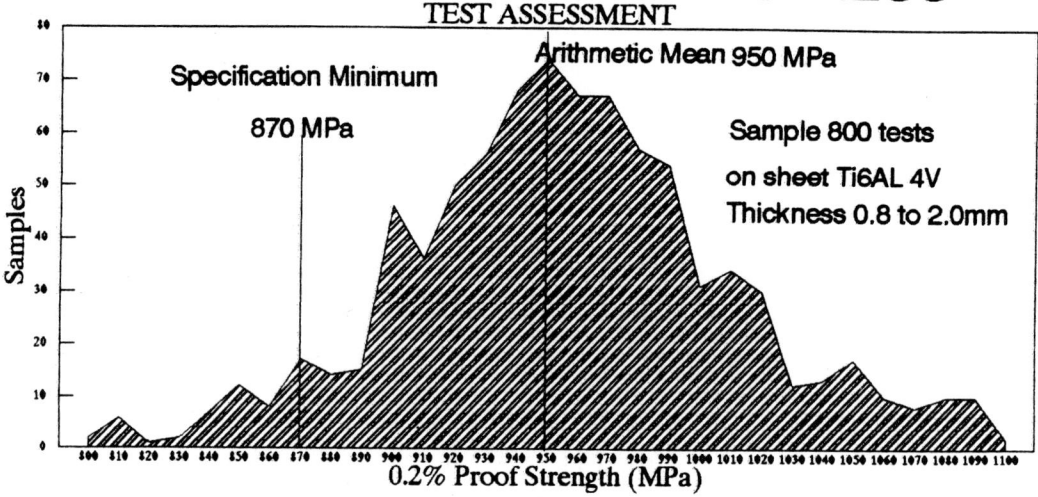

Fig 8   Post SPF Tensile Results

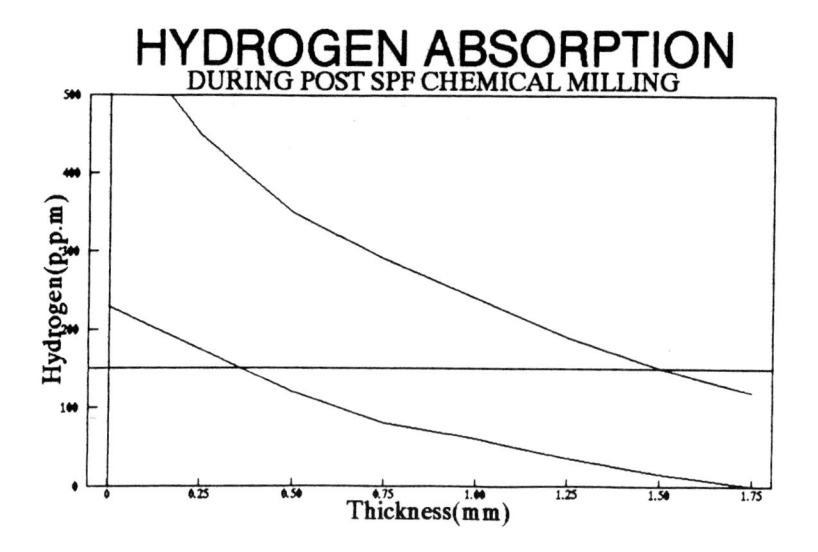

Fig 9   Post SPF & Chemical Milling Hydrogen Analysis Results

Application of SPF/DB Titanium Technology to Large Commercial Aircraft
Mr. C. F. Dressel
Senior Development Engineer
British Aerospace Airbus Ltd

Abstract

British Aerospace Airbus Ltd (and its predecessors) has been at the forefront of the development for application of superplastic forming and diffusion bonding (SPF/DB) of Titanium for some twenty years. This has resulted in a substantial portfolio of patents covering various structural forms including contoured multiple sheet cellular structures, now widely acknowledged as offering important design flexibility and efficiency.

The current paper describes the application of these technologies to large commercial aircraft, including a review of significant technology demonstrators and production experience to date, including development of an Airbus A320 spoiler. Much of the work done to establish the viability of this component has addressed issues such as quality control, process definition, and economics which is described here. This also serves to illustrate the rationale for embarking upon the development of larger structures and forms an essential link in BAe Airbus' continuing development strategy.

A vision for widespread application of superplastic forming (SPF) and diffusion bonding (DB) to future large commercial aircraft is presented. Realisation of this vision requires major extension of design and manufacturing capabilities, implying very significant investment, and therefore requires thorough justification. This is addressed in the later part of the paper.

1.0 Introduction

BAe Airbus began its involvement with SPF in the late 1960's related to the development of Concorde. The 1970's resulted in a diverse range of developments which covered, basic process investigations and material properties, simple structures (e.g. hemisphere, fluted struts, bellows for the CERN project) and developments exploring the ability of SPF to produce complex shapes (e.g. corrugated frame section, hat stiffened panels and warren girder type structures) This eventually led to the development of two sheet cellular structures and the first SPF/DB components.

Details of the production components and subsequent developments of SPF/DB by BAe Airbus are detailed in the following sections with cost and mass factors given in table 1.

2.0 Overview

The chronology of the developments described in this paper is presented in fig.1. It illustrates the sequential development programme from 2-sheet cellular technology to the current state-of-the-art, light weight, 4-sheet cellular structures.

Four sheet technology was developed through a Technology Demonstrator Programme based upon the 125 Business Jet Escape Hatch and Aileron, partially supported by the DTI. The technology was subsequently commercially exploited through its application to the Harrier AV8B Avionics Bay Door and licensing to IEP Structures Ltd.

This prolonged period of development has resulted in BAe Airbus owning significant intellectual property rights, the most substantial of which is related to material contouring to achieve balanced cell forming through the control of initial material thickness.

Applications such as the Escape Hatch, Aileron, Spoiler and Harrier Door have proven the applicability of SPF/DB to pressure fatigue, static strength, stiffness, weight efficiency and acoustic fatigue designed applications, using single and double curvatures, multi-stage forming, non-planner blanks and deep forming areas. Combined with the fact that BAe Airbus have considerable experience in the manufacture of UWAP and Jack Cans, with in excess of 17,000 flying components today, this work demonstrates a wide potential for SPF/DB Titanium technology for airframe manufacture.

In conjunction with the structural and component developments outlined we have also undertaken development in many other areas including; metallic tooling design, ceramic tooling, press concepts, SPF/DB process modelling (in conjunction with Swansea University), and material characterisation. The later has resulted in all SPF material being quality checked as standard, although it has been possible to relax or remove this requirement for very thin gauge material.

The application of new technologies to civil aircraft require long development times which has limited the exploitation of SPF/DB to date. However, it is clear that SPF/DB offers significant benefits due to:

> Simple flat starting blanks manufactured via a common route.
> High material utilisation.
> Reduced parts count.
> Multiple forming.
> Concurrent SPF/DB.
> Efficient structures and complex profiles.

3.0 Components & Developments

3.1 Jack Can (Fig.2)

Jack Cans were the first commercial SPF item, manufactured for the Airbus A310. Comprising part of the fuel tank boundary of the wing, the Jack Can is stability designed to resist implosive loads created by a crash fuel surge pressure.
A simple two sheet SPF part, it replaced a conventional warm formed fabrication. SPF has enabled a much better definition of the stiffeners resulting in a considerable improvement in the stability required by the crash case. Stability is assisted by a mechanically bonded flange, although the design relies purely on the peripheral TIG weld for integrity along the flange. This component was accompanied by the introduction of sheet material SPF quality checks in order to ensure the correct thickness distribution for the part.

3.2 Under Wing Access Panels (Figs.3 & 4)

The successful introduction of the A310 Jack Can was followed by the first 2-sheet SPF/DB cellular structure, the Airbus A310 Under Wing Access Panel. Introduced as a retro-fit component to replace a machined light alloy plate on the basis of weight saving for cost parity. It has subsequently become the design standard for all Airbus aircraft. There are 14 panels on the smaller A320 ranging up to 42 on the larger A330 & A340 aircraft. BAe now have considerable experience in the manufacture of Under Wing Access Panels and Jack Cans with in excess of 17,000 flying components today.

The forming process requires high deformation (~300%) with areas of local thickening to distribute loads from the attachment points. The very light weight hollow structure is foam filled to prevent impingement of molten metal on the inner fuel bearing surface during lightening strike. Additionaly, many parts need to be armour plated to resist tyre debris damage.

Diffusion bonding of the webs and armour plate takes place within an Argon purged tool cavity. Due to concerns regarding the quality of diffusion bonds achieved in non sealed environments, the component was designed to meet its design criteria without relying on bonding between the web walls. This was achieved by designing the inner fuel bearing surface with convolutions which direct the fuel load to compress the cell walls together. This overcame the concerns at the time and was proved by inhibiting the bonding on test parts. Armour plates are attached by forming the panel down onto a thin plate placed in the bottom of the tool. During the forming cycle the plate is adequately bonded to the panel for impact requirements.

3.3 Escape Hatch - 125 (Fig. 5)

This is a very significant development since it was the first major 4-sheet component produced. The technologies developed here, e.g. chemical etching to balance cell forming, have been applied to all subsequent structures.

Designed as a pressure cycle fatigue, single curvature part for mass parity and to be fully interchangeable with the conventional hatch; it employs complex chemical etching to balance the difficult cellular structure created by the non-symmetric bi-directional shear web pattern. This component achieves a large parts count reduction; from 90parts + 1100 fasteners to 25 parts + 230 fasteners.

3.4 Aileron - 125 (Fig.6)

Developed together with the Escape Hatch the Aileron comprises of a four sheet SPF/DB cellular construction made in three sections due to the size constraints of capital equipment. It has demonstrated the applicability of technology to such applications and required a high standard profile accuracy and repeatability, with many location features for fittings being incorporated into the tool.

Designed to be fully interchangeable with the conventional component, it is a mass balanced aileron control surface with bending, torsional stiffness and fatigue life equivalent to a light alloy version. Partial structural optimisation has resulted in a multiple spar design.

3.5 Harrier Avionics Bay Door (Fig.7)

A direct commercial exploitation of the 4-sheet SPF/DB technology developed on the 125 escape Hatch and Aileron it required the additional development of forming double curvature parts. This component is an important milestone in our development of SPF/DB, since it is our first and so far only production aircraft application of four sheet SPF/DB technology.

Designed for acoustic fatigue in a hot environment and for mass parity with the conventional fabrication, this component has provided a very large extension of the part life with significant cost savings. Parts count has reduced from 42 parts and 482 rivets to 4 sheets bonded to make 1 part.

The whole concept of 4-sheet SPF/DB cellular structures as developed by BAe Airbus is dependant upon using the core to skin bonding to prove the internal web structure. If manufacturing procedures are correctly followed to ensure the quality of the blank preparation and forming, this is directly representative of the bonding regime for the internal web structure. This has been accepted by the BAe & McDonnell Douglas authorities specifically for this part.

Part of the inner surface of the panel is removed to reduce weight, providing the added assurance of access to the internal structure of the component. However, the quality acceptance of the part is primarily based upon the above relation of skin to core bonding using ultrasonic C-scans.

Standard methods and tests used to prove the quality of all four sheet parts are:

1. X-ray to verify the internal structure of parts.
2. Ultrasonic C-scan to assess the degree of bonding of skins to core and consequently the degree of bonding in the webs.
3. Destructive tests on extractions from component to assess diffusion bonding, hydrogen content, alpha contamination.
4. Process control which is essential during manufacture in order to be able to reduce the levels of inspection to a practical minimum.

3.6 A320 Spoiler (Fig.8)

The A320 Spoiler represents the current state of the art of SPF/DB for BAe Airbus. It is a 4-sheet SPF/DB cellular component using extensive and complex etching (5 thickness) to achieve a light weight slender aerofoil section, bending and torsional stiffness designed for minimal mass in order to compete with the current carbon fibre component. It is a clear indication that significant cost benefits can be realised from SPF/DB, for some current carbon fibre applications, where weight parity is required.

3.7 Flap Track Beam (Fig.9)

This is a high load intensity structure with tough requirements relating to track loading, bending, twisting and the accommodation of shear loads and attachments.

Developed only as a series of test sections, it represents an entirely different concept of mixing thick and thin material sections for the SPF/DB process. These kinds of concepts are likely to be necessary for the extension of the technology for future products.

Such applications have raised higher levels of concern regarding the quality of the final product and have necessitated a change in the manner in which SPF/DB development is undertaken, ensuring that bond quality and the ability to apply current NDE techniques are built into the concepts from the outset. This will remain as a major emphasis of our future work.

Such test pieces have also been used to prove different forming concepts which cover the forming of flat blanks which incorporate thick material sections and part forming sections before incorporating preformed thick material sections prior to final SPF/DB.

4.0 Issues For The Future

4.1 Technical Application Of The Technology

The majority of recent SPF/DB technology development by BAe Airbus has focused upon the A320. Extending this technology to the larger aircraft in the product range requires further development to accommodate the physical sizes and the more demanding performance criteria. In considering the type of development necessary to extend the claimed benefits of SPF/DB to these large aircraft and the continued long term development of SPF/DB beyond this, it is possible to conceive a very large potential for the technology in main aircraft structures (Fig.10).

Realisation of SPF/DB is not yet possible on this scale but our continued development is very much in this direction. However, future development and applications will however be driven by prime order winning criteria i.e. first time cost and subsequently by other tangible benefits which affect cost of aircraft operation rather than as an unfocused extension of the technology. Our current work should shortly give us a clearer indication of the economic/technical benefits of this course of development.

Future realisation of SPF/DB will be dependant upon two factors:

1. Establishing a process capability for SPF/DB, that is accepted as a basis for design, making SPF/DB a 'mature' understood design/manufacturing option, with long term cost effectiveness against other competing technologies.
2. Achieving a concurrent working discipline in order to accommodate the compromises and synergy of the many elements which must be considered in the design of any component e.g. structural performance, reparability, inspection, formability, cost etc. This is particularly important for SPF/DB since all these aspects are achieved via a single manufacturing operation.

Extension of the technology to just a limited number of the potential areas identified (Fig.10), will require a major change in quality assurance; requiring extensive use of destructive and non-destructive testing during development, with carefully planned and monitored product introduction in order to establish a set of statistically based process parameters. Strict process control and comprehensive calibration of the whole component against critical inspection areas or specific self-proving features will allow minimal destructive and non-destructive inspection.

The quality techniques envisaged to prove such structures are not viewed as being particularly different from those currently employed. It is also not envisaged at this time that it would be possible to move away from the destructive examination of specimens taken from the body of the component. Inspection levels for the production of these parts could be the main factor determining whether the potential of the technology is sufficient to provide a satisfactory return on investment.

4.1 Business Application Of The Technology

The most significant issues that face this technology in the future are not technical but are concerned with the decision required to fully exploit it. Technology will never be enough in itself to satisfy the needs of industry unless it can be commercially exploited. Whether this is possible is dependant upon the drive determination and innovation of those concerned.

Based upon the benefits of the current range of production and development components it has been possible to make some scaling of the benefits to be expected from applying SPF/DB to future aircraft. This has in part been the basis for identifying the applications in figure 10.

Prior to initiation of a full scale demonstrator programme to prove the first of these developments, we are seeking to establish a cost effective option for the design so that we are confident that the objectives are both technically and economically feasible.

This aspect needs to be considered in the future development of all SPF and SPF/DB technology since in modern manufacturing industry the design activity is probably the biggest determinant of product costs and the benefits of any new technology development can be severely limited by manufacturing logistics. These in turn are primarily dependant upon product mix, manufacturing process variability, type and level of capital investment. One of the most significant developments in recent years affecting the cost assessment of technologies has been life cycle costing. Its importance has grown in many industries and will become particularly relevant to aerospace particularly with respect to energy and environmental issues.

The uptake of any new technology is significantly influenced by the cost of entry into the market place (or even the cost of exiting the old market). Additionally, all businesses have the need to make strategic technology and business decisions. It is essential to have the correct technology with a sufficient level of development to meet the business plan if the technology is to progress and this requires clear strategies for development and exploitation. It is the timeliness of the decision making process that most affects the success of any particular technology.

5.0 Acknowledgements

In thanking British Aerospace for permission to publish this paper the author emphasises that, although it is based on his professional work on the company's behalf, any views expressed are his own and do not necessarily represent those of The Company.

5.0 Tables & Figures

Table 1   Cost and Weight factors
Fig. 1   Chronology of SPF/DB Development
Fig. 2   A310 Jack Can
Fig. 3   Under Wing Access Panel - Inner fuel Bearing Surface
Fig. 4   Under Wing Access Panel - Outer Flying Surface
Fig. 5   Escape Hatch - 125
Fig. 6   Aileron (centre)
Fig. 7   AV8B Harrier Avionics Bay Door
Fig. 8   Spoiler - A320
Fig. 9   Flap Track Beam - Test Section
Fig. 10   Schematic Showing Potential SPF and SPF/DB Application to Commercial Aircraft

Table 1. Cost and Weight factors

Part	Conventional component Material	Cost Factor		Weight Factor	
		Conventional	SPF/DB	Conventional	SPF/DB
A310 Jack Can	Ti	1.00	0.66	1.00	0.65
UWAP	Al	1.00	1.00	1.00	0.60
Escape Hatch	Al	1.00	0.70	1.00	1.00
Aileron	Al	1.00	1.00	1.00	0.82
Spoiler	CFC	1.00	0.78	1.00	1.00
Harrier Bay Door	Ti/Al	1.00	0.65	1.00	0.78

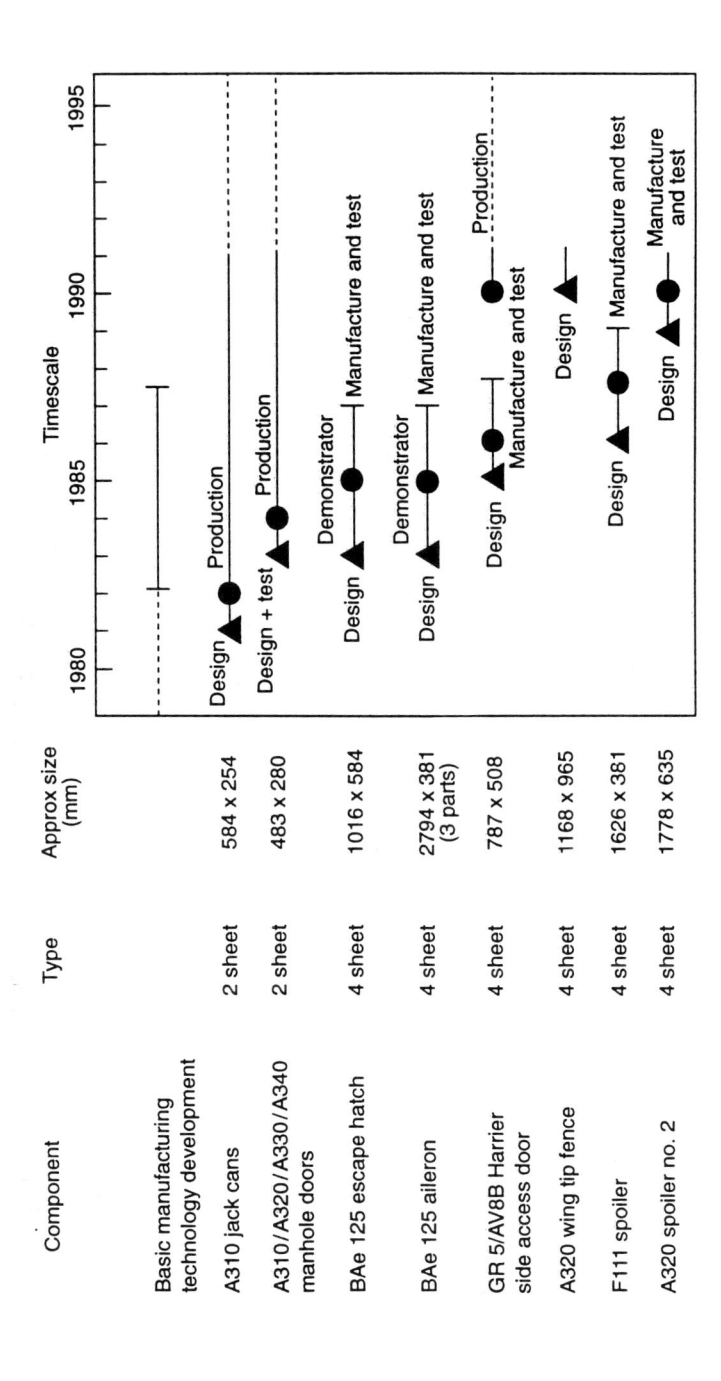

Component	Type	Approx size (mm)
Basic manufacturing technology development		
A310 jack cans	2 sheet	584 × 254
A310/A320/A330/A340 manhole doors	2 sheet	483 × 280
BAe 125 escape hatch	4 sheet	1016 × 584
BAe 125 aileron	4 sheet	2794 × 381 (3 parts)
GR 5/AV8B Harrier side access door	4 sheet	787 × 508
A320 wing tip fence	4 sheet	1168 × 965
F111 spoiler	4 sheet	1626 × 381
A320 spoiler no. 2	4 sheet	1778 × 635

Fig. 1 Chronology of SPF/DB Development

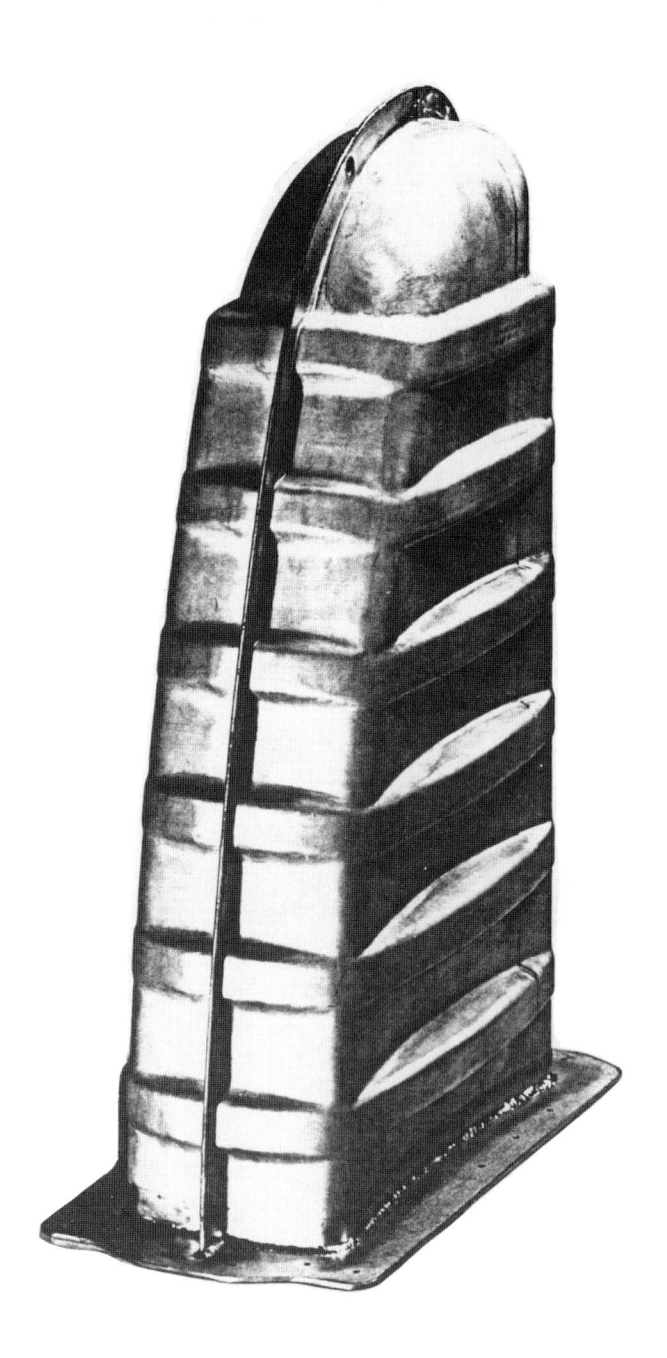

Fig. 2   A310 Jack Can

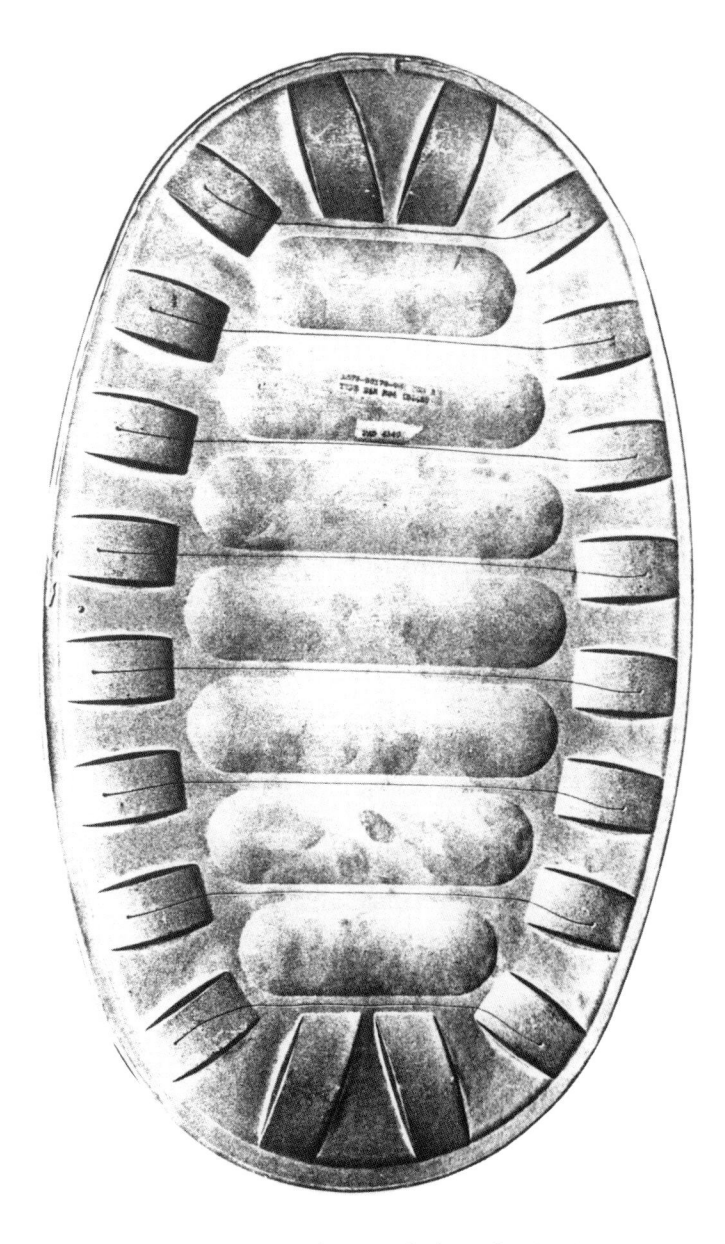

Fig. 3   Under Wing Access Panel - Inner fuel Bearing Surface

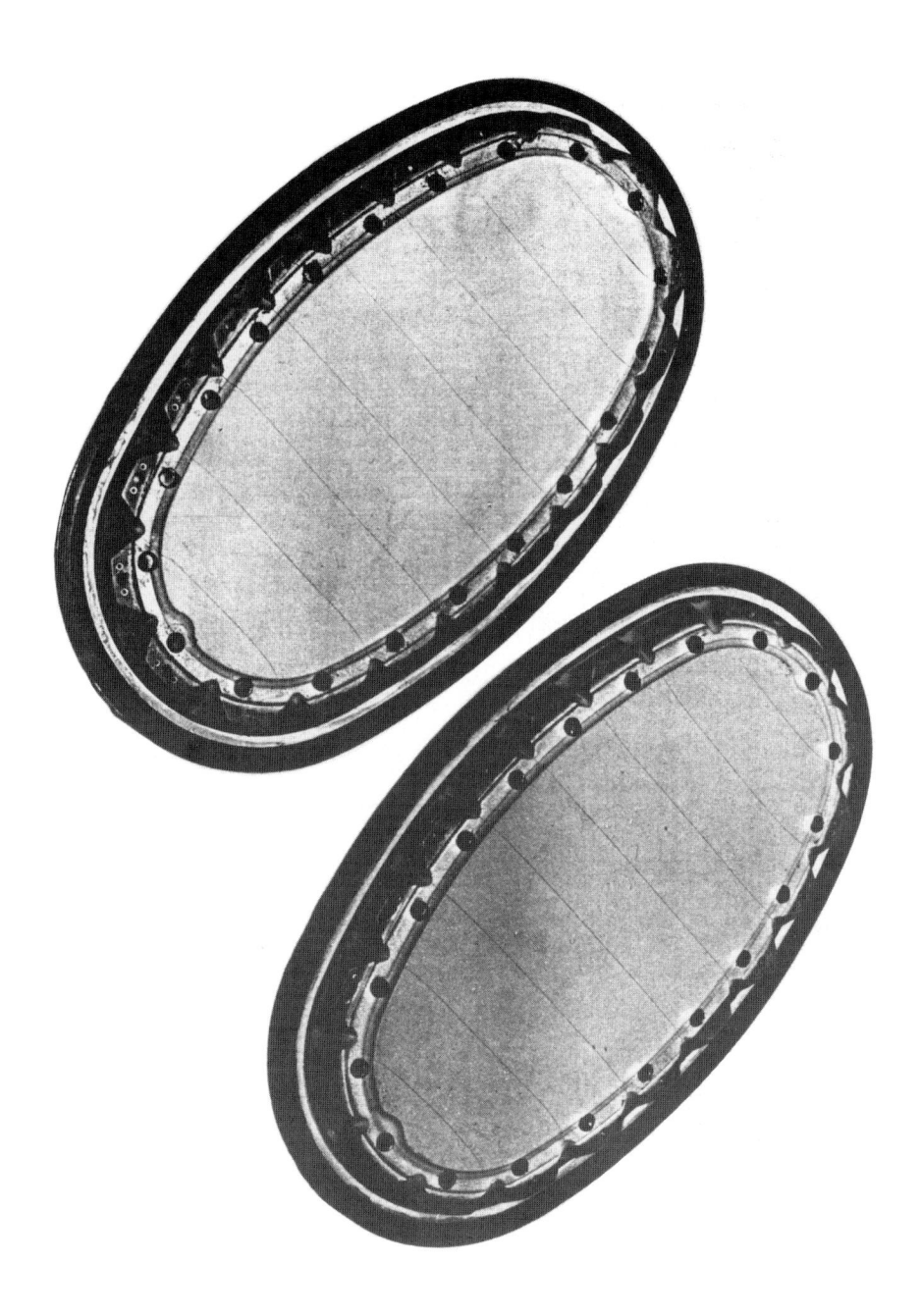

Fig. 4   Under Wing Access Panel - Outer Flying Surface

Superplasticity: 60 Years after Pearson 371

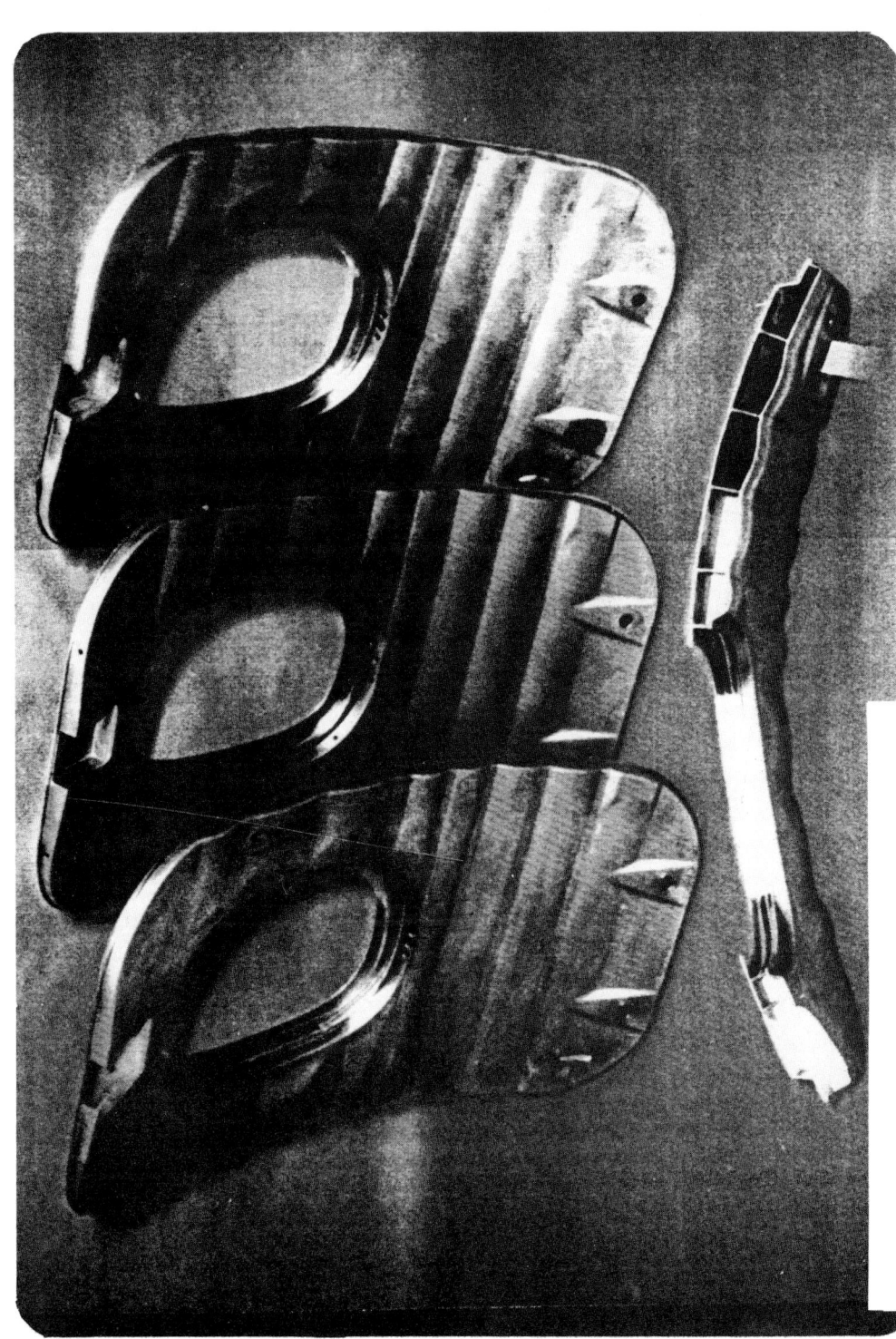

Fig. 5   Escape Hatch - 125

Fig. 6 Aileron (centre)

Fig. 7   AV8B Harrier Avionics Bay Door

Fig. 8   Spoiler - A320

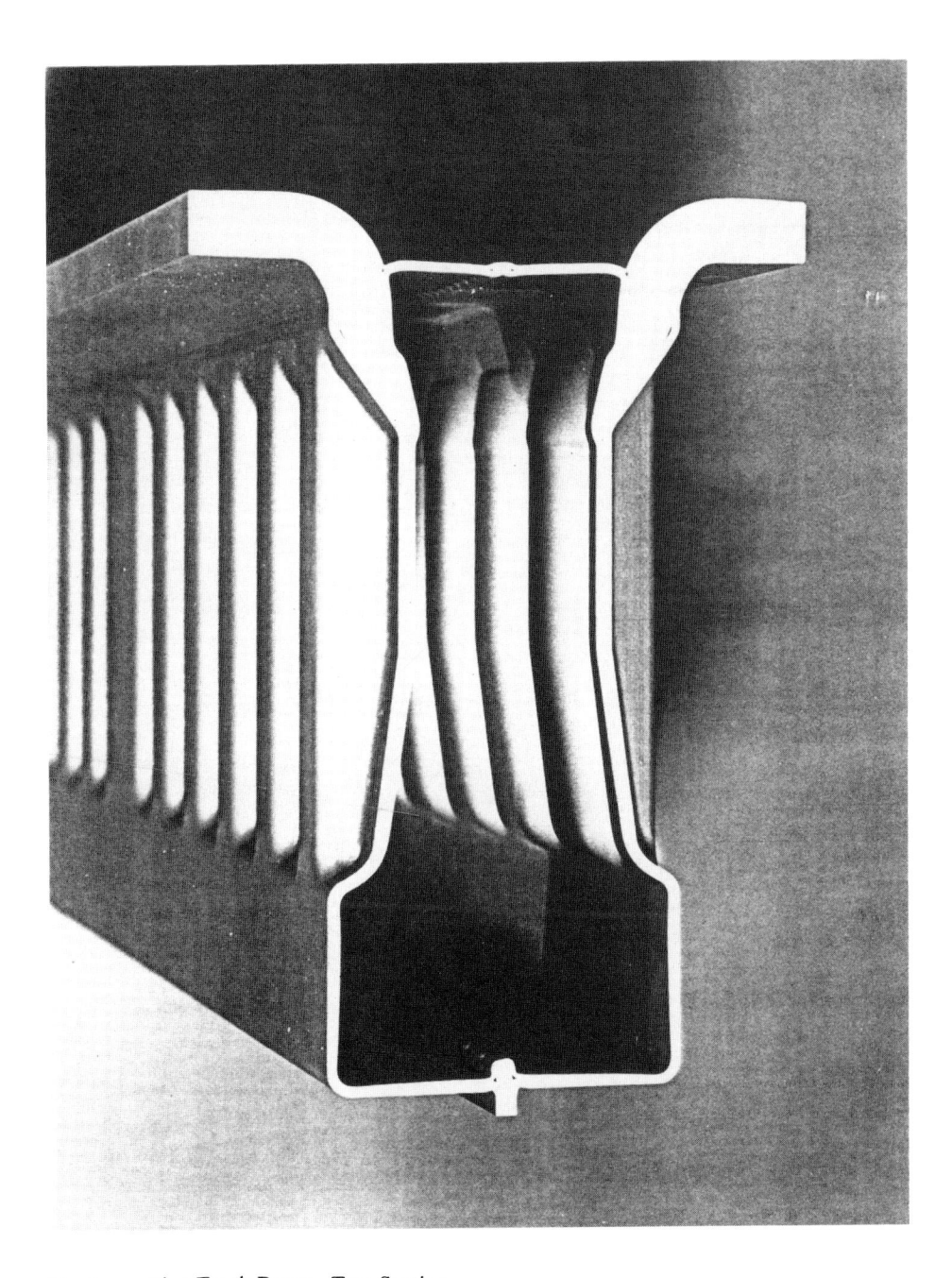

Fig. 9   Flap Track Beam - Test Section

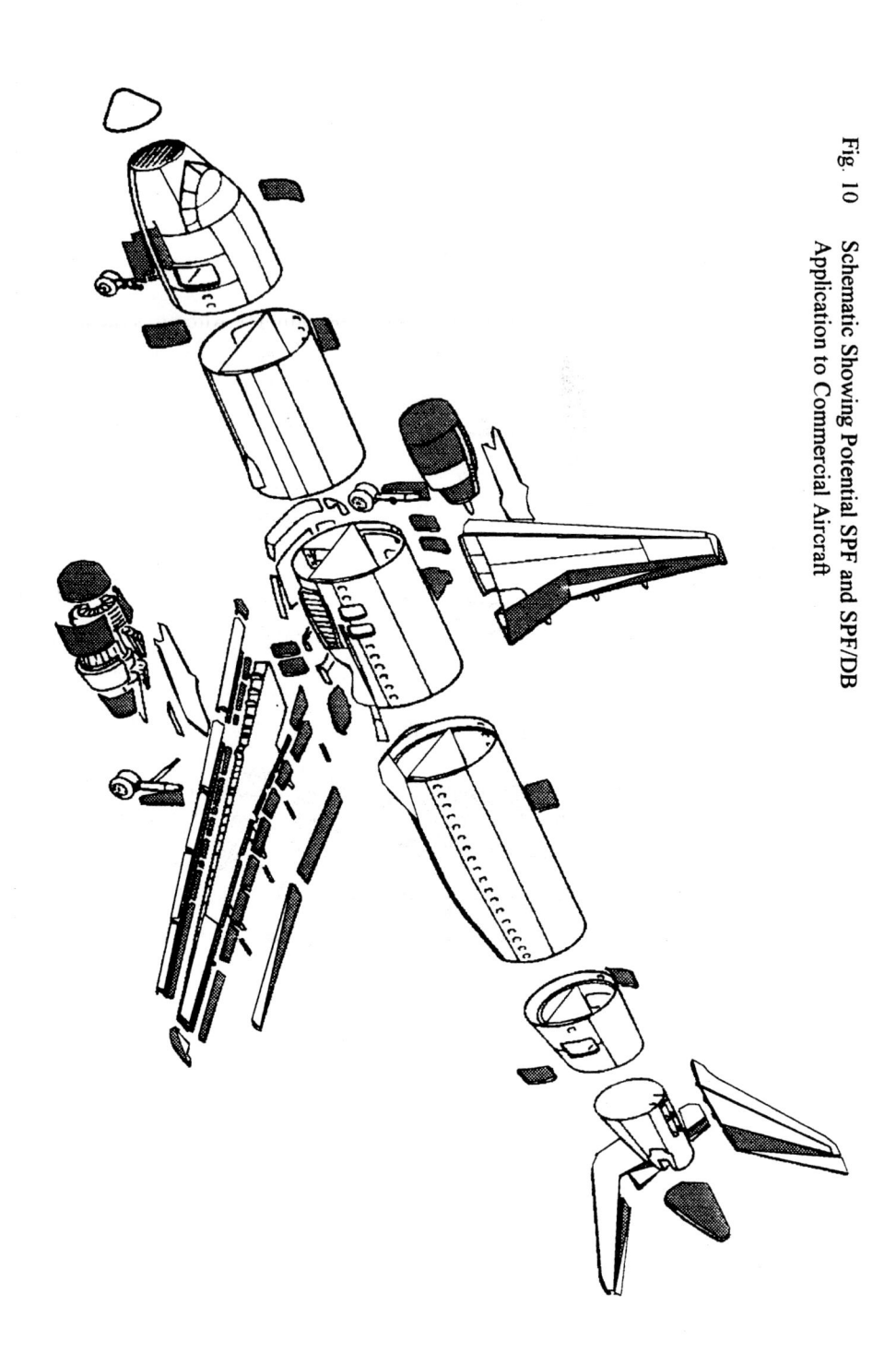

Fig. 10 Schematic Showing Potential SPF and SPF/DB Application to Commercial Aircraft

Industrialisation of SPF within BAe Military Aircraft

A.D. Collier, N. Jackson
Manufacturing Technology - British Aerospace (Defence) Ltd.

1.0 Introduction

British Aerospace first established the basic principles of Diffusion Bonding (DB) and Superplastic Forming (SPF) of Titanium at it's Filton site in the early 1970's. The benefits of applying the combined DB/SPF process to Military Aircraft structures were highlighted following a study in the early 1980's. On the recommendations of this study, a facility was specified and installed at Samlesbury in 1981. Following a substantial 10 year production/development phase, further plant has been installed within the facility to extend M.A.D.'s DB/SPF manufacturing capability.

This paper discusses the facilities currently in use for the production of DB/SPF components and possible further optimisation of the manufacturing cell.

2.0 Background

Figure 1 shows the range of development work carried out at Samlesbury over the past 20 years. As can be seen, a great deal of work has gone into the introduction of SPF and DB/SPF as production processes and to qualify their use for aircraft. This work included :-

(i) Basic manufacturing process development. Development of process parameters, tooling development, stop off development and Aluminium alloy forming development.

(ii) Material properties data base. Post formed properties, primary DB properties, effect of defects on DB joints and the effect of process parameter variables.

(iii) Structural elements programme. Manufacture and test of a wide range of panels with differing core configurations and geometry's to verify static and fatigue loading results and provide data concerning geometry tolerances, thinning and gas management techniques. These included 2 sheet stiffened panels, 3 sheet Truss cores, 4 sheet X cores and 4 sheet accordion cores.

(iv) Quality control. The generation of a data base concerning NDT techniques, allowable defects and quality control procedures.

3.0 Problems Associated with Industrialisation

Industrialisation of any novel process can be considered as the conversion of a laboratory technique to large scale component production. There are of course problems with the industrialisation of any process. Problems which were considered when Military Aircraft industrialised SPF and DB/SPF included : -

(i) Component Design. Structures must be designed for the process. It is unacceptable to attempt to design and manufacture components with no knowledge of the process benefits and limitations. The component material must also be taken into account, with care being taken to design out unwanted features, such as 'lock ons', deep forms where material may grip onto the tool surface as it cools.

(ii) Component Distortion. A database must be formulated of the acceptable levels of distortion. As the component size increases, so does the level of distortion. This is due to the removal technique applied to large components and can be overcome to some extent by the use of lifting frames and cool down fixtures.

(iii) Component Quality. A number of factors can affect component quality such as no-bonds, Alpha case contamination, Hydrogen contamination etc. These factors can be minimised by design and careful control of the process, which is more difficult in the production environment than the laboratory.

(iv) Tool Design. Consideration must be given to not only the tool design, but the type of tooling utilised on large scale components. Handling equipment must be used, as should tool load/unload fixtures, and as mentioned before, the use of cool down fixtures should be employed.

4.0 Facilities

Figure 2 shows the 600 tonne John Shaw press, installed in 1981. This press has a working size of 2000 mm by 1000 mm and is supported by a 900 °C tool pre-heat oven and tool loading equipment.

Figure 3 shows the 3000 tonne Chester Hydraulic press, installed in 1991. The press has a working size of 2500 mm by 1800 mm and again is supported by a 900 °C tool pre heat oven with tool loading facilities. This press also benefits from component load facilities.

Both presses are fully microprocessor controlled and the area is served by a dedicated clean room with silk screen print equipment and automated screen cleaning facility. Supercleaning of the Titanium sheet is carried out by an automated etch plant.

Pre and post form profiling is carried out by an abrasive water jet cutter mounted on a Cincinnati T3 786 robot. This is shown in figure 4.

5.0 Further Optimisation

Further optimisation of the manufacturing cell is possible through the inclusion of the following features : -

(i) Automated Chemical Etching. At present, all etching and associated processes are carried out by hand e.g. applying maskant, transferral of etch pattern, cutting, stripping, thickness surveys. This series of operations can be automated utilising robotic maskant spraying, laser maskant cutting, and automatic thickness comparison. This can then determine the re-etch time. This process is shown in figure 5.

(ii) Real Time X ray. Real time X ray is more suited to the non destructive testing of metallic structures than film X ray. The results of this technique can be downloaded onto video tape and remotely checked against known standards.

(iii) Process Improvement. The quality control data currently gathered within the SPF facility should be properly correlated and S.P.C. techniques applied to fully define process capability and set correct control/design limits.

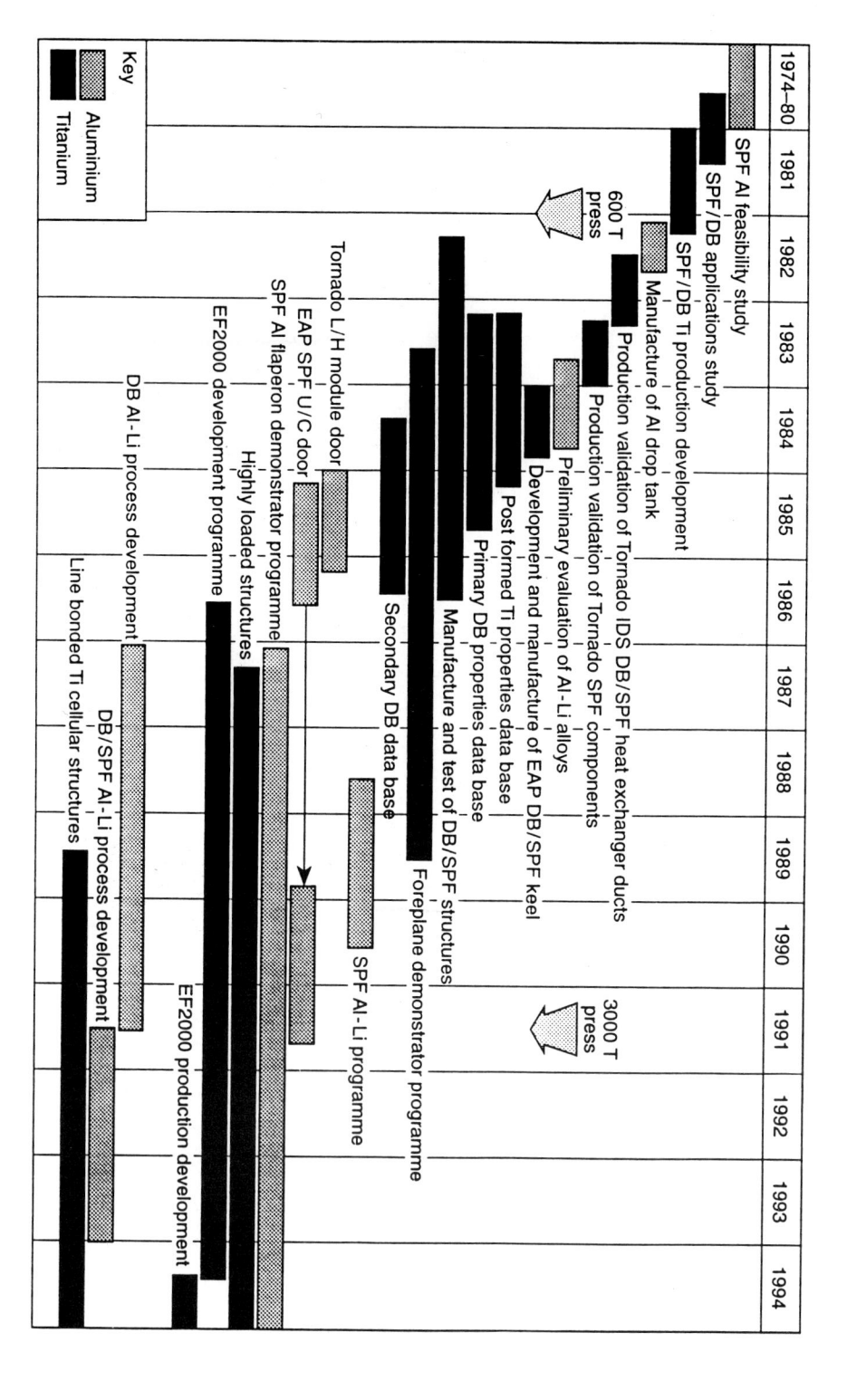

BAe MILITARY AIRCRAFT EXPERIENCE

Key: Aluminium (black), Titanium (shaded)

Timeline (1974–1994) of programmes:

- SPF Al feasibility study (1974–80)
- SPF/DB applications study (1981)
- SPF/DB Ti production development (1981–82)
- Manufacture of Al drop tank (1982)
- 600 T press (1982)
- Production validation of Tornado IDS DB/SPF heat exchanger ducts (1983–84)
- Production validation of Tornado SPF components (1984–85)
- Preliminary evaluation of Al-Li alloys (1984–85)
- Development and manufacture of EAP DB/SPF keel (1984–86)
- Post formed Ti properties data base (1984–86)
- Primary DB properties data base (1984–86)
- Tornado L/H module door (1984)
- EAP SPF U/C door (1984–85)
- SPF Al flaperon demonstrator programme (1984–86)
- EF2000 development programme (1985–91)
- Highly loaded structures (1985–87)
- DB Al-Li process development (1984–87)
- Secondary DB data base (1987–89)
- Manufacture and test of DB/SPF structures (1988–90)
- Foreplane demonstrator programme (1989–90)
- SPF Al-Li programme (1989–91)
- 3000 T press (1990)
- DB/SPF Al-Li process development (1988–93)
- Line bonded Ti cellular structures (1988–94)
- EF2000 production development (1991–94)

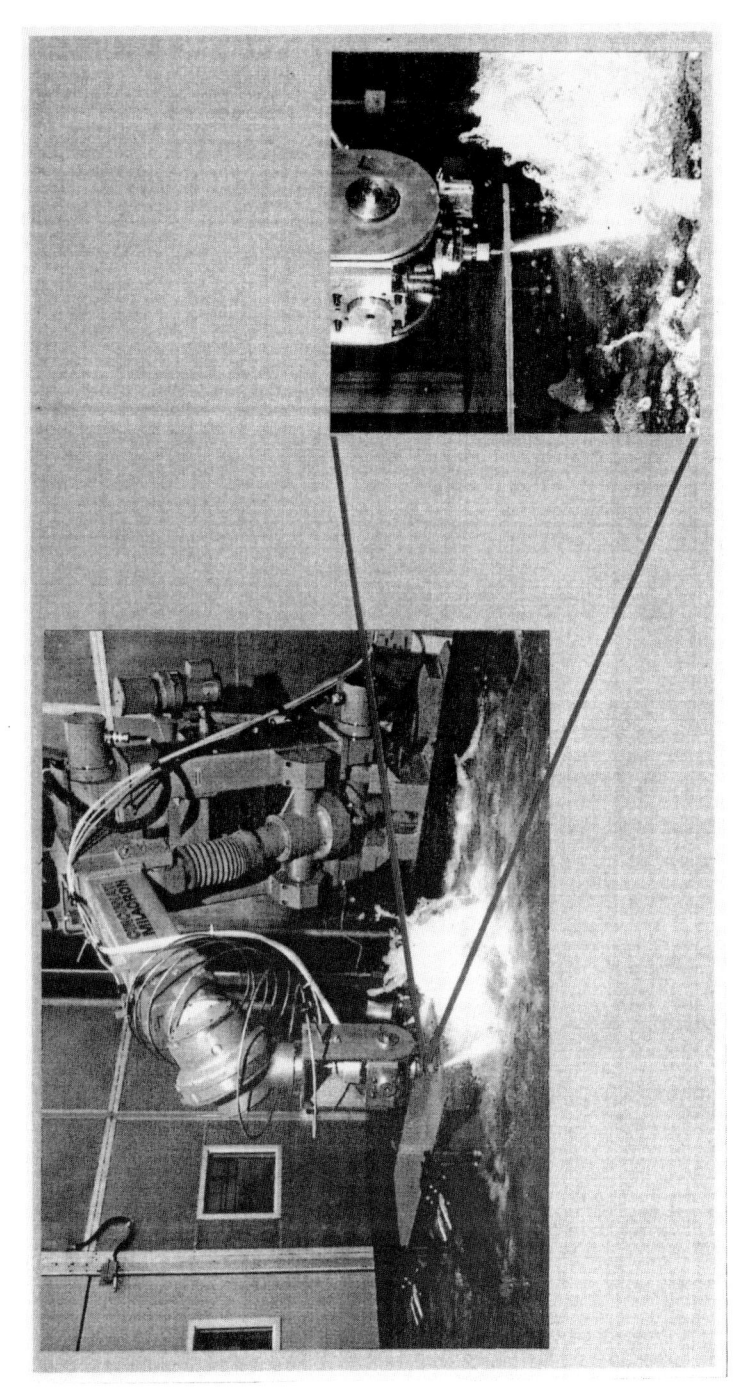

Superplasticity: 60 Years after Pearson 381

PROFILING Ti PLATE (7 MM)

3000 TONNE SPF - DB PRESS

Superplasticity: 60 Years after Pearson 383

600 TON - SPF / DB PRESS

CLOSING REMARKS

Where is Superplasticity 60 Years after Pearson?
M J Stowell
Alcan International Ltd
Banbury

These remarks are a response to the question asked by Dr Ridley in his opening account of the life and history of Professor Pearson. Before responding to his question, I feel it necessary to remind everyone that what we have seen at this conference is not the result of 60 years of continuous research and development in this field. The pioneering review by Underwood in 1962 opened the eyes of Western scientists to this fascinating field. He pointed out that, apart from some limited work on phase transformation superplasticity, little research had been done in the West on this topic since 1934 and that most interest had been shown in Russia. In fact, it was a Russian scientist, Bochvar, who coined the term "superplasticity" and it is very pleasing to me to note that his compatriots are again contributing vigorously to the field and are represented at this conference in Pearson's honour.

So to the question - where are we now? From today's presentations, it is obvious that superplasticity is now an established technology and business area which manufactures components for profit, having a range of commercially viable alloys and sophisticated machines and forming techniques available to operate with. All this has happened in the last 30 years and the field is still developing, albeit slowly.

Superplasticity is about making components and we have seen major advances in forming technology since the first simple experiments of Pearson in blowing up tubes. The first serious attempts to develop a superplastic forming technology were made nearly 30 years ago by Davis Fields at IBM; he was probably the first to adopt an engineering approach to this problem. Since then we have developed female, male and snap-back forming, introduced back-pressure forming to overcome cavitation problems and combined superplastic forming with diffusion bonding. All of these allow savings to be made in the manufacture of components of complex shape and all are designed to accommodate the low deformation rates traditionally associated with the phenomenon.

Pearson worked with what I call "useless" materials, those for which commercial application is zero or strictly limited. Many of us here, including myself, have worked on such materials - they have excellent superplastic properties, are often easily fabricated in the laboratory and allow one to explore the basics of the phenomenon. Since the late 60s we have developed an impressive collection of commercially viable materials, including steels, aluminium-, nickel-, titanium- and zinc-based alloys. A survey of these alloys reveals an interesting and highly relevant message:-
> in order to obtain relatively rapid acceptance, it is important to develop materials that are close in composition to those already proven in the market place and which have "service" properties which match those required for application.

We have heard in this conference of the impressive advances that have recently been made in two areas; (i) superplastic ceramics and (ii) high strain rate superplastics. There is significant potential for commercial benefit from these developments but, in order to obtain early benefit from them, the message above needs to be taken to heart.

Another message voiced at the conference is that, although much progress has already been made in educating engineers and designers on both the materials and forming capabilities of superplasticity, much more needs to be done in order to get maximum benefit from the technology. The solution to this problem will require the concerted efforts of both the industry and academia.

Turning now to what has been achieved in understanding superplasticity since Pearson's time, it is interesting to consider what Pearson's reactions might have been on hearing Terry Langdon confirm yesterday that grain boundary sliding is the principal mechanism of superplastic deformation. Of course, most of us have felt that this is the case for a long time but it has taken much scientific effort to prove the point and elaborate the mechanistic details; there probably are people who will continue to contest the conclusion. But, even without a thorough understanding of the mechanism of the phenomenon, we have been able successfully to design microstructures that give rise to superplasticity and to develop the processes for commercially manufacturing sheet although, as we heard this morning, we still have some way to go to get fully acceptable consistency of the starting material for superplastic forming. This illustrates that there are very relevant metallurgical problems to be solved in this field which do not hang on being able to understand completely the mechanisms but which do influence significantly the ability of businesses to be commercially successful.

We have heard that excellent progress is being made in developing mathematical models of the superplastic forming process, some of which have been industry driven; surely Pearson would never have imagined his effect being described by a set of equations ! We also heard that the modellers are not necessarily producing what industry wants; this message should not be ignored ! Progress is also being made in modelling the development of microstructure during superplastic forming.

In conclusion, we have come a very long way since Pearson's beautiful experiments of 60 years ago. Having been in this field since 1966, I see that we are now much more realistic about the commercial opportunities for this technology and are still generating new, exciting scientific challenges. The field was composed mainly of metallurgists when I entered it and it has gained enormous benefit from the injection of ideas from other disciplines. More "alloying" is needed and I urge the academics to get even closer to the industrialists to unearth further demanding problems and the superplasticity industry to make greater efforts to share their problems with academia.

Finally may I, on behalf of the attendees, thank Dr Ridley and his committee for organising an excellent conference and wish Norman a happy, long and fruitful retirement.

Subject index

A

ABAQUS finite element modelling 218, 219, 238
accommodation helpers 93, 101
accommodation process 60, 64, 100, 101, 105, 110, 126, 144
activation energy 98–100, 139, 156
aerospace applications
    aluminium alloys 274, 284
    diffusion bonding 296, 347, 359–376
    superplastic forming 218, 235, 296, 347–350, 351–358, 359–376
    titanium alloys 129, 223, 305, 324, 347, 351, 357–376
aircraft safety 347
alloy additions 161, 308
alloy steel hot platens 255
alloys
    consitutive description 207
    microstructure 51–59
    positive exponent superplasticity 93–102
alumina 70–72, 86, 121, 123, 147
aluminium
    addition to titanium alloy 161
    lattice diffusion 195
    matrix composites 75–84
    superplastic, market areas 284–295
aluminium alloys
    aerospace applications 284
    cavitation 66, 67, 75–84
    continuous nucleation 64
    diffusion bonding 296
    elongated cavities 67
    fine-grained 93, 105
    flow hardening 195, 197, 198
    grain sizes 195, 196

hydrogen outgassing 68
marine vehicle applications 284
melting point 97
microstructure 173–182, 176
nanocrystalline powders 93, 95
rail vehicle applications 284
recrystallization 193–198
rheology 193–198
roll bonding 296–303
sheet forming 273–276
stress–strain relationships 195, 196
superplastic forming 274, 296–303
superplasticity 94, 183–192, 193–198, 277–283
thermomechanical processing 173–182
types in current use 273
unrecrystallized 193
aluminium composites
    grain size 103
    silicon carbide content 106
aluminium oxide, surface layer 296
aluminium sheet
    aerospace applications 274
    rail vehicle applications 274–275
    shaped parts 296
aluminium titanate 133
aluminium–lithium alloys 183, 218, 219, 222
aluminium–magnesium alloys 277–281
aluminium–zinc eutectoid alloy 297
anelasticity 37
annealing 68, 69, 166, 197, 338
architectural applications, aluminium alloys 284
automobile industry, superplastic forming 201
AWE 338

B

B-spline methods 201
band theory of solids 141
Bezier surfaces 222
Bingham flow law 45
bird ingestion 347, 349
boat hulls, aluminium 282–285
boron micro-alloy 151
boundary annihilation 180
boundary ledges 64
brass orientation 184
brasses, flow softening 197
British Aerospace Airbus Ltd 359
British Aerospace (Defence) Ltd 377
bulk transport applications 284, 290–291
Burgers vector 12, 34, 36, 38, 44, 52, 186, 309

C

Cauchy stress 238
cavitation
    in aluminium alloy 75–84
    in ceramics 70–72
    characteristics 75–84
    control 67–69, 235
    directionality 89
    effects of pressure and temperature 280–281
    and electrical fields 69
    intergranular 123
    in intermetallic alloys 127
    and maximum elongation 88
    in metal matrix composites 69–70
    in metals 63–69
    resistance 147
    strain induced 51, 77–80
    in superplastic aluminium composite 75–84
    in superplastic ceramics 85–90
    in superplastic materials 63–74
    in titanium alloys 305–306

cavities
    coalescence 67, 82
    directionality 85–87, 90
    effect of impurities 64
    elongation 88
    formation mechanisms 75
    nucleation and growth 56–57, 63–67, 71, 82, 88, 109
    volume fraction 64, 66, 70, 76, 77, 85
ceramic alloys 135, 136
ceramic hot platens 255
ceramics
    cavitation 70–72, 85–90
    fabrication processes 121, 128–129
    gas-pressure forming 129
    grain boundary structure 121, 125, 133
    lattice defects 144
    mechanical properties 122
    'metallic' electronic behaviour 139–150
    microstructure 125
    shaped parts 121, 129, 130, 133
    stress–strain relationship 124, 139
    superplastic forming 128–129
    superplasticity 13–15, 121–132, 133–138, 139–150, 144
    tensile ductility 123
    ternary 136
ceria stabilised zirconia 143
chemical etching, automated 379
chemical milling 352
Chester Hydraulics 253, 378
cobalt additions 151–158, 161
Coble creep 36, 37, 39, 46, 126
cold working 177–179
commercial practice 68, 72, 93, 103, 274, 305, 387
composite boundary 60
composites *see* matrix composites

compression testing 133, 152, 153–154, 260, 263, 264
compressive processes 178–179, 260
computational costs
    finite element modelling 220, 222, 223, 224
    genetic algorithms 211
computer controlled press 257, 378
constitutive equations, for titanium alloys 237
contact algorithms 201, 202, 222
continuous recrystallization 180, 181, 183
copper, in aluminium alloys 173, 180, 185
copper alloys, cavities 66, 67
cracking 66, 82, 123, 139, 143
creep tests 144
cross-head measurements 133, 167, 211, 261
crystalline phase, metastable 113
crystallographic techniques 325

D

defect structure 53
defects, prediction in finite element modelling 218, 220–222
deformation gradient 209
deformation mechanisms 30–31, 35, 93, 100
deformation time 55
design, industrial 304, 348, 364, 378
diffractometry, in textural analysis 164
diffusion, and superplastic flow 122
diffusion bonding
    aerospace applications 347, 359–376
    of aluminium alloys 296
    equipment 253
    numerical methods 218–234
    and superplastic forming 296, 304
    of titanium alloys 220–221, 296, 305, 324, 327, 348, 359–377
diffusion creep *see* Coble creep
diffusion growth model, for cavitation 66–67

diffusive accommodation 64
discretization 208–209
dislocation creep 60, 167, 168
dislocation loops 142–143
dislocation motion 60
dislocation networks 107
dislocation slip 25
dislocations, in yttria stabilised tetragonal zirconia 143
dispersion strengthening 115
displacement loading history 235
distortion, acceptable levels 378
domes, bulge formed 85, 90
ductility 101, 102, 123, 147
duplex materials 64, 127
dynamic precipitation, and grain size 109
dynamic recrystallization 180

E

economic considerations
    superplastic ceramics and intermetallics 130
    superplastic forming 349, 351, 364
    titanium alloys 305, 309, 324
effective stress 36
elastic after-effect 37
elastic-viscoplastic model 237
electrical fields, and cavitation control 69
electron backscattered patterns (EBSP) 174, 183, 184, 306, 324, 326
electron tunnelling 141, 145
elongation
    and cavitation 88
    superplastic 94, 278–279
elongation to failure 98–99
energy dissipation method 207
engineering components, superplastic forming 235–236
equilibrium equations 209

eutectoid alloys
    aluminium–zinc 297
    tin–lead 37
    titanium 307

F

fatigue stress 347–350
finite element modelling
    ABAQUS 218, 219, 238
    applications 215–216
    computational costs 220, 222, 223, 224
    defect prediction 218, 220–222
    of strain rates 241
    of superplastic forming 201–205
    validation 238–239
fire containment 351
flow hardening, aluminium alloys 195, 197, 198
flow softening, brasses 197
flow stress
    ceramic alloys 136
    and elongation 98
    titanium 309
forming temperatures, low 308, 310, 324
forming tools 274
four-sheet technology 360, 362
fracture initiation 83
free bulging 297
furnace enclosure 254

G

gas management systems 256
gas-pressure forming 129
gas-pressure source 296, 297
gauge length strain rate 241–243
genetic algorithms 207, 211–215
Gifkins core–mantle model 139
glassy phase, grain boundary 125, 144

grain boundary
    behaviour 43–44
    cavitation site 64
    defect structure 53
    dislocations 28
    glassy phase 125, 144
    ledges 64
    liquid phase 44, 80–81, 93, 98
    migration 35, 54, 133
    misorientation 106
    separation 121
    structure 33–42, 125
    triple junction *see* triple junction
grain boundary sliding
    accomodation process 101, 144
    angle 180, 183
    cooperative 29–31, 32, 37–38
    and grain size 235
    and intragranular dislocation slip 45–46
    as mechanism of superplasticity 11, 12, 25–32, 33, 60, 93, 107, 117, 388
    rate 43
    and stress concentration 82
    in zinc bicrystals 25–29
grain growth
    kinetics 197, 235
    in mechanically alloyed materials 96
    rate 54, 185
    and strain 38, 51, 186
    in titanium alloy 166
grain refinement 105–109
grain rotation 11, 37, 186
grain size
    in aluminium alloys 195, 196
    in aluminium–magnesium alloys 278–279
    and cavity nucleation 64
    in magnesium alloys 113
    refinement 93, 102

and superplasticity 109, 117, 122, 133–138
  in titanium alloys 308
growth microcavities 51

H

hardness testing 114
high-temperature behaviour, of titanium alloys 307
high-temperature deformation, activation energy 156
high-temperature strain, and microstructure 180
HIP *see* pressure
hot extrusion 107
hot platen presses 254–255, 352
Hubbard model 141–142
hydraulic presses 253–259
hydrogen
  removal 67, 68, 352
  as stabiliser 308
hydroxyapatite 121

I

impurities, and cavity nucleation 64
incremental flow formulation 206
industrialisation, problems 377–378
insulator–metal transitions 141, 143
intergrain shear, resistance 43
intergranular deformation 43–45
intergranular fracture 100
intermetallics
  cavitation 63, 64
  stress–strain relationship 127, 128
  superplastic 121–132
  titanium aluminide 307, 310
intragranular deformation 51
intragranular dislocation 11, 43, 45–46
ionic and electric conductors 143

iron
  addition to aluminium alloys 277
  addition to aluminium–magnesium alloys 279
  addition to nickel alloys 151–158
  addition to titanium alloys 161
iron–iron carbide 121

J

Jaumann stress rate 238
jump tests 134

K

Kikuchi patterns 174
kinetics
  of cavity nucleation 72
  of defect accumulation 51, 52
  of grain growth 54–57, 197, 235
  of lattice dislocation 44
Kirchhoff stress 238

L

lattice diffusion 195
lattice dislocations 27, 28, 31, 33, 36, 43–45, 47, 51, 98, 144
liquid phase, grain boundary 44, 100, 103–105, 110
lithium 68, 183, 185
loading transients 36
low-temperature performance 353

M

macrostructure development, in superplasticity 183–192
magnesium 97, 173, 180
magnesium alloys 66, 113–117
magnesium aluminate spinel 144
manufacturing processes 206, 235, 352, 360, 377

marine vehicle applications, aluminium alloys 284
martensite structure 151, 152
material parameter evaluation 206–217
mathematical modelling *see* numerical methods
matrix composites
    aluminium 75–84
    cavitation 69–70
    microstructural design 103–112
    positive exponent superplasticity 93–102
    superplastic properties 100
matrix–reinforcement interface 83
mechanical alloys 93, 96, 97, 100, 103, 307
medical applications 284, 288
melting, partial 111
melting point
    of aluminium alloys 97
    incipient 100
metal matrix composites *see* matrix composites
'metallic' nature, of ceramics 88, 139–150
metals
    cavitation 63–69
    superplasticity 10–13
microcracks 56, 64
microduplex materials 63, 64, 194
microstructure
    of aluminium alloys 173–182
    of ceramics 125
    control 67–69
    evolution mechanism 180
    of extruded materials 107
    of matrix composites 103–112
    of mechanical alloys 96
    of nickel alloys 153–154
    of superplastic alloys and ceramics 51–59
    and superplasticity 183–192, 235
    of titanium alloy 161–172, 306
misorientations 106, 173, 185

modelling *see* numerical methods
multiaxial constitutive equations 235

N

Nabarro–Herring creep 126
nanophase materials 66, 69–70, 93, 95, 96, 100
necking 75, 123, 155, 235, 325
Neopral 194
neutron diffraction analysis 114
Newton–Raphson procedure 208, 211, 215
nickel additions 154, 161
nickel alloys
    microstructures 153–154
    stress-strain relationships 153–154, 157
    superplastic behaviour 151–158
nickel-based intermetallics 125
nonuniform rational B-spline surfaces 201–205
nucleation 51, 64
nucleation rate 109
numerical methods
    diffusion bonding 218–234
    finite element modelling 201–205, 206–217
    modelling of cavity growth 66
    modelling of superplastic forming 206–217
    modelling of superplasticity 43–50, 60, 260, 340, 388
    predictive possibilities 33, 38
    specimen testing 235–250
    superplastic forming 218–234

O

optical microscopy 75, 76, 85, 133, 152, 174
orientation misfit dislocations 43–44, 51
oxides, structural 121

P

Pearson, findings 183
Pearson, C.E. 1–5, 9, 387–388

peel strength 298
Peierls stress 143
phase shape and distribution, of titanium 306
phases, mutually insoluble 133
5-phenyltetrazole 297–298
physical vapour deposition 93, 103
pinned dislocations 107
plastic deformation, non-uniformity 51
plastic matrix deformation 65
platens, heating elements 255–256
platinum, and ceramic metallization 145–146
porosity, localized 67
positive exponent superplasticity 93–102
post-forming properties 281–282
powder metallurgy 93, 100, 103, 307
press equipment 254, 378
pressure
    applied 68, 72
    HIPing 69, 85
    hydrostatic 76, 89
pressure cycle 207, 222
pressure forming 348
process improvement 379
process modelling tools 304
profiled blanks 340
pseudo-brittle fracture 63, 70

Q

quality control 363, 364, 377, 378
quantitative metallography 66
quasi-single phase materials 63, 64

R

rail vehicle applications 274–275, 284, 292–294
rapid solidification processing 113–117
rare earth elements, chemistry 142
recrystallization 108, 193–198

reinforcements
    effect on ductility 123
    and recrystallization 106–108
relaxation times 44–45
rheology 43–50, 193–198
road vehicle applications 284
roll bonding/superplastic forming 296–303
Rolls Royce plc 347

S

scanning electron microscopy 75, 76, 85, 164, 184
sculptured surfaces 202
secondary grain boundary dislocations 34, 35
segregation, control 111
shaped parts
    aluminium sheet 296
    aluminium–lithium alloy 219
    ceramic 121, 129, 130, 133
    finite element modelling 215–216
    formation 85
    from sheet metal 273–276
    superplastic forming 201, 253, 359
    surface treatment 282–283
    thickness 218, 281
    titanium alloys 305, 351, 352
    titanium plate 338
    trimming 274, 378
shear texture 185
shear-induced metallization 140
silicon
    in aluminium alloys 276
    in aluminium–magnesium alloys 279
    shear-induced metallization 140
silicon carbide reinforcements 69–70, 75–76, 106, 121, 123
silicon nitride reinforcements 75–76, 95, 96
simulation *see* numerical methods

sliding processes, accommodation 110
slip, limited 186
slip-casting 85
specimen testing, numerical methods 235–250
-spodumene glass ceramics 121
stacking fault energy 185
stainless steel 67
stopping-off 296, 297
strain anisotropy 327
strain gradients 108
strain rate
   and cavitation 77–80
   dependence 55
   finite element modelling 241
   and necking 75
   and rheology 46
   and superplasticity 93, 94, 103
strain rate sensitivity 76, 98, 99, 155, 260–272
strain rate sensitivity index 33, 260
strain-assisted continuous recrystallization 183, 183–184
strain-induced grain growth 38, 54–57, 186
stress–strain relationship
   aluminium alloys 107, 195, 196
   aluminium–magnesium alloys 278
   ceramics 123, 124, 139
   intermetallics 127, 128
   magnesium alloys 116
   nickel alloys 153–154, 157
   steady-state 36
   titanium alloys 162, 164, 325
structural ceramics, experimental studies 56
structural materials, superplastic properties 100
superplastic alloys and ceramics, microstructure evolution 51–59
superplastic aluminium alloys 277–283
   new market areas 284–295
   roll bonding/superplastic forming 296–303

superplastic aluminium composite, cavitation characteristics 75–84
superplastic ceramics
   cavitation 85–90
   microstructure evolution 51–59
   potential applications 121–132
superplastic compression tests 260–272
superplastic diffusion growth 65
superplastic flow 9–24, 43–50, 63, 107, 122
superplastic forming
   aerospace applications 218, 235, 347–350, 351–358, 359–376
   of aluminium alloys 274–274
   of aluminium–lithium alloy 218, 222
   of ceramics 128–129
   and diffusion bonding 296, 304
   economic considerations 349, 351, 364
   of engineering components 235–236
   equipment 253–59
   from sheet metal 273–276
   of intermetallics 129–130
   numerical methods 201–205, 206–217, 218–234
   of shaped parts 201, 253, 359
   thermal cycle 338, 339
   of titanium alloys 220, 305–323, 338–346, 351–358, 359–376, 377
superplastic intermetallics 121–132
superplastic materials, cavitation 63–74
'superplastic partition', ceramics 139–150
superplasticity
   in aluminium alloys 183–192, 193–198
   in ceramics 13–15, 133–138, 144
   commercial importance 387
   composite model 60
   definition 9
   flow characteristics 10–11
   and grain boundary sliding 11, 12, 25–32, 33, 60, 93, 107, 117, 388

and grain boundary structure 35–42

historical aspects 1–5, 9, 25

and macrostructure 183–192

in magnesium alloys 113–117

in metals 10–13

and microstructure 43, 109, 117, 133–138, 183–192, 235

in nickel-based alloys 151–158

numerical methods 43–50, 60, 260, 340, 388

optimum 102

positive exponent 93–102

and strain rate 93, 103

in structural materials 100

and temperature 103

in titanium alloys 161–172, 235–250, 236–237, 305–306, 324–337

in yttria-stabilized zirconia 14–15, 121, 141

in zinc alloys 263

in zinc bicrystals 25–32

Supral 173, 179, 273

surface treatment 282–283

T

telecommunications applications 129, 130, 284, 286–288

temperature
    and deformation 179–180
    and microstructure 176
    and superplasticity 103

tensile ductility 121, 123

tensile elongation 9, 146

tensile properties
    of superplastic materials 93
    of titanium alloys 306, 325, 326

tensile testing 95, 114, 157, 162–163, 264–265

textural analysis, in diffractometry 164

textural changes, in titanium alloy 161–172

texture, random 186

theoretical results, agreement with experiment 47, 67

thermal cycle, in superplastic forming 338, 339

thermal stability, of mechanical alloys 97

thermomechanical treatment 103, 105–109, 173–182

thickness, of shaped parts 218, 281

thinning, local 236

tin, additions to titanium alloy 161

tin alloys, Pearson's observations 2, 9

tin–lead eutectic alloy 37

titanium alloys
    aerospace applications 223, 304, 305, 324, 347, 351, 357–376
    alloying metals 308
    annealing 166
    cavitation 63, 64, 305–306
    commercial practice 305
    constitutive description 207
    constitutive equations 237
    cost advantages 305, 309, 324
    diffusion bonding 296, 305, 324, 327, 348, 359–377
    electron back scattered patterns 306
    eutectoid alloys 307
    flow stress 309
    high temperature behaviour 307
    materials data 236
    mechanical properties 339
    microstructure 28, 161–172, 306, 308
    shaped parts 305, 351, 352
    stress–strain relationship 162, 164, 325
    superplastic forming 220, 305–323, 338–346, 351–358, 359–376, 377
    superplasticity 161–172, 235–250, 236–237, 305–306, 324–337
    tensile properties 306, 325, 326
    textural changes 161–172
    two-phase 168

titanium aluminide 129, 307, 310
titanium matrix composites 307
titanium-based intermetallics 127, 129, 307, 310
TKR International 351
tool design 352, 378
transition metal chemistry 141–147
transmission electron microscopy 82, 95, 114, 144
triple junction 37–38, 51, 56, 64, 70, 82, 83, 100, 125

U

uniaxial constitutive equations 236–237
USSR (former), work on superplasticity 9–10, 387

V

vacancy diffusion 65, 68
vacuum degassing 352
vapour quenched alloys 100
variational equations 209
Von Mises equivalent strain 208

W

weld strength 282
Wigner–Seitz cell 144

X

X-ray methods 114, 125, 152, 185, 379

Y

yttria-stabilized tetragonal zirconia (YTZP) 57
   cavitation 70–72, 85
   creep tests 144
   dislocations 143
   flow stress 133
   metallic nature 88
   reinforced 123
   superplasticity 121, 141
   transmission electron microscopy 144
   see also ceramics

Z

Zener–Hollomon parameter 139, 140
zinc
   in aluminium alloys 105, 173, 180
   bicrystals 25–32
zinc alloys, superplasticity 263
zinc–aluminium eutectoid 36
zirconium, in aluminium alloys 173, 179

Contributors

Alalykin, A.A. 193—198

Baldo, O. 218—234

Bate, P.S. 183—192

Baudelet, Bernard 60

Blackwell, P.L. 183—192

Bonet, J. 206—217

Bowen, A.W. 324—337

Butler, R.G. 273—276, 284—295

Collier, A.D. 377—383

Cullen, E. 173—182

Davies, D.P. 324—337

Davies, T.J. 85—90, 139—150

Diaz, J. 218—234

Dressel, C.F. 357—376

Dunne, F.P.E. 235—250

Fitzpatrick, G.A., 347—350

Furihata, Satoshi 296—303

Geary, B. 324—337

Hammond, C. 161—172

Higashi, K. 75—84, 93—102, 103—112, 113—117

Humphreys, F.J. 173—182

Inoue, A. 113—117

Iwasaki, H. 75—84

Jackson, N. 377—383

Kaibyshev, O.A. 25—32

Katramados, I. 235—250

Kearns, M.W. 305—323

Kobayashi, M. 151—158

Langdon, Terence G. 9—24

Larin, Sergey A. 43—50

Lian, Jianshe 60

Mabuchi, M. 75—84, 103—112

Masumoto, T. 113—117

Matsuo, M. 277—283

Moran, F.J. 338—346

Mori, T. 75—84

Nieh, T.G. 121—132

Novikov, I.I. 193—198

Ochiai, S. 151—158

Ogwu, A.A. 139—150

Ohsawa, H. 260—272, 296—303

Pavlov, V.I. 193—198

Payne, James 133—138

Perevezentsev, Vladimir N. 43—50, 51—59

Pilling, John 133—138

Portnoy, V.K. 193—198

Purcell, Zachary 201—205

Ridley, N. 1—5, 63—74, 85—90, 139—150, 173—182

Sadeghi, Reza 201—205

Shibata, T. 113—117

Stephen, D. 304

Stowell, M.J. 387—388

Stracey, R.J. 273—276, 284—295

Swale, W. 351—358

Todd, R.I. 33—42

Tuffs, M. 161—172

Ubhi, H.S. 324—337

Uoya, A. 113—117

Wadsworth, J. 121—132

Wang, Z.C. 63—74, 85—90, 139—150

Ward-Close, C.M. 324—337

Whittingham, Roy 253—59

Williams, B.C. 324—337

Wisbey, A. 305—323, 324—337

Wood, R.D. 206—217

Yamaguchi, Y. 113—117

Yoshizawa, M. 260—272